Rank-Deficient
and Discrete
Ill-Posed Problems

SIAM Monographs on Mathematical Modeling and Computation

About the Series

In 1997, SIAM began a new series on mathematical modeling and computation. Books in the series develop a focused topic from its genesis to the current state of the art; these books

- present modern mathematical developments with direct applications in science and engineering;
- describe mathematical issues arising in modern applications;
- develop mathematical models of topical physical, chemical, or biological systems;
- present new and efficient computational tools and techniques that have direct applications in science and engineering; and
- illustrate the continuing, integrated roles of mathematical, scientific, and computational investigation.

Although sophisticated ideas are presented, the writing style is popular rather than formal. Texts are intended to be read by audiences with little more than a bachelor's degree in mathematics or engineering. Thus, they are suitable for use in graduate mathematics, science, and engineering courses.

By design, the material is multidisciplinary. As such, we hope to foster cooperation and collaboration between mathematicians, computer scientists, engineers, and scientists. This is a difficult task because different terminology is used for the same concept in different disciplines. Nevertheless, we believe we have been successful and hope that you enjoy the texts in the series.

Joseph E. Flaherty

Per Christian Hansen, *Rank-Deficient and Discrete Ill-Posed Problems: Numerical Aspects of Linear Inversion*

Michael Griebel, Thomas Dornseifer, Tilman Neunhoeffer, *Numerical Simulation in Fluid Dynamics: A Practical Introduction*

Khosrow Chadan, David Colton, Lassi Päivärinta and William Rundell, *An Introduction to Inverse Scattering and Inverse Spectral Problems*

Charles K. Chui, *Wavelets: A Mathematical Tool for Signal Analysis*

Editor-in-Chief
Joseph E. Flaherty
Rensselaer Polytechnic Institute

Editorial Board

Ivo Babuska
University of Texas at Austin

H. Thomas Banks
North Carolina State University

Margaret Cheney
Rensselaer Polytechnic Institute

Paul Davis
Worcester Polytechnic Institute

Stephen H. Davis
Northwestern University

Jack J. Dongarra
University of Tennessee at Knoxville and Oak Ridge National Laboratory

Christoph Hoffmann
Purdue University

George M. Homsy
Stanford University

Joseph B. Keller
Stanford University

J. Tinsley Oden
University of Texas at Austin

James Sethian
University of California at Berkeley

Barna A. Szabo
Washington University

Rank-Deficient and Discrete Ill-Posed Problems

Numerical Aspects of Linear Inversion

Per Christian Hansen
Technical University of Denmark
Lyngby, Denmark

Society for Industrial and Applied Mathematics
Philadelphia

Copyright ©1998 by the Society for Industrial and Applied Mathematics.

10 9 8 7 6 5 4 3 2 1

All rights reserved. Printed in the United States of America. No part of this book may be reproduced, stored, or transmitted in any manner without the written permission of the publisher. For information, write to the Society for Industrial and Applied Mathematics, 3600 University City Science Center, Philadelphia, PA 19104-2688.

Library of Congress Cataloging-in-Publication Data

Hansen, Per Christian.
   Rank-deficient and discrete ill-posed problems : numerical aspects of linear inversion / Per Christian Hansen.
      p. cm. -- (SIAM monographs on mathematical modeling and computation)
   Includes bibliographical references and index.
   ISBN 0-89871-403-6 (pbk.)
   1. Equations, Simultaneous--Numerical solutions. 2. Iterative methods (Mathematics) 3. Sparse matrices. I. Title. II. Series.
QA218.H38 1997
512. 9'42--dc21   97-32066

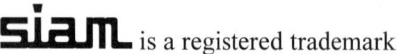 is a registered trademark.

Contents

Preface   ix

Symbols and Acronyms   xiii

CHAPTER 1. Setting the Stage   1
  1.1 Problems with Ill-Conditioned Matrices 1
  1.2 Ill-Posed and Inverse Problems 4
      1.2.1 The Singular Value Expansion 6
      1.2.2 The Smoothing Property of the Kernel 8
      1.2.3 The Picard Condition and the Instability of the Solution   9
  1.3 Prelude to Regularization 10
  1.4 Four Test Problems 13
      1.4.1 Signal Processing: Sinusoids in Noise 13
      1.4.2 Computation of the Second Derivative (deriv2) 14
      1.4.3 One-Dimensional Image Restoration Model (shaw) . . . 15
      1.4.4 A Problem with a Discontinuous Solution (wing) 16

CHAPTER 2. Decompositions and Other Tools   19
  2.1 The SVD and Its Generalizations 19
      2.1.1 The (Ordinary) SVD 19
      2.1.2 The GSVD 22
      2.1.3 What the SVD and GSVD Look Like 25
      2.1.4 Other SVDs 26
      2.1.5 Algorithms and Software 29
  2.2 Rank-Revealing Decompositions 29
      2.2.1 Rank-Revealing QR and LU Decompositions 31
      2.2.2 Rank-Revealing URV and ULV Decompositions 33
      2.2.3 What RRQR and URV Decompositions Look Like . . . 35
      2.2.4 Generalized Rank-Revealing Decompositions 36
      2.2.5 Algorithms and Software 37
  2.3 Transformation to Standard Form 38

	2.3.1	Explicit Transformation	40
	2.3.2	Implicit Transformation	41
	2.3.3	GSVD Computations Based on Standard-Form Transformation	42
2.4	Computation of the SVE		43

CHAPTER 3. Methods for Rank-Deficient Problems 45

3.1	Numerical Rank		45
3.2	Truncated SVD and GSVD		48
	3.2.1	Rank-Deficient Systems of Equations	49
	3.2.2	Truncated Total Least Squares	52
	3.2.3	Matrix Approximations	53
	3.2.4	Perturbation Bounds	56
3.3	Truncated Rank-Revealing Decompositions		58
	3.3.1	The Use of Rank-Revealing QR Decompositions	58
	3.3.2	The Use of UTV Decompositions	60
3.4	Truncated Decompositions in Action		63
	3.4.1	Subset Selection by SVD and RRQR	63
	3.4.2	Minimum-Norm Solutions and Null Spaces by SVD and UTV	63
	3.4.3	Numerical Ranges by SVD and GSVD	66

CHAPTER 4. Problems with Ill-Determined Rank 69

4.1	Characteristics of Discrete Ill-Posed Problems		69
4.2	Filter Factors		71
4.3	Working with Seminorms		74
4.4	The Resolution Matrix, Bias, and Variance		78
4.5	The Discrete Picard Condition		81
4.6	L-Curve Analysis		83
4.7	Random Test Matrices for Regularization Methods		88
4.8	The Analysis Tools in Action		90
	4.8.1	Standard-Form Regularization of deriv2 and shaw	90
	4.8.2	General-Form Regularization of deriv2 and wing	94
	4.8.3	The Importance of the Relative Decay Rate	97

CHAPTER 5. Direct Regularization Methods 99

5.1	Tikhonov Regularization		99
	5.1.1	Formulations and Algorithms	100
	5.1.2	Perturbation Bounds	103
	5.1.3	Least Squares with a Quadratic Constraint	105
	5.1.4	Inequality Constraints	106
	5.1.5	Related Methods	106
5.2	The Regularized General Gauss–Markov Linear Model		108
5.3	Truncated SVD and GSVD Again		109

- 5.4 Algorithms Based on Total Least Squares 111
  - 5.4.1 Truncated TLS Again 112
  - 5.4.2 Regularized TLS 114
- 5.5 Mollifier Methods 115
  - 5.5.1 The Target Function Method 116
  - 5.5.2 The Backus–Gilbert Method 118
- 5.6 Other Direct Methods 120
- 5.7 Characterization of Regularization Methods 123
- 5.8 Direct Regularization Methods in Action 126
  - 5.8.1 A Geometric Perspective 126
  - 5.8.2 From Oversmoothing to Undersmoothing 128
  - 5.8.3 Six Direct Solutions to shaw 130
  - 5.8.4 The Backus–Gilbert Solution to shaw 130
  - 5.8.5 Discontinuous Solutions to wing 131

CHAPTER 6. Iterative Regularization Methods — 135
- 6.1 Some Practicalities 136
- 6.2 Classical Stationary Iterative Methods 138
- 6.3 Regularizing CG Iterations 141
  - 6.3.1 Implementation Issues 142
  - 6.3.2 The Regularizing Effects of CG Iterations 145
- 6.4 Convergence Properties of Regularizing CG Iterations 149
  - 6.4.1 Convergence of the Ritz Values 149
  - 6.4.2 Convergence Rates for the CG Solution 153
  - 6.4.3 Filter Factors for the CG Solution 154
- 6.5 The LSQR Algorithm in Finite Precision 157
- 6.6 Hybrid Methods 162
- 6.7 Iterative Regularization Methods in Action 164
  - 6.7.1 Error Histories 164
  - 6.7.2 Four Iterative Solutions to shaw 165
  - 6.7.3 Ritz Plots for Lanczos Bidiagonalization 166
  - 6.7.4 The Influence of Reorthogonalization in LSQR ... 171
  - 6.7.5 A Hybrid Method in Action 172

CHAPTER 7. Parameter-Choice Methods — 175
- 7.1 Pragmatic Parameter Choice 176
- 7.2 The Discrepancy Principle 179
- 7.3 Methods Based on Error Estimation 181
- 7.4 Generalized Cross-Validation 184
- 7.5 The L-Curve Criterion 187
  - 7.5.1 Distinguishing Signal From Noise 188
  - 7.5.2 Computational Aspects 190
  - 7.5.3 Other Aspects 192

7.6	Parameter-Choice Methods in Action		193
7.7	Experimental Comparisons of the Methods		195
	7.7.1	Inverse Helioseismology	199
	7.7.2	Image Deblurring	206

CHAPTER 8. Regularization Tools   **209**

Bibliography   **215**

Index   **243**

Preface

This research monograph describes the numerical treatment of certain linear systems of equations which we characterize as either *rank-deficient problems* or *discrete ill-posed problems*. Both classes of problems are characterized by having a coefficient matrix that is very ill conditioned; i.e., the condition number of the matrix is very large, and the problems are effectively underdetermined.

Given a very ill conditioned problem, the advice usually sounds something like "do not trust the computed solution, because it is unstable and most likely dominated by rounding errors." This is good advice for general ill-conditioned problems, but the situation is different for rank-deficient and discrete ill-posed problems. These particular ill-conditioned systems can be solved by numerical regularization methods in which the solution is stabilized by including appropriate additional information. Since the two classes of problems share many of the same regularization algorithms, it is natural to discuss the numerical aspects of both problem classes in the same book.

The intended audience of this book is applied mathematicians and engineers studying rank-deficient and linear inverse problems that need to be solved numerically. Inverse problems arise, e.g., in applications in astronomy and medical imaging where one needs to improve unsharp images or get pictures from raw measurements. Rank-deficient and inverse problems arise in many other areas of science and engineering as well, such as geophysics, heat conduction, and signal processing.

The present book gives a survey of the state-of-the-art numerical methods for solving rank-deficient problems and linear discrete ill-posed problems. The goal of the book is to present and analyze new and existing numerical algorithms, to obtain new insight into these algorithms, and to describe numerical tools for the analysis of rank-deficient and discrete ill-posed problems.

The book is based on the author's research in numerical regularization methods. This research has focused on two main issues: *algorithms* and *insight*. Regarding the first issue, the emphasis is on reliability and efficiency, both of which are important as the size and complexity of the computational problems grow and the demand for advanced real-time processing increases. Regarding the second issue, the goal of the book is to characterize the regularizing effect of various methods in a common framework.

Although there are many algorithmic details in this book, it is not a "cookbook" of numerical regularization software. The development of robust and efficient implementations of numerical algorithms is a huge topic which lies outside the scope of this book. Implementation details of all the numerical algorithms discussed here can be found elsewhere in the literature, and adequate references are always given. We emphasize, though, that many of the algorithms are based upon standard "building blocks" in numerical linear algebra (such as QR factorizations and singular value decompositions (SVDs)) which are available in very efficient implementations on most high-performance computers.

In each chapter, the numerical algorithms and tools are illustrated by several numerical examples. The purpose of these examples is to illustrate the various methods in action by applying them to realistic small-scale test problems. All of these computations are performed in MATLAB.

The book focuses on the *numerical aspects* of regularization, and we assume that the reader is already familiar with the underlying theory of rank-deficient and inverse problems. One should not read the book without some background in numerical linear algebra and matrix computations.

Chapter 1 sets the stage for the remaining chapters. We introduce rank-deficient and discrete ill-posed problems as two important subclasses of the general class of linear ill-conditioned problems, and we emphasize that the algorithms presented in this book are relevant for these problems only. We also briefly summarize the main features of the underlying continuous ill-posed and inverse problems in the form of Fredholm integral equations of the first kind, and we motivate the incorporation of prior information in order to stabilize the computed solution to these problems. Finally, we introduce four test problems that are used throughout the book to illustrate the behavior of the numerical algorithms and tools.

Chapter 2 describes the decompositions and transformations that are the most important tools in numerical linear algebra for treating rank-deficient and discrete ill-posed problems. These tools serve two purposes: one is to provide matrix decompositions for reliable and powerful analysis of the involved linear systems of equations, and the other is to provide fast and numerically stable methods for computing the desired stabilized solutions. First, we focus on the SVD and its generalizations, and then we describe alternative rank-revealing decompositions. We also discuss numerical methods for transforming a regularization problem in general form into one in standard form, thus simplifying the solution procedure, and we give an algorithm for computing the singular value expansion (SVE) of a first-kind Fredholm integral operator.

Chapter 3 surveys the concept of numerical rank, and a number of stable methods for treating numerically rank-deficient matrices are presented. In all of these algorithms, matrix approximations of lower rank play a central role—either as the desired result itself or as a means for defining stabilized solutions. Some of the methods presented here are based on the SVD and its generalizations, while others are based on rank-revealing decompositions which are almost as reliable but faster to compute and/or update. In addition, we present some relevant perturbation bounds related to the matrix approximations and the corresponding stabilized solutions.

Chapter 4 surveys a number of important concepts and tools for problems with ill-determined numerical rank. All of them give insight into how the stabilizing or regularizing effects of a particular regularization method influence the solution, and how the solution depends on the regularization parameter. We briefly discuss the influence of the noise on the amount of information that can be extracted from the given problem. Then we introduce filter factors, resolution matrices, and other quantities associated with regularization methods, and we show how these quantities characterize the regularized solution. We also introduce the L-curve, which is a parametric plot of the solution's size versus the corresponding residual norm, and we discuss the information that this curve gives about the regularized solution. Finally, we describe an algorithm for generating random test matrices for discrete ill-posed problems.

The next two chapters describe direct and iterative regularization methods for discrete ill-posed problems. Direct regularization methods can be characterized as methods which are based on some kind of "canonical decomposition," such as the QR factorization or the SVD. Chapter 5 presents a number of these direct methods—some well known and some new—along with an analysis of the regularizing properties of the methods. The following methods are discussed: Tikhonov's classical method and related methods, the regularized general Gauss–Markov linear model, the truncated SVD and its generalizations, methods based on total least squares, mollifier methods (i.e., the Backus–Gilbert method and related methods), and some methods based on the use of other norms than the 2-norm. The similarities and differences between the methods are discussed, and we also include a general characterization of a broad range of regularization methods.

Chapter 6 first surveys some classical stationary iterative regularization methods. The remaining part of the chapter is devoted to a detailed description of the regularizing effects of the conjugate gradient (CG) method applied to the normal equations, with particular emphasis on the CGLS implementation (based on conjugate gradients) and the LSQR implementation (based on Lanczos bidiagonalization). Special attention is given to the filter factors associated with the CG method. This theory is still under development, and we present new, hitherto unpublished, results about the CG filter factors. The behavior of regularizing CG iterations in finite-precision arithmetic is also discussed. Finally, we discuss hybrid methods in which Lanczos bidiagonalization is combined with an "inner regularization" scheme in each iteration step.

Chapter 7 describes a number of algorithms for choosing the regularization parameter to be used in the direct and iterative regularization methods. The focus here is on the "pragmatic" aspects of parameter-choice methods when the problem's data and size are fixed—in contrast to the more theoretical aspects associated with the convergence of the regularized solution as the error in the right-hand side tends to zero. After a discussion of some general aspects of parameter-choice methods, we describe methods that require a good estimate of the norm of the errors in the right-hand side, namely, the discrepancy principle and some variants of this method. Then follows a discussion of several methods that do not require information about the error norm: methods based on error estimates, the generalized cross-validation method, and the L-curve method. Finally, we present the numerical results of an extensive comparison of these methods applied to two realistic test problems in helioseismology and image deblurring.

The last chapter is an overview of the public-domain package REGULARIZATION TOOLS, which consists of 53 MATLAB routines for analysis and solution of discrete ill-posed problems, and which complements this book. Most of the numerical results presented here are computed by means of routines from the REGULARIZATION TOOLS package.

Acknowledgments

I wish to thank all the people I have collaborated with throughout the past 10 years for inspiration and motivation, as well as access to real data sets. These collaborations have shaped this book. I also wish to acknowledge my present and former colleagues at the Department of Mathematical Modelling, at UNI•C, and at the Astronomical Observatory, for providing stimulating working conditions. Special thanks go to Prof. Åke Björck for many insightful discussions and for detailed comments on an earlier version of the manuscript. Also thanks to Susan Ciambrano, Kristen Kirchner, and Deborah Poulson from SIAM and Beth Gallagher for an enjoyable collaboration on this book.

Finally, I thank my wife Nancy with all my heart for her endless love and tremendous support throughout so many years.

Per Christian Hansen
Virum, Denmark
September 1997

Symbols and Acronyms

The page and equation references given below point to the location where the particular symbol or acronym is introduced or defined.

Symbol	Name	Page	Equation
A	$m \times n$ matrix	12	–
$\bar{A}$	standard-form matrix	38	(2.35)
$A^{\dagger}$	pseudoinverse of A	21	(2.3)
$A^{\#}$	regularized inverse	78	(4.17)
A_k	truncated SVD matrix	49	(3.10)
A_k^{alg}	"alg"-based approximation to A_k	–	–
$A_{L,k}$	truncated GSVD matrix	55	(3.28)
$A_{L,k}^{\#}$	truncated GSVD regularized inverse	100	(5.32)
$A_{\lambda}^{\#}$	Tikhonov regularized inverse	100	(5.3), (5.33)
b	right-hand side vector	12	–
$\bar{b}$	standard-form right-hand side	38	(2.35)
b^{exact}	exact right-hand side	–	–
$\bar{B}$	bidiagonal standard-form matrix	101	(5.5)
B_k	bidiagonal $(k+1) \times k$ Lanczos matrix	143	(6.15), (6.17)
$\text{cond}(A)$	condition number of A	22	(2.5)
$\text{diag}(\cdot)$	diagonal matrix	–	–
e	perturbation vector	–	–
$e_i^{(n)}$	ith unit vector of length n	–	–
E	perturbation matrix	–	–
$\mathcal{E}(\cdot)$	expected value	–	–
f_i	filter factor	72	–
$f(t)$	solution function	5	(1.1), (1.2)
$\tilde{f}(t)$	approximate solution function	11	–
F	diagonal filter matrix	72	–
$g(s)$	right-hand side function	5	(1.1), (1.2)

xiii

Symbol	Name	Page	Equation
$\mathcal{G}(\lambda)$	GCV function	184	(7.17)
I_n	identity matrix of order n	–	–
$J(t_0, t)$	Backus–Gilbert criteria function	118	–
k	discrete regularization parameter	–	–
$k_i(t)$	kernel	6	(1.2)
$K(s, t)$	kernel	5	(1.1)
$\mathcal{K}(t_0, t)$	averaging kernel	116	(5.48)
$\mathcal{K}_k(\cdot, \cdot)$	Krylov subspace	145	(6.19)
L	$p \times n$ regularization matrix	12	(1.14)
	also lower triangular matrix	–	–
L_1, L_2	approx. to 1st and 2nd deriv. operators	13	(1.15), (1.16)
$L_A^\dagger$	A-weighted pseudoinverse of L	39	(2.32), (2.34)
$L_\mathcal{Q}$	modified regularization matrix	77	(4.16)
$\mathcal{L}_i^{(k)}(\lambda)$	Lagrange polynomial in CG analysis	151	–
m	row dimension of matrix A	12	–
M	diagonal $p \times p$ GSVD matrix	22	(2.6)
n	column dimension of matrix A	12	–
N	length of sampled signal	14	–
$\mathcal{N}(\cdot)$	null space	–	–
$\mathcal{N}_k(\cdot)$	numerical null space	48	(3.7)
$\mathcal{P}_k(\theta)$	CG polynomial	148	(6.26)
Q	left $m \times n$ QR-factor	31	(2.21)
$\mathcal{Q}(\lambda)$	quasi-optimality function	182	(7.14)
$r^{(k)}$	residual vector in iterative methods	138	–
r_ϵ	numerical ϵ-rank	46	(3.2), (3.3)
R	upper triangular matrix	31	–
$\mathcal{R}(\cdot)$	range	–	–
$\mathcal{R}_k(\cdot)$	numerical range	48	(3.8)
$\mathcal{R}_k(\theta)$	Ritz polynomial	146	(6.21)
$\mathcal{T}(\lambda)$	effective number of degrees of freedom	181	(7.10)
$T_\gamma(t_0, t)$	target function	116	–
u_i	left singular vector	19	–
	also left GSVD vector	22	–
$u_i^T b$	Fourier coefficient of b	–	–
$u_i(s)$	left singular function	7	(1.4)
$\hat{u}_1$	initial left Lanczos vector	143	–
U	left $m \times n$ singular matrix	19	(2.1)
	also left $m \times n$ GSVD matrix	22	(2.6)
	also left matrix in UTV decomposition	–	–
$\hat{U}_{k+1}$	left $m \times (k+1)$ Lanczos matrix	143	(6.15)
v_i	right singular vector	19	–
	also left GSVD vector	22	–
$v_i(t)$	right singular function	7	(1.4)

Symbols and Acronyms

Symbol	Name	Page	Equation
V	right $n \times n$ singular matrix	19	(2.1)
	also $p \times p$ left GSVD matrix	22	(2.6)
	also right matrix in UTV decomposition	–	–
$\hat{V}_k$	right $n \times k$ Lanczos matrix	143	(6.15)
$\mathcal{V}(\lambda)$	scaled squared residual	180	(7.9)
$w^T x$	linear functional on x	122	–
W_{r_ϵ}	approximate null-space basis in RRQR	31	–
x	solution vector	12	–
x^*	a priori estimate of solution vector	13	–
$\bar{x}$	standard-form solution vector	38	–
x_0	solution component in $\mathcal{N}(L)$	39	(2.33), (2.34)
x^{exact}	exact solution vector	–	–
x_i	right GSVD vector	22	–
x_k	truncated SVD solution	50	(3.11)
$\bar{x}_k$	truncated TLS solution	52	(3.22)
x_k^{alg}	"alg"-based approximation to x_k	–	–
x_k^{basic}	basic solution	50	(3.13)
$x^{(k)}$	iteration vector	135	–
x_{LS}	least squares solution	21	(2.4)
$x_{L,k}$	truncated GSVD solution	51	(3.15)
$\hat{x}_{L,k}$	modified truncated SVD solution	51	(3.16), (3.17)
$x_{L,\lambda}$	Tikhonov regularized solution	100	(5.3)
x_{reg}	regularized solution	72	–
x_λ	standard-form Tikhonov solution	100	–
X	right $n \times n$ GSVD matrix	22	(2.6)
Y_{r_ϵ}	approximate null-space basis	31	(2.22)
α	relative decay rate	82	(4.30)
γ_i	generalized singular value	22	(2.7)
δ_0, δ_∞	extreme residual norms	85	(4.36)
δ_e	upper bound in discrepancy principle	179	(7.5)
δ_E	upper bound in generalized discrep. princ.	179	(7.6)
$\delta_i^{(k)}$	ith column of Δ_k	150	–
Δ_k	matrix in CG analysis	150	(6.29)
η_{res}	effective resolution limit	71	(4.1), (4.2)
$\theta_j^{(k)}$	Ritz value	146	–
Θ	subspace angle	45	(3.1)
$\kappa(\lambda)$	curvature of L-curve	189	(7.24)
λ	continuous regularization parameter	–	–
μ_i	singular value of kernel	7	(1.4)
	also diagonal GSVD element	22	–
Ξ	resolution matrix	78	(4.20)

Symbol	Name	Page	Equation
$\rho(\lambda)$	sum of filter factors	178	(7.2)
σ_i	singular value of matrix	19	–
	also diagonal GSVD element	22	–
Σ	diagonal $n \times n$ matrix of singular values	19	(2.1)
	also $p \times p$ diagonal GSVD matrix	22	(2.6)
τ_A	error level for singular values	70	–
τ_b	error level for Fourier coefficients	70	(1.14), (1.17)
$\omega(f)$	smoothing norm of function f	11	–
$\Omega(x)$	discrete smoothing norm of vector x	12	–
(ϕ, ψ)	inner product of functions	7	(1.6)
$\|x\|_2$	2-norm of vector x	–	–
$\|\phi\|_2$	2-norm of function ϕ	10	–
$\|A\|_2$	2-norm of matrix A	–	–
$\|A\|_F$	Frobenius norm of matrix A	–	–
$\|K\|_2$	2-norm of kernel K	6	(1.3)
$\|x\|_{\underline{p}}$	p-norm in SVD basis of vector x	124	(5.73)

Acronym	Full name	Page	Equation
CG	Conjugate gradient	141	–
CGLS	CG applied to normal equations	143	(6.14)
GCV	Generalized cross-validation	184	(7.17)
GSVD	Generalized SVD	22	(2.6)
LS	Least squares	–	–
LSQR	Least squares via Lanczos bidiag.	143	(6.15)–(6.18)
MTSVD	Modified truncated SVD	51	(3.16)
PP-TSVD	Piecewise polynomial TSVD	121	(5.63)
RRQR	Rank-revealing QR	31	(2.21)
RSVD	Restricted SVD	27	(2.17)
R-TLS	Regularized TLS	114	(5.41)
SVD	Singular value decomposition	19	(2.1)
SVE	Singular value expansion	6	(1.4)
TGSVD	Truncated GSVD	51	(3.15)
TLS	Total least squares	52	–
TSVD	Truncated SVD	50	(3.11), (3.12)
T-TLS	Truncated TLS	52	(3.22)
TV	Total variation	120	(5.61), (5.62)
ULV	Two-sided orthogonal decomposition	33	(2.29)
ULLV	Generalized ULV decomposition	36	(2.30), (2.31)
URV	Two-sided orthogonal decomposition	33	(2.28)
UTV	Either ULV or URV	33	–

1

Setting the Stage

A very large condition number of the coefficient matrix A in a linear system of equations $Ax = b$ implies that some (or all) of the equations are numerically linearly dependent. Hence, the standard advice to avoid solving such systems numerically is not bad. Indeed, it is sometimes the case that the large condition number is caused by an incorrect mathematical model which should be modified before one attempts to compute a numerical solution. Numerical "tools," such as the singular value decomposition (SVD) (see §2.1), can identify the linear dependencies and thus help to improve the model and lead to a modified system with a better-conditioned matrix. This modified system can then be solved by standard numerical techniques [36], [154], [230].

However, there are classes of problems for which the coefficient matrix is *correctly* very ill conditioned, i.e., where this property is part of the formulation of the problem. Then the standard linear algebra techniques no longer apply, and the numerical treatment often becomes more difficult.

This book gives a survey of advanced numerical methods for solving such problems with ill-conditioned matrices.

1.1. Problems with Ill-Conditioned Matrices

The numerical treatment of very ill conditioned linear systems of equations is more complicated than the treatment of well-conditioned systems, for the following two reasons.

- The user should know what **kind of ill conditioning** to expect and how to deal with it. Is the problem rank deficient or ill posed? Is it possible to regularize, i.e., include additional information to stabilize the solution? What additional information is available and is it suited for stabilization purposes?

- The user should also know which **numerical regularization method** should be used to treat the problem efficiently and reliably on a computer. Are both the analysis and the solution phases important? Should one prefer a direct or an iterative method? How much stabilization should be added?

In other words, one cannot expect to deal satisfactorily with ill-conditioned problems without both theoretical and numerical insight.

It is beyond the goal of this book to give a thorough treatment of the theoretical aspects of mathematical models that lead to ill-conditioned systems of equations. Moreover, this topic has been extensively studied during the last few decades, and literature at any level of sophistication can be found. The same is true for the mathematical aspects of regularization. For the reader's convenience, the most important aspects are summarized in §1.2 below.

In contrast to the above, descriptions of *numerical regularization methods* are scattered in the literature. Only few attempts have been taken to unify the work and to characterize the methods and their interrelations [37], [161, §5], [171], [260], despite the fact that many of the techniques have been widely used by scientists and engineers.

The goal of this book is to give a unified treatment of efficient and reliable numerical methods that are suited for regularization of problems with an ill-conditioned coefficient matrix. With this insight in hand, the reader should be able to understand the details of various regularization methods, to distinguish between different methods, and, ultimately, to select the best algorithm(s) for a given problem. No regularization method is superior to the other methods. Rather, each method has its advantages, depending on the application in which it is used.

Any discussion of ill-conditioned matrices requires knowledge of the SVD of the matrix A; see, e.g., [154] and §2.1. In particular, the condition number of A is defined as the ratio between the largest and the smallest singular values of A. The numerical treatment of systems of equations with an ill-conditioned coefficient matrix depends on the type of ill-conditioning of A. There are two important classes of problems to consider, and many practical problems belong to one of these two classes.

Rank-deficient problems are characterized by the matrix A having a cluster of small singular values, and there is a well-determined gap between large and small singular values. This implies that one or more rows and columns of A are nearly linear combinations of some or all of the remaining rows and columns. Therefore, the matrix A contains almost redundant information, and the key to the numerical treatment of such problems is to extract the linearly independent information in A, to arrive at another problem with a well-conditioned matrix.

Discrete ill-posed problems arise from the discretization of ill-posed problems such as Fredholm integral equations of the first kind. Here all the singular values of A, as well as the SVD components of the solution, on the average, decay gradually to zero, and we say that a discrete Picard condition (see §4.5) is satisfied. Since there is no gap in the singular value spectrum, there is no notion of a numerical rank for these matrices. For discrete ill-posed problems, the goal is to find a balance between the residual norm and the size

1.1. PROBLEMS WITH ILL-CONDITIONED MATRICES

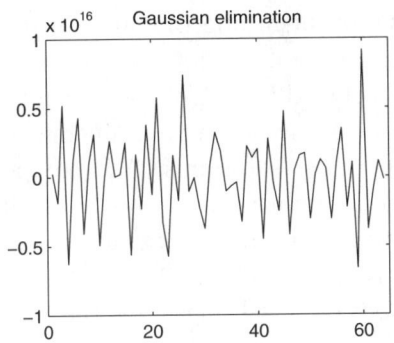 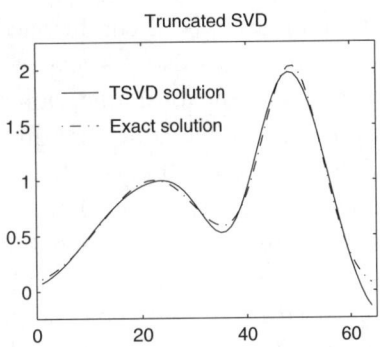

FIG. 1.1. *Two numerical solutions to a 64×64 discretization $A x = b$ of an inverse problem. The left part shows the solution computed by means of Gaussian elimination with partial pivoting. The right part shows the TSVD solution (solid line), obtained by retaining the seven largest SVD components, together with the exact solution (dash-dotted line).*

of the solution that matches the errors in the data as well as one's expectations to the computed solution. Here, "size" should be interpreted in a rather broad sense; e.g., size can be measured by a norm, a seminorm, or a Sobolev norm.

Due to the large condition number of A, both classes of problems are effectively underdetermined, and therefore many of the regularization methods described in this book can be used for both classes of problems. Moreover, in both classes of problems, there is a strong relation between the amount of extracted linearly independent information and the norm of the solution and the corresponding residual. Nevertheless, it is often advantageous to keep in mind the basic difference between the two problem classes, namely, a gap in the singular value spectrum versus an overall decay.

Figure 1.1 shows numerical results for a 64×64 discretization $A x = b$ of a Fredholm integral equation of the first kind (test problem shaw from §1.4.3). The left part of the figure shows the solution x_{GE} to $A x = b$ computed via Gaussian elimination with partial pivoting. This solution is dominated by oscillations with very large amplitude, and the norm of the solution is huge: $\|x_{\text{GE}}\|_2 = 2.6 \cdot 10^{16}$. The right part of the figure shows the exact solution x^{exact} (dash-dotted line) plus a regularized solution (solid line), the so-called truncated SVD (TSVD) solution x_7 (cf. §3.2), in which only the seven largest SVD components are retained. The TSVD solution is a fair approximation to the exact solution, with a reasonable balance between the relative error norm $\|x^{\text{exact}} - x_7\|_2 / \|x^{\text{exact}}\|_2 = 0.054$ and the relative residual norm $\|b - A x_7\|_2 / \|b\|_2 = 0.014$. The TSVD method has its origin in rank-deficient problems, and the example illustrates that it can also be suited for solving discrete ill-posed problems.

If the matrix is ill conditioned and the problem does not belong to either of the two classes listed above, then regularization cannot produce a suitable solution, and the best one can do is to solve the problem as accurately as possible—without regularization—by means of, say, iterative refinement or extended precision software. If, in particular, the solution's SVD components increase, on the average, then the condition number can be overly pessimistic as a measure of the solution's sensitivity to perturbations of b, and such problems are therefore characterized in [54] as effectively well conditioned. As emphasized in [57, §4] the condition number of A still measures the solution's sensitivity to rounding errors. See [60] for an example in boundary element methods.

Our experience is that it is fairly simple to understand why some mathematical problems lead to rank-deficient matrices. For example, if A is a Hankel matrix (cf. (1.18)) or a Toeplitz matrix derived from oversampling of k (damped) sinusoids—which is typical in many signal processing applications [92], [250], [343]—then the numerical rank of A is never greater than $2k$ (see §1.4.1). Rank-deficiency is also common in statistics where the term "collinearity" is used; see [321] and the references therein.

The treatment of discrete ill-posed problems, on the other hand, usually requires a deeper insight into the underlying mathematical model. For this reason, we include below a very brief introduction to ill-posed problems in order to point out some fundamental concepts. Other introductions are given in [171, §3.1], [185, §2], [187, §2], and [296, Chapter 1], while more complete treatments can be found in, e.g., [10], [21], [28], [107], [111], [142], [160], [161], and [226]. Newcomers to this field are encouraged to read Wing and Zahrt's primer [373] on first-kind integral equations.

1.2. Ill-Posed and Inverse Problems

The concept of well-posed and ill-posed problems goes back to Hadamard at the beginning of this century; cf., e.g., [164]. Hadamard essentially defined a problem as *ill posed* if the solution is not unique or if it is not a continuous function of the data—i.e., if an arbitrary small perturbation of the data can cause an arbitrarily large perturbation of the solution. Hadamard believed that ill-posed problems were "artificial" in that they would not describe physical systems.

He was mistaken, though, and today ill-posed problems arise in the form of inverse problems in many areas of science and engineering. *Inverse problems* arise quite naturally if one is interested, say, in determining the internal structure of a physical system from the system's measured behavior, or in determining the unknown input that gives rise to a measured output signal (in contrast to direct problems where the interest is in the system's behavior given

1.2. ILL-POSED AND INVERSE PROBLEMS

the input or internal structure). Some examples are acoustics [302], astrometry [65], computerized tomography [259], continuation problems [45], early vision [30], electromagnetic scattering [202], geophysics [248], [278], inverse geo- and helioseismology [59], [263], [269], [277], mathematical biology [72], [229], optics and image restoration [9], [26], [29], [85], remote sensing [19], inverse scattering theory [62], [226, Chapter 5], signal processing [335], and statistics [340]. Other examples can be found in [21], [107], [110], [111], [161], [223], [211], [256], [338], and [373].

This book is devoted to real linear ill-posed problems (the extensions to the complex case are obvious). Nonlinear ill-posed problems constitute a much broader area, and their numerical treatment is often specialized to the particular application. For this reason, nonlinear problems are not discussed in this book. Discussions of some general methods for nonlinear ill-posed problems can be found in, e.g., [107], [111], [212], [358], and the references therein.

We focus on linear inverse problems that can be formulated in the following very general form:

$$\int_\Omega \text{input} \times \text{system} \, d\Omega = \text{output} .$$

In this formulation, the direct problem is to compute the output, given the input and the mathematical description of the system. The goal of the inverse problem is to determine either the input or the system that gives rise to the (noisy) measurements of the output. For example, in astronomical image deblurring the "input" is the night sky, the blurring "system" consists of the telescope and the atmosphere, and the "output" is the recorded blurred image. The goal is to reconstruct the "input," i.e., the unblurred image, given a mathematical description of the blurring effects of the telescope and the atmosphere. Another example is computerized tomography, where the "input" is, say, an X-ray source, the "system" is the object being scanned (often the brain), and the "output" is the measured damping of the X-rays. The goal here to reconstruct the "system," i.e., the scanned object, from information about the locations of the X-ray sources and measurements of their damping.

The classical example of a linear ill-posed problem is a Fredholm integral equation of the first kind[1] with a square integrable kernel [160], which can always be written in the generic form

$$\int_0^1 K(s,t) \, f(t) \, dt = g(s) , \qquad 0 \leq s \leq 1 , \qquad (1.1)$$

where the *right-hand side* g and the *kernel* K are known functions, at least in principle, while f is the unknown, sought solution. In many—but not all—practical applications of (1.1) the kernel K is given exactly by the underlying

---

[1]Volterra integral equations of the first kind, $\int_0^s K(s,t) \, f(t) \, dt = g(s)$, are generally less difficult to deal with, as discussed in [17, §6.10].

mathematical model, while the right-hand side g typically consists of measured quantities; i.e., g is only known with a certain accuracy and only in a finite set of points $s_1, \ldots, s_m$.

An important special case of (1.1) is the first-kind Fredholm integral equation with a discrete right-hand side, which takes the following generic form:

$$\int_0^1 k_i(t)\, f(t)\, dt = b_i\,, \qquad i = 1, \ldots, m\,. \tag{1.2}$$

Here, we are given m functionals (or kernels) k_i on an unknown function f, and (1.2) can be obtained from (1.1) with $k_i(t) = K(s_i, t)$ and $b_i = g(s_i)$. The problem (1.2) is therefore continuous in only one variable t. Both forms (1.1) and (1.2) give rise to ill-conditioned systems of linear algebraic equations.

The difficulties with the integral equation (1.1) are inseparably connected with the compactness of the operator which is associated with the kernel K [227, Chapter 15]. In physical terms, the integration with K in (1.1) has a "smoothing" effect on f in the sense that high-frequency components, cusps, and edges in f are "smoothed out" by the integration. To illustrate this damping of high-frequency components, let

$$f(t) = \sin(2\pi p t)\,, \qquad p = 1, 2, \ldots\,.$$

Then the corresponding right-hand side g is given by

$$g(s) = \int_0^1 K(s, t)\, \sin(2\pi p t)\, dt, \qquad p = 1, 2, \ldots\,,$$

and the Riemann–Lebesgue lemma[2] states that $g \to 0$ as $p \to \infty$. We can therefore expect that the reverse process, i.e., that of computing f from g, will tend to amplify any high-frequency components in g. As we shall illustrate below, this is indeed the case for both (1.1) and (1.2).

1.2.1. The Singular Value Expansion

The superior analytical tool for analysis of first-kind Fredholm integral equations (1.1) with square integrable kernels is the *singular value expansion* (SVE) of the kernel. A kernel K is square integrable if the norm

$$\|K\|^2 \equiv \int_0^1 \int_0^1 K(s,t)^2\, ds\, dt \tag{1.3}$$

is bounded. By means of the SVE, any square integrable kernel K can be written as the following infinite sum:[3]

$$K(s,t) = \sum_{i=1}^{\infty} \mu_i\, u_i(s)\, v_i(t) \tag{1.4}$$

---

[2] The Riemann–Lebesgue lemma can be formulated as follows: if the function ψ has limited total fluctuation in the interval $(0,1)$, then, as $\lambda \to \infty$, $\int_0^1 \psi(\theta)\, \sin(\lambda \theta)\, d\theta$ is $\mathcal{O}(\lambda^{-1})$.

[3] The equality signs in (1.4) and (1.7) hold "almost everywhere."

1.2. ILL-POSED AND INVERSE PROBLEMS

(for degenerate kernels, the ∞ should be replaced by the rank of the kernel). The functions u_i and v_i are termed the *singular functions* of K. They are orthonormal with respect to the usual inner product, i.e.,

$$(u_i, u_j) = (v_i, v_j) = \begin{cases} 1 & \text{if } i = j, \\ 0 & \text{if } i \neq j, \end{cases} \tag{1.5}$$

where $(\cdot, \cdot)$ is defined by

$$(\phi, \psi) \equiv \int_0^1 \phi(t)\, \psi(t)\, dt\,. \tag{1.6}$$

The numbers μ_i are the *singular values* of K; they are nonnegative and they can always be ordered in nonincreasing order such that

$$\mu_1 \geq \mu_2 \geq \mu_3 \geq \cdots \geq 0\,.$$

The singular values satisfy the relation $\sum_{i=1}^{\infty} \mu_i^2 = \|K\|^2$, showing that the μ_i must decay faster than $i^{-1/2}$.

The triplets $\{\mu_i, u_i, v_i\}$ are related to the following two eigenvalue problems associated with the kernel K: $\{\mu_i^2, u_i\}$ are the eigensolutions of the symmetric kernel $\int_0^1 K(s,x)\,K(t,x)\,dx$, while $\{\mu_i^2, v_i\}$ are the eigensolutions of $\int_0^1 K(x,s)\,K(x,t)\,dx$. This illustrates that the triplets $\{\mu_i, u_i, v_i\}$ are characteristic and essentially unique for the given kernel K. For more details, see [28], [111, §2.2], [227, §15.4], and [314, §8]. Numerical methods for computing the SVE are treated in §2.4.

Perhaps the most important relation between the singular values and functions is the following fundamental relation:

$$\int_0^1 K(s,t)\, v_i(t)\, dt = \mu_i\, u_i(s)\,, \qquad i = 1, 2, \ldots\,, \tag{1.7}$$

which shows that any singular function v_i is mapped onto the corresponding u_i, and that the singular value μ_i is the amplification of this particular mapping. If this relation, together with the SVE (1.4), is inserted into the integral equation (1.1), then we obtain the equation[4]

$$\sum_{i=1}^{\infty} \mu_i\, (v_i, f)\, u_i(s) = \sum_{i=1}^{\infty} (u_i, g)\, u_i(s), \tag{1.8}$$

which, in turn, leads to the following expression for the solution to (1.1):

$$f(t) = \sum_{i=1}^{\infty} \frac{(u_i, g)}{\mu_i}\, v_i(t)\,. \tag{1.9}$$

---

[4]Equations (1.8) and (1.9) hold with relatively uniform absolute convergence [314, §2.4 and Theorem 8.3.2], which implies mean convergence [314, p. 55]. Uniform convergence holds if K is continuous [314, p. 147].

We stress that f only exists if the right-hand side of (1.9) indeed converges, which is equivalent to requiring that g belong to $\mathcal{R}(K)$, the range of K. From (1.9) we see that f is expressed in terms of the singular functions v_i and the corresponding expansion coefficients $\mu_i^{-1}(u_i, g)$. One can therefore completely characterize the solution f by an analysis of the coefficients $\mu_i^{-1}(u_i, g)$ and the functions v_i.

There is also a system of m triplets $\{\mu_i, u_i, v_i\}$ associated with the operator in the Fredholm integral equation (1.2) with a discrete right-hand side. In this case, u_i are m orthonormal functions while v_i are m orthonormal vectors of length m. See [28] for more details.

1.2.2. The Smoothing Property of the Kernel

The overall behavior of the singular values μ_i and the singular functions u_i and v_i is by no means "arbitrary"; their behavior is strongly connected with the properties of the kernel K. The following holds.

- The "smoother" the kernel K, the faster the singular values μ_i decay to zero (where "smoothness" is measured by the number of continuous partial derivatives of K). If the derivatives of order $0, \ldots, q$ exist and are continuous, then σ_i is approximately $\mathcal{O}(i^{-p-1/2})$. The precise result is proved in [313] and summarized in [79].

- The smaller the μ_i, the more oscillations (or zero-crossings) there will be in the singular functions u_i and v_i. This property is perhaps impossible to prove in general, but it is often observed in practice. It is related to the Riemann–Lebesgue lemma mentioned above.

The practical implication of the above properties of the triplets $\{\mu_i, u_i, v_i\}$ is that the expression (1.9) for f can be regarded as a spectral expansion in which the coefficients $\mu_i^{-1}(u_i, g)$ describe the spectral properties of the solution f. We see from (1.8) that the integration with K indeed has a smoothing effect: the higher the spectral components in f, the more they are damped in g by the multiplication with μ_i. Moreover, Eq. (1.9) shows that the inverse problem, that of computing f from g, indeed has the opposite effect on the oscillations in g, namely, an amplification of the spectral components (u_i, g) with a factor μ_i^{-1}. This, of course, amplifies the high-frequency components.

The decay rate of the singular values μ_i is so fundamental for the behavior of ill-posed problems that it makes sense to use this decay rate to characterize the degree of ill-posedness of the problem, as mentioned in, e.g., [111, p. 40] and [364]. Hofmann [211, Definition 2.42] gives the following definition: if there exists a positive real number α such that the singular values satisfy $\mu_i = \mathcal{O}(i^{-\alpha})$, then α is called the *degree of ill-posedness*, and the problem is characterized as mildly or moderately ill posed if $\alpha \leq 1$ or $\alpha > 1$, respectively.

1.2. ILL-POSED AND INVERSE PROBLEMS

On the other hand, if $\mu_i = \mathcal{O}(e^{-\alpha i})$, then the problem is termed severely ill posed.

1.2.3. The Picard Condition and the Instability of the Solution

With this behavior in mind, it is obvious that not every right-hand side g will lead to a "smooth" solution f due to the amplification factors μ_i^{-1}. In effect, the right-hand side g must be somewhat "smoother" than the desired function f, in order that the right-hand side in (1.9) actually converges to f. The following Picard condition is therefore essential (see [111, §2.2], [160, §1.2], and [227, Theorem 15.18] for more details).

The Picard Condition. In order that there exist a square integrable solution f to the integral equation (1.1), the right-hand side g must satisfy

$$\sum_{i=1}^{\infty} \left(\frac{(u_i, g)}{\mu_i} \right)^2 < \infty . \qquad (1.10)$$

The Picard condition says that from some point in the summation in (1.9), the absolute value of the coefficients (u_i, g) must decay faster than the corresponding singular values μ_i in order that a square integrable solution exists. For g to be square integrable the coefficients (u_i, g) must decay faster than $i^{-1/2}$, but the Picard condition puts a stronger requirement on g, in that the coefficients (u_i, g) must decay faster than $\mu_i\,i^{-1/2}$. Since the Picard condition is so essential in connection with Fredholm integral equations of the first kind, it should—whenever possible—be checked before one attempts to solve the integral equation. In §2.4 we return to the numerical aspects of computing the quantities in (1.10).

The requirement (1.10) in the Picard condition is identical to the requirement that the right-hand side g belong to $\mathcal{R}(K)$, the range of K. If g has any, arbitrarily small, component outside $\mathcal{R}(K)$, then there is no square integrable solution. Consider a $g \notin \mathcal{R}(K)$, and let g_k denote the approximation to g obtained from truncating its SVE expansion after k terms,

$$g_k(s) = \sum_{i=1}^{k} (u_i, g)\, u_i(s) .$$

This g_k clearly satisfies the Picard condition (1.10) for all $k = 1, 2, \ldots$, and the corresponding approximate solution f_k is given by

$$f_k(t) = \sum_{i=1}^{k} \frac{(u_i, g)}{\mu_i}\, v_i(t) .$$

We conclude that as $k \to \infty$ we have $g_k \to g$, but

$$\|f_k\|_2 \equiv (f_k, f_k)^{1/2} \to \infty \quad \text{as} \quad k \to \infty .$$

It is exactly this lack of stability of f that makes the integral equation (1.1) ill posed.

Unfortunately, in practical situations we will typically only have access to an approximate right-hand side g which is contaminated with unavoidable errors:

$$g = g^{\text{exact}} + \eta\,, \qquad g^{\text{exact}} \in \mathcal{R}(K)\,, \qquad \|\eta\|_2 \lesssim \|g^{\text{exact}}\|_2\,.$$

Here, g^{exact} denotes the unknown, exact right-hand side and η denotes the perturbation. Ideally, we want to compute $f^{\text{exact}} = \sum_{i=1}^{\infty} \mu_i^{-1} (u_i, g^{\text{exact}})\, v_i$. We cannot expect the errors η to satisfy the Picard condition, and hence $g \notin \mathcal{R}(K)$. Any naive approach that tries to compute f^{exact} via the infinite sum $\sum_{i=1}^{\infty} \mu_i^{-1} (u_i, g)\, v_i$ will therefore usually diverge or return a useless result with extremely large norm, no matter how small the perturbation η is. Instead, it is necessary to use a regularization method that replaces the original problem (1.1) by a regularized problem with a stable solution which approximates the desired f^{exact}. If the perturbation is too big, compared to g^{exact}, then it is impossible to compute an approximation to f^{exact}; hence, the assumption $\|\eta\|_2 \lesssim \|g^{\text{exact}}\|_2$. We return to regularization methods in §4.2 and Chapters 5–6.

We mention in passing that there are also important examples of first-kind Fredholm integral equations, such as the inverse Laplace transformation, for which the associated operator is not compact. These operators have a continuum of singular values instead of countably many singular values, but the above discussion still applies with the sums and expansion coefficients replaced by integrals and distribution functions; see [159].

1.3. Prelude to Regularization

As we have seen in the previous section, the primary difficulty with ill-posed problems is that they are practically underdetermined due to the cluster of small singular values of K. Hence, it is necessary to incorporate further information about the desired solution in order to stabilize the problem and to single out a useful and stable solution. This is the purpose of *regularization*.

Although many types of additional information about the solution f to (1.1) are possible in principle—see, e.g., the survey in [84]—the dominating approach to regularization is to allow a certain residual associated with the regularized solution, with residual norm

$$\rho(f) = \left\| \int_0^1 K(s,t)\, f(t)\, dt - g(s) \right\|_2,$$

and then use one of the following four schemes.

1.3. Prelude to Regularization

1. Minimize $\rho(f)$ subject to the constraint that f belongs to a specified subset, $f \in \mathcal{S}_f$.

2. Minimize $\rho(f)$ subject to the constraint that a measure $\omega(f)$ of the "size" of f is less than some specified upper bound δ, i.e., $\omega(f) \leq \delta$.

3. Minimize $\omega(f)$ subject to the constraint $\rho(f) \leq \alpha$.

4. Minimize a linear combination of $\rho(f)^2$ and $\omega(f)^2$:

$$\min \left\{ \rho(f)^2 + \lambda^2 \, \omega(f)^2 \right\} ,$$

where λ is a specified weighting factor.

Here, α, δ, and λ are known as regularization parameters, and the function ω is sometimes referred to as the "smoothing norm." The fourth scheme is the well-known Tikhonov regularization scheme [336], [337]. The underlying idea in all four schemes is that a regularized solution having a suitably small residual norm and satisfying the additional constraint will (hopefully) be not too far from the desired, unknown solution to the underlying unperturbed problem. We return to these issues in Chapters 4–6. The choice of the regularization parameter is also a very important topic, which is treated in Chapter 7.

From a statistical point of view, the introduction of regularization decreases the size of the solution's covariance matrix at the cost of adding bias to the solution [246] (see also §4.4).

In practice, we must somehow discretize the regularization problem in order to solve it numerically. There are many ways to discretize integral equations, and we do not attempt to cover this topic here. Instead, we refer to [17], [80], [81], and [226, Chapter 3]. Suffice it here to mention that in connection with integral equations, there are essentially two main classes of methods, namely, quadrature methods and Galerkin methods. Both methods compute an approximation $\tilde{f}$ to f. In the *quadrature method*, a quadrature rule with abscissas $t_1, \ldots, t_n$ and corresponding weights $w_1, \ldots, w_n$ is used to approximate an integral as

$$\int_0^1 \phi(t) \, dt \approx \sum_{j=1}^n w_j \, \phi(t_j) ,$$

and when this rule is applied to the integral equation (1.1) for m distinct values $s_1, \ldots, s_m$, then we obtain an $m \times n$ matrix A and a right-hand side b with elements given by

$$a_{ij} = w_j \, K(s_i, t_j) , \qquad b_i = g(s_i) . \tag{1.11}$$

The solution vector is $(\tilde{f}(t_1), \ldots, \tilde{f}(t_n))^T$, i.e., samples of $\tilde{f}$. In the *Galerkin method*, one chooses two sets of basis function ϕ_i and ψ_j, and then the matrix and right-hand side elements are given by

$$a_{ij} = \int_0^1 K(s,t)\,\phi_i(s)\,\psi_j(t)\,ds\,dt\;, \qquad b_i = \int_0^1 g(s)\,\phi_i(s)\,ds\;. \qquad (1.12)$$

Solving the linear system of equations $A\,\xi = b$ for the vector ξ, we obtain

$$\tilde{f}(t) = \sum_{i=1}^n \xi_i\,\psi_i(t)\;. \qquad (1.13)$$

Some collocation methods are special cases of the Galerkin method with delta functions as the ϕ_i basis functions, $\phi_i(s) = \delta(s - s_i)$. If K is symmetric and $\phi_i = \psi_i$, then A is symmetric and Galerkin's method is called the Rayleigh–Ritz method.

When a rank-deficient or ill-posed problem is discretized, then the inherent difficulties carry over to the discrete problem in the sense that the coefficient matrix will also have either a cluster of small singular values or singular values that decay gradually to zero. Hence, the discrete problem will also be effectively underdetermined (unless the discretization is so coarse that the small singular values do not show up in the matrix). This is true for any discretization method; see, for example, [373, Chapter 6]. Hence, some kind of regularization is also required to solve the discretized problem.

The most natural approach to adding regularization to the discrete problem is to discretize the underlying, regularized problem. That is, we advocate the strategy "first regularize, then discretize." In this way, we have the best control of which additional information we actually enforce on the regularized solution. Thus, if the discretization leads to a square system,

$$A\,x = b\;, \qquad A \in \mathbb{R}^{n \times n}\;,$$

or an overdetermined system,

$$\min \|A\,x - b\|_2\;, \qquad A \in \mathbb{R}^{m \times n}\;, \qquad m > n\;,$$

where the vector x represents the function f, then in analogy with the above four regularization schemes we will need either some subset $\mathcal{S}_x$ or some measure of "size" $\Omega(x) \approx \omega(f)$. For example, the discrete Tikhonov regularization scheme becomes

$$\min \left\{ \|A\,x - b\|_2^2 + \lambda^2\,\Omega(x)^2 \right\}\;.$$

The function Ω is termed the *discrete smoothing norm*, and it is often—but not always (see §§5.6 and 5.7)—of the form

$$\Omega(x) = \|L\,x\|_2\;, \qquad (1.14)$$

where the matrix L is typically either the identity matrix, a diagonal weighting matrix, or a $p \times n$ discrete approximation of a derivative operator (e.g., $\Omega(x) \approx$

$\omega(f) = \|f''\|_2$), in which case L is a banded matrix with full row rank. For example, for certain discretizations, the matrices

$$L_1 = \begin{pmatrix} 1 & -1 & & \\ & \ddots & \ddots & \\ & & 1 & -1 \end{pmatrix} \in \mathbb{R}^{(n-1) \times n} \qquad (1.15)$$

and

$$L_2 = \begin{pmatrix} 1 & -2 & 1 & & \\ & \ddots & \ddots & \ddots & \\ & & 1 & -2 & 1 \end{pmatrix} \in \mathbb{R}^{(n-2) \times n} \qquad (1.16)$$

are scaled approximations to the first and second derivative operators. When $p < n$ then $\|L \cdot \|_2$ is said to be a *seminorm*; i.e., there exist vectors $x \neq 0$ (in the null space of L) such that $\|L x\|_2 = 0$. We stress that even when $\omega(f) = \|f\|_2$, the matrix L need not be the identity matrix—this will depend on the discretization scheme used.

If an a priori estimate x^* of the desired regularized solution is available, then this information can be taken into account by including x^* in the discrete smoothing norm $\Omega(x)$, which then takes the form

$$\Omega(x) = \|L(x - x^*)\|_2 . \qquad (1.17)$$

In this way, the regularized solution will be "biased" towards the a priori estimate x^*, and the simple version of $\Omega(x)$ in (1.14) corresponds to the choice $x^* = 0$.

1.4. Four Test Problems

We end this chapter with a description of four test problems which, throughout the rest of the book, will serve to illustrate the various regularization algorithms and techniques. The first test problem comes from signal processing and leads to a numerically rank-deficient matrix; the three other test problems come from the discretization of Fredholm integral equations of the first kind, and they lead to discrete ill-posed problems and matrices with ill-determined numerical rank. The problems are so simple that they can easily be understood by nonexperts, yet they have the characteristic features of inverse problems. The last three test problems are included in the REGULARIZATION TOOLS [186] package for MATLAB as functions deriv2, shaw, and wing, respectively.

1.4.1. Signal Processing: Sinusoids in Noise

Hankel and Toeplitz matrices arise frequently in signal processing [347], where they play a central role in, e.g., linear prediction and adaptive filters [203]. In many of these applications, the matrices are numerically rank deficient. Given a sampled signal of length N,

$$s_1, s_2, \ldots, s_N,$$

it is common to form the associated $m \times n$ Hankel matrix

$$A = \begin{pmatrix} s_1 & s_2 & \cdots & s_n \\ s_2 & s_3 & \cdots & s_{n+1} \\ \vdots & \vdots & & \vdots \\ s_m & s_{m+1} & \cdots & s_{m+n-1} \end{pmatrix} \quad \text{with} \quad m+n-1 = N. \quad (1.18)$$

The numerical rank of this matrix depends on the properties of the signal.

Consider first a simple example where the signal is a sampled sinusoid,

$$s_i = \sin(i\,\omega), \quad i = 1, \ldots, N,$$

where ω is a constant. As long as ω is not a multiple of π (in which case $A = 0$), the first two columns of A are linearly independent. Moreover, from standard trigonometry we easily obtain the relation

$$s_i + s_{i+2} = 2\cos(\omega)\,s_{i+1}, \quad i = 1, \ldots, N-2,$$

showing that any two neighboring columns of A are linearly dependent. Hence, we conclude that in this case A has rank 2.

In this test problem the signal is a weighted sum of two sinusoids with different frequencies and phases, plus additive white noise:

$$s_i = a_1 \sin(i\,\omega_1 + \phi_1) + a_2 \sin(i\,\omega_2 + \phi_2) + \eta_i, \quad i = 1, \ldots, N,$$

where η_i are normally distributed numbers with zero mean and standard deviation σ_{noise}. The matrix A is therefore a sum of two rank-2 Hankel matrices plus a random Hankel matrix. If $\omega_1 \neq \omega_2$ and neither are multiples of π, if $\sigma_{\text{noise}} \ll |a_1| + |a_2|$, and if $m \geq n > 4$, then A has numerical rank 4, i.e., four large singular values and $n - 4$ small singular values of order σ_{noise}.

1.4.2. Computation of the Second Derivative (deriv2)

This is a classical example of an ill-posed problem, and it is used in numerous papers on regularization algorithms. The kernel K of the integral equation (1.1) is Green's function for the second derivative:

$$K(s,t) = \begin{cases} s(t-1), & s < t, \\ t(s-1), & s \geq t, \end{cases}$$

1.4. FOUR TEST PROBLEMS

and both integration intervals are $[0, 1]$. The symmetric kernel K is not differentiable across the line $s = t$. The singular values and functions for $i = 1, 2, \ldots$ are given by (see, e.g., [227, Example 15.13])

$$\begin{aligned} \mu_i &= (i\pi)^{-2}, \\ u_i(s) &= \pm\sqrt{2}\sin(i\pi s), \\ v_i(t) &= \mp\sqrt{2}\sin(i\pi t). \end{aligned}$$

Since the singular values are proportional to i^{-2}, the problem is moderately ill posed. The right-hand side g and the corresponding solution f are given by

$$g(s) = (s^3 - s)/6, \qquad f(t) = t.$$

This particular choice of f and g is from [80, p. 315].

The symmetric $n \times n$ matrix A, the solution vector x^{exact}, and the corresponding right-hand side b^{exact} are computed from K, f, and g by means of the Galerkin method (cf. Eq. (1.12)) with orthonormal ⊓ basis functions defined on the uniform mesh $[0, \frac{1}{n}, \frac{2}{n}, \ldots, 1]$. This leads to the following relations for $i = 1, \ldots, n$:

$$\begin{aligned} a_{ii} &= h^2\left(h\left(i^2 - i + \tfrac{1}{4}\right) - \left(i - \tfrac{2}{3}\right)\right), \\ a_{ij} &= h^2\left(j - \tfrac{1}{2}\right)\left(h\left(i - \tfrac{1}{2}\right) - 1\right), \qquad j < i, \\ b_i^{\text{exact}} &= \tfrac{1}{6} h^{3/2}\left(i - \tfrac{1}{2}\right)\left(\tfrac{1}{2} h^2\left(i^2 - (i-1)^2\right) - 1\right), \\ x_i^{\text{exact}} &= h^{3/2}\left(i - \tfrac{1}{2}\right), \end{aligned}$$

where $h = \frac{1}{n}$.

1.4.3. One-Dimensional Image Restoration Model (shaw)

In image restoration problems, one seeks to compensate for the deblurring effects of the optical system in the recorded image. Unfortunately, these two-dimensional problems get very large and are therefore not always suitable as test problems.

This test problem from [309, Eq. (5.1)] uses a first-kind Fredholm integral equation to model a one-dimensional image restoration situation. Both integration intervals are $[-\pi/2, \pi/2]$, and the kernel K and solution f are given by

$$\begin{aligned} K(s,t) &= (\cos(s) + \cos(t))^2 \left(\frac{\sin(u)}{u}\right)^2, \\ u &= \pi\left(\sin(s) + \sin(t)\right), \\ f(t) &= a_1 \exp\left(-c_1(t - t_1)^2\right) + a_2 \exp\left(-c_2(t - t_2)^2\right). \end{aligned}$$

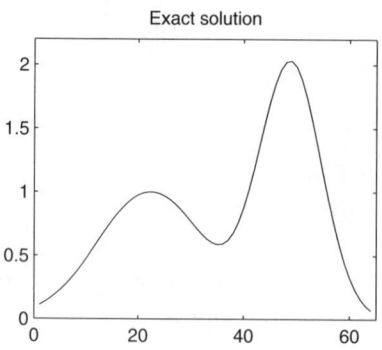 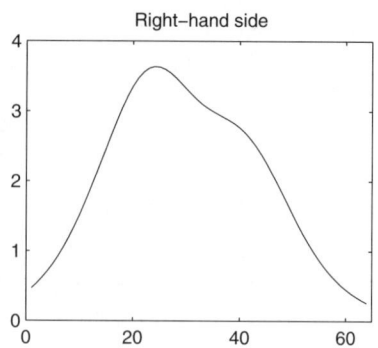

FIG. 1.2. *The exact solution x^{exact} and the corresponding right-hand side $b^{\text{exact}} = A\,x^{\text{exact}}$ in the one-dimensional image restoration test problem* shaw.

This symmetric kernel is very smooth, and its singular values are roughly proportional to e^{-2i}; i.e., the problem is severely ill posed.

The kernel K is the point spread function for an infinitely long slit of width one wavelength, f is the source's light intensity as a function of the angle t of incidence, and g is the observed light intensity in the image produced on the other side of the slit. The parameters a_1, a_2, c_1, c_2, t_1, t_2 are constants that determine the intensity, width, and position of the two light sources; here we use $a_1 = 2$, $a_2 = 1$, $c_1 = 6$, $c_2 = 2$, $t_1 = .8$, $t_2 = -.5$, giving an f with two distinct "humps." As shown in Figure 1.2, these two "humps" are smeared out in g by the integration with K.

The kernel K and the solution f are discretized by simple collocation with n collocation points $t_i = (i - 0.5)\,\pi/n$, $i = 1, \ldots, n$, to produce the symmetric $n \times n$ matrix A and solution vector x^{exact}. Then the corresponding discrete right-hand side is produced as $b^{\text{exact}} = A\,x^{\text{exact}}$.

1.4.4. A Problem with a Discontinuous Solution (wing)

This test problem appears as problem VI.10 in [373], and it is different from the above two problems in that the solution is discontinuous. Although the solutions to inverse problems are often continuous, discontinuous solutions also arise in practice, e.g., in image processing (across edges in the image) and in seismic deconvolution (across layer boundaries), so it makes sense to include such a test problem here.

Both integration intervals are $[0, 1]$, and the kernel, the solution, and the right-hand side are given by

$$K(s,t) = t\,\exp(-s\,t^2)\,,$$

1.4. Four Test Problems

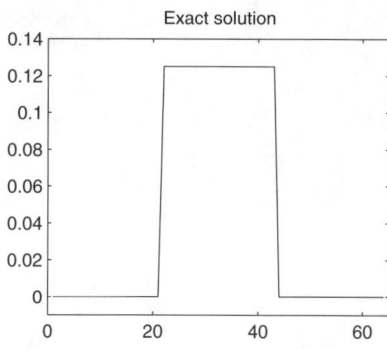 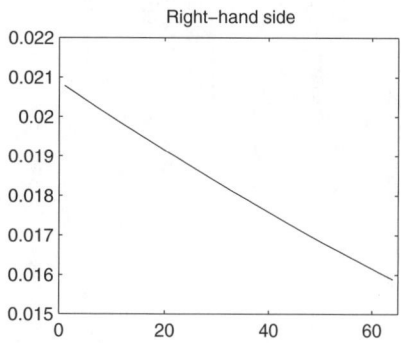

FIG. 1.3. *The exact solution x^{exact} and the corresponding right-hand side b^{exact} in the* wing *test problem.*

$$f(t) = \begin{cases} 1, & \text{for } t_1 < t < t_2, \\ 0, & \text{elsewhere}, \end{cases}$$

$$g(s) = \frac{\exp(-s\,t_1^2) - \exp(-s\,t_2^2)}{2\,s}.$$

Here, t_1 and t_2 are constants satisfying $0 < t_1 < t_2 < 1$, and we use the values $t_1 = 1/3$ and $t_2 = 2/3$. The functions f and g are shown in Fig. 1.3. The singular values of K decay very fast (approximately as $e^{-4.5i}$), and the problem is severely ill posed.

As in the first test problem, the $n \times n$ matrix A and the corresponding vectors x^{exact} and b^{exact} are obtained by discretization by means of the Galerkin method (cf. (1.12)) with orthonormal ⊓ basis functions defined on the uniform mesh $[0, \frac{1}{n}, \frac{2}{n}, \ldots, 1]$. The numerical integrations in (1.12) are done by the midpoint rule.

2

Decompositions and Other Tools

Before we turn to the numerical regularization methods in the following chapters, it is convenient to summarize the "canonical decompositions" which are important theoretical as well as computational tools in connection with rank-deficient and discrete ill-posed problems. Along with this discussion, it is natural to briefly summarize methods for transforming regularization problems in general form into standard form. We also describe how the singular value expansion (SVE) of a kernel can be computed by means of the singular value decomposition (SVD).

2.1. The SVD and Its Generalizations

The superior numerical "tools" for analysis of rank-deficient and discrete ill-posed problems are the (ordinary) SVD of A and its generalization to two matrices, the generalized SVD (GSVD) of the matrix pair (A, L); see [154, §§2.5.3–4 and 8.7.3]. The SVD reveals all the difficulties associated with the ill-conditioning of the matrix A, while the GSVD of (A, L) yields important insight into regularization problems involving both the matrix A and the regularization matrix L, such as in (1.14).

The use of the SVD and the GSVD in the analysis of discrete ill-posed problems goes back to Hanson [201] and Varah [352], [353]. The SVD has many similarities with the SVE discussed in §1.2.1, and the early history of both is described in [324].

2.1.1. The (Ordinary) SVD

Let $A \in \mathbb{R}^{m \times n}$ be a rectangular or square matrix, and assume for ease of presentation that $m \geq n$. Then the SVD of A is a decomposition of the form

$$A = U \Sigma V^T = \sum_{i=1}^{n} u_i \sigma_i v_i^T, \qquad (2.1)$$

where $U = (u_1, \ldots, u_n) \in \mathbb{R}^{m \times n}$ and $V = (v_1, \ldots, v_n) \in \mathbb{R}^{n \times n}$ are matrices with orthonormal columns, $U^T U = V^T V = I_n$, and where the diagonal matrix $\Sigma =$

$\mathrm{diag}(\sigma_1,\ldots,\sigma_n)$ has nonnegative diagonal elements appearing in nonincreasing order such that
$$\sigma_1 \geq \sigma_2 \geq \cdots \geq \sigma_n \geq 0 \ .$$
The numbers σ_i are called the *singular values* of A while the vectors u_i and v_i are the *left and right singular vectors* of A, respectively. The decomposition in (2.1) is called the "thin SVD" in [154, §2.5.4] because U is rectangular when $m > n$. The SVD is defined for any m and n; if $m < n$, simply apply (2.1) to A^T and interchange U and V.

Geometrically speaking, the SVD of A provides two sets of orthonormal basis vectors, namely, the columns of U and V, such that the matrix becomes diagonal when transformed into these two bases.

The singular values are always well conditioned with respect to perturbations: if A is perturbed by a matrix E then the norm $\|E\|_2$ is an upper bound for the absolute perturbation of each singular value. The classical perturbation bounds for singular values and vectors can be found, e.g., in [36, Theorem 1.2.7], and a variant of the perturbation bound for singular values is given in [317].

The singular values σ_i can also be defined as the stationary values of $\|Ax\|_2/\|x\|_2$; cf. [154, Theorem 8.6.1]. An alternative SVD, in which $U^T S U = V^T T V = I_n$ and σ_i are the stationary values of $\|Ax\|_S/\|x\|_T$, with $\|x\|_M = \|x^T M x\|_2$, is briefly discussed by Van Loan in [350].

From the relations $A^T A = V \Sigma^2 V^T$ and $A A^T = U \Sigma^2 U^T$ we see that the SVD of A is strongly linked to the eigenvalue decompositions of the symmetric semidefinite matrices $A^T A$ and $A A^T$. This shows that the SVD is unique for a given matrix A, up to a sign change in the pair (u_i, v_i)—except for singular vectors associated with multiple singular values, where only the spaces spanned by the vectors are unique. In connection with discrete ill-posed problems, two characteristic features of the SVD are very often found.

- The singular values σ_i decay gradually to zero with no particular gap in the spectrum. An increase of the dimensions of A will increase the number of small singular values.

- The left and right singular vectors u_i and v_i tend to have more sign changes in their elements as the index i increases, i.e., as σ_i decreases.

Both features are consequences of the fact that the SVD of A is closely related to the SVE of the underlying kernel K. In fact, the singular values σ_i of A are in many cases approximations to the singular values μ_i of K. This property has been studied by Wing and his coworkers in a series of papers [1], [2], and [372], and by this author in [178]. See also [293]. Moreover, in some situations the singular vectors u_i and v_i yield information about the singular functions of K; see [114] and [178]. The results from [178] are summarized in §2.4. If

2.1. THE SVD AND ITS GENERALIZATIONS

both $A^T A$ and $A A^T$ are totally positive then the ith left and right singular vectors have exactly $i - 1$ sign changes [189].

To see how the SVD gives insight into the ill-conditioning of A, consider the following relations which follow directly from (2.1):

$$\left.\begin{array}{rclrcl} A\,v_i & = & \sigma_i\,u_i\,, & \|A\,v_i\|_2 & = & \sigma_i \\ A^T u_i & = & \sigma_i\,v_i\,, & \|A^T u_i\|_2 & = & \sigma_i \end{array}\right\} \quad i = 1,\ldots,n\,. \tag{2.2}$$

We see that a small singular value σ_i, compared to $\sigma_1 = \|A\|_2$, means that there exists a certain linear combination of the columns of A, characterized by the elements of the right singular vector v_i, such that $\|A\,v_i\|_2 = \sigma_i$ is small. The same holds for u_i and the rows of A. In other words, a situation with one or more small σ_i implies that A is nearly rank deficient, and the vectors u_i and v_i associated with the small σ_i are the numerical null vectors of A^T and A, respectively. From this property and the above-mentioned characteristic features of the SVD of A, we conclude that the matrix in a discrete ill-posed problem is always highly ill conditioned, and its numerical null space is spanned by vectors with many sign changes.

The SVD also gives important insight into another aspect of discrete ill-posed problems, namely, the smoothing effect typically associated with a square integrable kernel K as discussed in §1.2. Notice that as σ_i decreases, the singular vectors u_i and v_i become more and more oscillatory. Consider now the mapping $A\,x$ of an arbitrary vector x. Using the SVD, we get

$$x = \sum_{i=1}^{n} (v_i^T x)\, v_i \quad \text{and} \quad A\,x = \sum_{i=1}^{n} \sigma_i\,(v_i^T x)\, u_i\,.$$

These relations clearly show that due to multiplication with the σ_i, the high-frequency components of x are more damped in $A\,x$ than the low-frequency components. Moreover, the inverse problem, namely, that of computing x from $A\,x = b$ or $\min \|A\,x - b\|_2$, must have the opposite effect: it amplifies the high-frequency oscillations in the right-hand side b.

Yet another use of the SVD is in connection with least squares problems, possibly rank deficient. If A is invertible, then its inverse is given by $A^{-1} = \sum_{i=1}^{n} v_i\,\sigma_i^{-1} u_i^T$, and therefore the solution to $A\,x = b$ is $x = \sum_{i=1}^{n} \sigma_i^{-1}(u_i^T b)\,v_i$. Otherwise, the *pseudoinverse* [36, §1.2.5], [154, §5.5.3] (or Moore–Penrose generalized inverse) $A^\dagger$ is given by

$$A^\dagger \equiv \sum_{i=1}^{\mathrm{rank}(A)} v_i\,\sigma_i^{-1} u_i^T, \tag{2.3}$$

and the *least squares solution* x_{LS} to the least squares problem $\min \|A\,x - b\|_2$ (with minimum 2-norm, if $\mathrm{rank}(A) < n$) is given by

$$x_{\mathrm{LS}} = A^\dagger b = \sum_{i=1}^{\mathrm{rank}(A)} \frac{u_i^T b}{\sigma_i}\,v_i\,. \tag{2.4}$$

Note that $x = A^{-1}b$ is merely a special case of (2.3)–(2.4). It is the division by the small singular values in the expressions for the solutions x and x_{LS} that amplifies the high-frequency components in b.

The sensitivity of the solutions x and x_{LS} to perturbations of A and b can be measured by the 2-norm *condition number* of A, defined by

$$\mathrm{cond}(A) \equiv \|A\|_2 \, \|A^\dagger\|_2 = \sigma_1/\sigma_{\mathrm{rank}(A)} \; . \tag{2.5}$$

This definition includes the invertible case where $\mathrm{cond}(A) = \|A\|_2 \, \|A^{-1}\|_2$. See, e.g., [36, §1.4], [154, §§2.7 and 5.3.7], [230, Chapter 9], [326, Chapter III] for details on the sensitivity of x and x_{LS}.

2.1.2. The GSVD

The GSVD of the matrix pair (A, L) is a generalization of the SVD of A in the sense that the generalized singular values of (A, L) are essentially the square roots of the generalized eigenvalues of the matrix pair $(A^T A, L^T L)$. In order to keep our exposition simple, we assume that the dimensions of $A \in \mathbb{R}^{m \times n}$ and $L \in \mathbb{R}^{p \times n}$ satisfy $m \geq n \geq p$, which is always the case in connection with discrete ill-posed problems. We also assume that $\mathcal{N}(A) \cap \mathcal{N}(L) = \{0\}$ and that L has full row rank. Then the GSVD is a decomposition of A and L in the form

$$A = U \begin{pmatrix} \Sigma & 0 \\ 0 & I_{n-p} \end{pmatrix} X^{-1} , \qquad L = V(M, 0) X^{-1} . \tag{2.6}$$

The columns of $U \in \mathbb{R}^{m \times n}$ and $V \in \mathbb{R}^{p \times p}$ are orthonormal, $U^T U = I_n$, and $V^T V = I_p$; $X \in \mathbb{R}^{n \times n}$ is nonsingular with columns that are $A^T A$-orthogonal (see (2.8) below); and Σ and M are $p \times p$ diagonal matrices: $\Sigma = \mathrm{diag}(\sigma_1, \ldots, \sigma_p)$, $M = \mathrm{diag}(\mu_1, \ldots, \mu_p)$. Moreover, the diagonal elements of Σ and M are nonnegative and ordered such that

$$0 \leq \sigma_1 \leq \cdots \leq \sigma_p \leq 1 , \qquad 1 \geq \mu_1 \geq \cdots \geq \mu_p > 0 ,$$

and they are normalized such that

$$\sigma_i^2 + \mu_i^2 = 1 , \qquad i = 1, \ldots, p .$$

Then the *generalized singular values* γ_i of (A, L) are defined as the ratios

$$\gamma_i = \sigma_i/\mu_i , \qquad i = 1, \ldots, p , \tag{2.7}$$

and they obviously appear in nondecreasing order. For historical reasons, this ordering is the opposite of the ordering of the ordinary singular values of A. Since

$$X^T A^T A X = \begin{pmatrix} \Sigma^2 & 0 \\ 0 & I_{n-p} \end{pmatrix} \quad \text{and} \quad X^T L^T L X = \begin{pmatrix} M^2 & 0 \\ 0 & 0 \end{pmatrix} , \tag{2.8}$$

2.1. The SVD and Its Generalizations

we see that (γ_i^2, x_i) are the generalized eigensolutions of the pair $(A^T A, L^T L)$ associated with the p finite generalized eigenvalues.

Similar to the SVD, the pairs (σ_i, μ_i) are well conditioned with respect to perturbations of A and B. This follows from the general perturbation bounds for the GSVD in [234].

If L has more rows than columns and if $L = QR$ is the QR factorization of L, then we can replace L with R in (2.6) such that $A = U \Sigma X^{-1}$, $R = V M X^{-1}$, and $L = (QV) M X^{-1}$.

Similarly with the SVD, the GSVD of (A, L) provides three new sets of linearly independent basis vectors (the columns of U, V, and X) such that the two matrices A and L become diagonal when transformed into these new bases. The two bases associated with the columns of U and V are orthonormal.

For $p < n$ the matrix $L \in \mathbb{R}^{p \times n}$ always has a nontrivial null space $\mathcal{N}(L)$. For example, if L is an approximation to the second derivative operator given by (1.16), then $\mathcal{N}(L)$ is spanned by the vectors $(1, 1, \ldots, 1)^T$ and $(1, 2, \ldots, n)^T$. In the GSVD, the last $n - p$ columns x_i of X satisfy

$$L x_i = 0, \quad i = p+1, \ldots, n, \tag{2.9}$$

and they are therefore basis vectors for the null space $\mathcal{N}(L)$.

There is a slight notational difficulty here because the matrices U, Σ, and V in the GSVD of (A, L) are *different* from the matrices with the same symbols in the SVD of A. However, in this presentation it will always be clear from the context which decomposition is being used. When L is the identity matrix I_n, then the U and V of the GSVD are identical to the U and V of the SVD, and the generalized singular values of (A, L) are identical to the singular values of A, except for the reverse ordering of the singular values and vectors.

In the special case where L is square and nonsingular, we have $L = V M X^{-1}$ and thus

$$A L^{-1} = U (\Sigma M^{-1}) V^T, \tag{2.10}$$

from which we see that in this case, the SVD of the "matrix quotient" $A L^{-1}$ simply consists of part of the GSVD of the pair (A, L)—hence the alternative name "quotient SVD" (QSVD) (see [91]).

There is no simple connection between the generalized singular values and vectors and the ordinary singular values and vectors. The ratios between the ordinary singular values of A and L, and the quantities σ_i and μ_i in the GSVD of (A, L), are bounded as follows.

Theorem 2.1.1. [179, Theorem 2.4]. *Let $\psi_i(A)$ and $\psi_i(L)$ denote the ordinary singular values of A and L, and let σ_i and μ_i denote the diagonal elements in the GSVD of (A, L). Then, for all $\sigma_i \neq 0$ and all μ_i,*

$$\left\| \begin{pmatrix} A \\ L \end{pmatrix}^\dagger \right\|_2^{-1} \leq \frac{\psi_{n-i+1}(A)}{\sigma_i} \leq \left\| \begin{pmatrix} A \\ L \end{pmatrix} \right\|_2 \tag{2.11}$$

and

$$\left\|\begin{pmatrix}A\\L\end{pmatrix}^\dagger\right\|_2^{-1} \leq \frac{\psi_i(L)}{\mu_i} \leq \left\|\begin{pmatrix}A\\L\end{pmatrix}\right\|_2. \qquad (2.12)$$

Regarding the condition of X, the following general result holds.

Theorem 2.1.2. [179, Theorem 2.3]. *Let $P_{\mathcal{N}(L)}$ denote the projection matrix onto the null space of L, let $\inf(A P_{\mathcal{N}(L)})$ denote the smallest nonzero singular value of $A P_{\mathcal{N}(L)}$, and define ν_p by*

$$\nu_p = \begin{cases} \|L^{-1}\|_2, & p = n, \\ \max\{\|L^\dagger\|_2, \inf(A P_{\mathcal{N}(L)})^{-1}\}, & p < n. \end{cases} \qquad (2.13)$$

Then

$$\|X^{-1}\|_2 = \left\|\begin{pmatrix}A\\L\end{pmatrix}\right\|_2 \leq \|A\|_2 + \|L\|_2, \qquad \|X\|_2 = \left\|\begin{pmatrix}A\\L\end{pmatrix}^\dagger\right\|_2 \leq \nu_p, \qquad (2.14)$$

and therefore $\operatorname{cond}(X) = \|X\|_2 \|X^{-1}\|_2 \leq (\|A\|_2 + \|L\|_2)\, \nu_p$.

The latter theorem implies that if ν_p is not large and if $\|A\|_2 \approx \|L\|_2$, then X is guaranteed to be well conditioned and there is no unit vector z such that $\|A z\|_2$ and $\|L z\|_2$ are small simultaneously.

For discrete ill-posed problems we can actually say more about the SVD–GSVD connection. First, we can show that if the singular vectors of A have the oscillation property described in §2.1.1, then this property carries over to the generalized singular vectors x_i; see Theorem 3.2 and the discussion following it in [179]. This result, in turn, is used in [179, §3] to show that if L is a reasonably well conditioned matrix scaled such that $\|L\|_2 \approx \|A\|_2$, then the quantity $\inf(A P_{\mathcal{N}(L)})^{-1}$ in Theorem 2.1.2 is smaller than $\|L^\dagger\|_2$, such that ν_p is equal to $\|L^\dagger\|_2$. We conclude that for discrete ill-posed problems we *usually* have

$$\|X\|_2 \leq \|L^\dagger\|_2 \quad \text{and} \quad \operatorname{cond}(X) \leq (\|A\|_2 + \|L\|_2)\|L^\dagger\|_2. \qquad (2.15)$$

That is, the matrix X is *usually* approximately as well conditioned as L.

Due to Theorem 2.1.1 it then follows that the diagonal matrix Σ must display the ill-conditioning of A. Since $\gamma_i = \sigma_i (1 - \sigma_i^2)^{-1/2} \approx \sigma_i$ for small σ_i, the generalized singular values must therefore decay gradually to zero as the ordinary singular values do. As a consequence, for a discrete ill-posed problem, the following three characteristic features of the GSVD, similar to those of the SVD, are usually found.

- The generalized singular values γ_i decay to zero with no gap in the spectrum. An increase of the dimensions of A increases the number of small generalized singular values.

2.1. THE SVD AND ITS GENERALIZATIONS

- The singular vectors u_i, v_i, and x_i have more sign changes in their elements as the corresponding γ_i decreases.

- If L approximates a derivative operator, then the last $n - p$ columns x_i of X have very few sign changes, since they are the null vectors of L.

We emphasize again that the second of these features is very difficult—perhaps impossible—to prove in general, but that it is still observed in many discrete ill-posed problems arising from applications.

The GSVD is a much more recent tool in numerical linear algebra than the SVD; it was originally introduced by Van Loan [350] in 1976, and later developed by Paige and Saunders in [272]. The present notation is based on [36, Theorem 4.2.2]. A similar definition of a generalized SVE of a compact operator/differential operator pair was introduced by Hanke [167] in 1992. The relationships between the GSVD and the CS decomposition, along with the latter's history, is given in [275].

2.1.3. What the SVD and GSVD Look Like

We now illustrate the features of the SVD and the GSVD of matrices involved in discrete ill-posed problems by a numerical example. The model problem used here is the one-dimensional image restoration problem **shaw** from §1.4.3, discretized with $m = n = 32$. The matrix L is an approximation to the second derivative operator (cf. (1.16)), and the condition numbers of L and X (of the GSVD) are 183 and 53, respectively, illustrating that the matrix X is typically as well conditioned as L.

The discretization method used for this model problem, namely, simple collocation, is a special case of the Galerkin method with orthonormal basis functions. Hence, the singular values σ_i of A are approximations to the largest singular values of the underlying operator, and the singular vectors u_i and v_i are approximations to "sampled" versions of the corresponding singular functions of the operator; cf. §2.4. This means that inspection of the SVD yields insight into the SVE of the underlying operator.

As mentioned in §1.4.3, the singular values decay fast to zero. This agrees with Fig. 2.1, whose left part shows the computed singular values σ_i of A. The singular values decay until they settle at a level which is approximately equal to σ_1 times the machine precision.

The left singular vectors u_i corresponding to the largest eight singular values are shown in the top part of Fig. 2.2. Due to the symmetry of A, the right singular vectors v_i look the same. We clearly see the oscillation property mentioned in §1.2.2; in fact, the ith singular vector has exactly $i - 1$ sign changes.

Next, we illustrate that the general properties of the singular triplets carry over to the generalized singular values and vectors of (A, L). The generalized

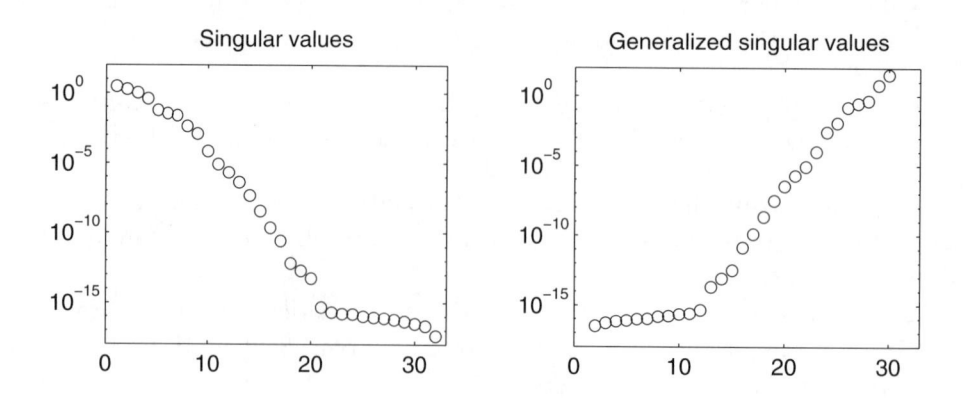

FIG. 2.1. *The 32 singular values σ_i of A and the 30 generalized singular values γ_i of (A, L), for the* shaw *test problem with $L = L_2$.*

singular values γ_i (cf. (2.7)) are shown in the right part of Fig. 2.1; recall the reverse ordering and that there are "only" p of them. The decay rates of the σ_i and the γ_i are approximately the same.

Eight left generalized singular vectors u_i, as well as eight GSVD vectors x_i, are shown in the bottom part of Fig. 2.2 for $i = 32, 31, \ldots, 25$. The two vectors u_{31} and u_{32} correspond to the submatrix $I_{n-p} = I_2$ in (2.6). The same is true for x_{31} and x_{32}, which span the null space of L. The remaining six vectors shown in Fig. 2.2 correspond to the six largest generalized singular values. As expected from the discussion in §2.1.2, the generalized singular vectors are very similar to the ordinary singular vectors. The oscillation properties also carry over to the generalized singular vectors v_i, which are not shown here.

2.1.4. Other SVDs

After the introduction of the GSVD (or QSVD), it took some years before other generalizations of the SVD were introduced. The next to appear was the *product SVD* (PSVD) of (A, L), which was introduced in [118]:

$$A = \hat{U}\,\hat{\Sigma}\,\hat{X}^T, \qquad L = \hat{V}\,\hat{M}\,\hat{X}^{-1}. \qquad (2.16)$$

The name of this decomposition comes from the fact that if L is square and nonsingular, then $A L^T = \hat{U}\,(\hat{\Sigma}\,\hat{M})\,\hat{V}^T$; i.e., the SVD of $A L^T$ is made up of matrices from the PSVD. The PSVD is useful, e.g., in connection with the orthogonal Procrustes problem: given A and B, the orthogonal matrix Q that minimizes $\|A - B Q\|_F$ is given by $Q = \hat{U}\,\hat{V}^T$ where $B^T A = \hat{U}\,\hat{\Sigma}\,\hat{V}^T$ (see also [205]). The PSVD is also useful in connection with rank-deficient problems in signal processing; see [87] and [118] for details and applications.

2.1. THE SVD AND ITS GENERALIZATIONS

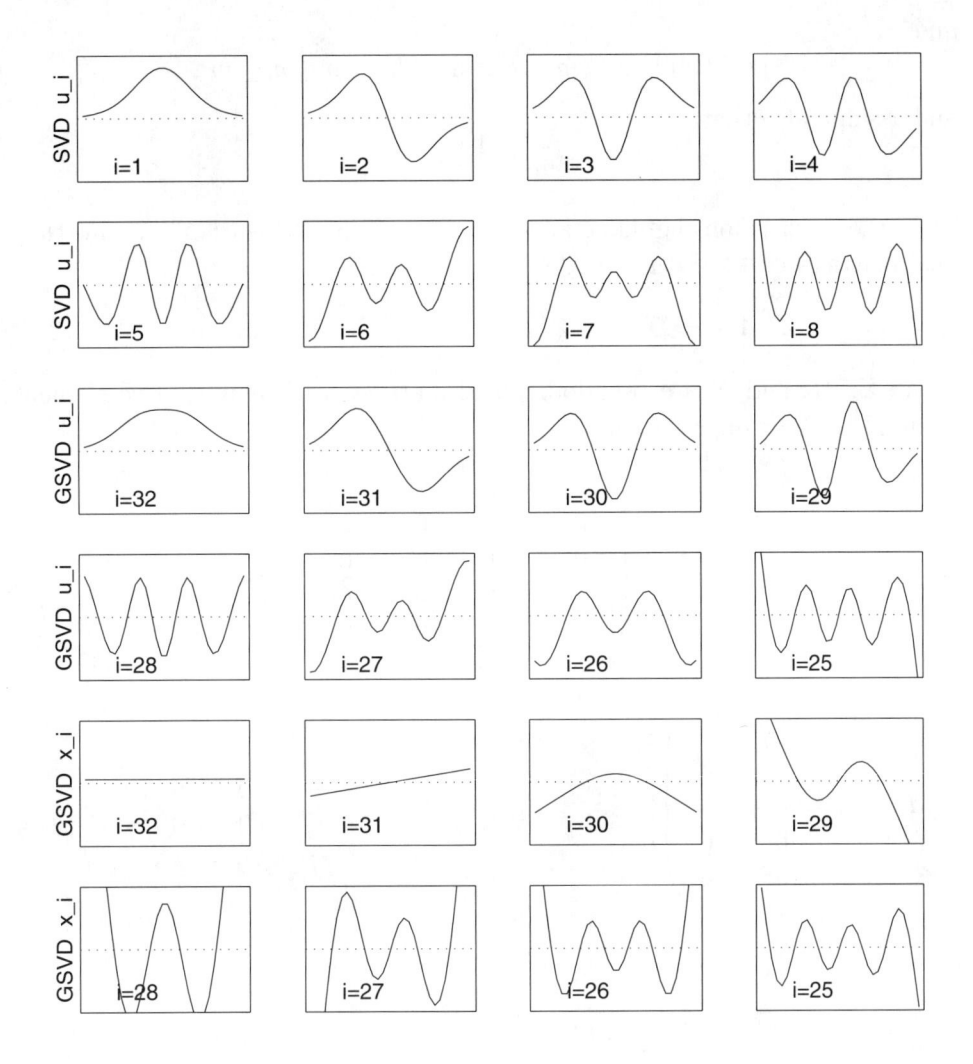

FIG. 2.2. *Eight left singular vectors u_i of A and eight left generalized singular vectors u_i of (A, L), as well as eight GSVD vectors x_i, for the 32×32 shaw test problem with $L = L_2$. The horizontal dotted lines represent zero.*

The next member of the family of generalized SVDs to be introduced was the *restricted SVD* (RSVD) [374] (see also [113] for a less general version of this decomposition). The RSVD provides a simultaneous diagonalization of three matrices. A complete discussion of the RSVD can be found in [90] and [374]; here, we summarize those aspects that are important in connection with discrete ill-posed problems as described in §5.2. Let the triplet of matrices (A, C, L) satisfy

$$A \in \mathbb{R}^{m \times n}, \quad C \in \mathbb{R}^{m \times q}, \quad L \in \mathbb{R}^{p \times n}$$

and
$$\text{rank}(C) = q \leq m \,, \qquad \text{rank}(L) = p \leq n \leq m \,,$$
and assume also that
$$\text{rank}\begin{pmatrix} A \\ L \end{pmatrix} = n \,.$$
Then there exist nonsingular matrices $X \in \mathbb{R}^{n \times n}$ and $Z \in \mathbb{R}^{m \times m}$, and orthogonal matrices $U \in \mathbb{R}^{q \times q}$ and $V \in \mathbb{R}^{p \times p}$ such that
$$Z^T A X = \Sigma \,, \qquad Z^T C U = M \,, \qquad V^T L X = N \,, \qquad (2.17)$$
where Σ, M, and N are pseudodiagonal matrices with nonnegative elements having the following structure:

$$\Sigma = \begin{pmatrix} \Sigma_A & 0 & 0 & 0 \\ 0 & I_j & 0 & 0 \\ 0 & 0 & I_k & 0 \\ 0 & 0 & 0 & I_\ell \\ 0 & 0 & 0 & 0 \\ t & j & k & \ell \end{pmatrix} \begin{matrix} s \\ j \\ k \\ \ell \\ u \end{matrix} \,, \qquad (2.18)$$

$$M = \begin{pmatrix} I_s & 0 \\ 0 & 0 \\ 0 & I_k \\ 0 & 0 \\ 0 & 0 \\ s & k \end{pmatrix} \begin{matrix} s \\ j \\ k \\ \ell \\ u \end{matrix} \,, \qquad N = \begin{pmatrix} I_t & 0 & 0 & 0 \\ 0 & I_j & 0 & 0 \\ t & j & k & \ell \end{pmatrix} \begin{matrix} t \\ j \end{matrix} \,, \qquad (2.19)$$

and where
$$\Sigma_A = \text{diag}(\sigma_1, \ldots, \sigma_{\min(s,t)}) \in \mathbb{R}^{s \times t} \,, \qquad \sigma_1 \geq \cdots \geq \sigma_{\min(s,t)} \geq 0 \,. \qquad (2.20)$$
The dimensions of the submatrices are $j = r_{ac} + p - r_{ac\ell}$, $k = n + q - r_{ac\ell}$, $\ell = r_{ac\ell} - p - q$, $s = r_{ac\ell} - n$, $t = r_{ac\ell} - r_{ac}$, and $u = m - r_{ac}$, where
$$r_{ac} = \text{rank}(A, C) \qquad \text{and} \qquad r_{ac\ell} = \text{rank}\begin{pmatrix} A & C \\ L & 0 \end{pmatrix} \,.$$

Again, we emphasize that the quantities U, V, etc. with the same names as those in the SVD and the GSVD are different from these quantities. The name RSVD originates in applications where the goal is to find the matrix X of minimal norm that reduces the rank of $A + CXL$, i.e., by *restricting* the modifications to the column space of C and the row space of L.

An important feature of the RSVD is that if A is ill conditioned and both C and L are well conditioned, then the ill-conditioning of A is reflected in the matrix Σ, while both X and Z are well conditioned. See [378, §2] for details. The use of the RSVD is discussed in §5.2.

After the introduction of the RSVD it was realized that the PSVD, QSVD, and RSVD are members of a tree of SVD generalizations to an arbitrary number of matrices. These aspects are treated in [91]. Apart from the QSVD and the RSVD, we are not aware of any applications of these generalized SVDs in connection with discrete ill-posed problems. A discussion of some applications of these generalized SVDs in signal processing can be found in [86].

2.1.5. Algorithms and Software

The classical algorithm for computing the SVD of a dense matrix is due to Golub, Kahan, and Reinsch [148], [151]. The algorithm consists of two main stages. In the first stage, A is transformed into upper bidiagonal form B by means of a finite sequence of alternating left and right Householder transformations. In the second, iterative, stage, the shifted QR algorithm is applied implicitly to the matrix $B^T B$, and consequently B converges to Σ. The left and right orthogonal transformations, if accumulated, produce the matrices U and V.

Surveys of algorithms for computing the SVD of dense matrices can be found in many references; see, e.g., [23], [82], [154], [165], and the references therein. References to algorithms for computing the GSVD can be found in [4]. An algorithm for the PSVD is given in [118], and an algorithm for the RSVD can be found in [376] (see also [41] for the case where C and L are nonsingular). Finally, an algorithm for computing a PP-SVD (one of the further generalizations of the SVD) is given in [375]. We are not aware of any other algorithms for SVD generalizations.

Algorithms for computing singular triplets of large sparse matrices are usually based on (block) Lanczos bidiagonalization. See [23], [24], [71, Chapter 5], and [150] for details. An extension of the Lanczos bidiagonalization algorithm for computing the largest generalized singular values and associated vectors when L is banded is described in [193]. A general algorithm is described in [377].

Subroutines for computing the SVD of a dense rectangular matrix are available in most mathematical software libraries today, and there are also a few subroutines available for large sparse matrices; see Table 2.1 for a list of software. Regarding the GSVD, the situation is different: only the LAPACK Library includes a subroutine _GGSVD, based on a Jacobi-type method [16], for computing the GSVD. We are not aware of software for other generalized SVDs.

2.2. Rank-Revealing Decompositions

The main reason for the success of the SVD and the GSVD as analysis tools for rank-deficient and discrete ill-posed problems is that they provide new coordinate systems, determined by the columns of U and V (and X), such that

TABLE 2.1. *Software for computing the SVD.*

Software package	Subroutines
Dense matrices	
ACM TOMS [48]	HYBSVD
EISPACK [136]	SVD
IMSL [216]	LSVRR
LAPACK [4]	_GESVD
LINPACK [93]	_SVDC
NAG [257]	F02WEF
Numerical Recipes [287]	SVDCMP
Sparse matrices	
LANCZOS [71]	LSVAL, LSVEC
SVDPACK [23]	BLS, LAS, SIS, TMS

the matrices become diagonal when transformed into these coordinate systems. The price we pay for this feature is a fairly high computational cost; for example, the Golub–Kahan–Reinsch SVD algorithm for dense matrices needs $14mn^2 + 8n^3$ flops to compute the full SVD. Moreover, both decompositions are difficult to update[5] efficiently when rows are appended to A or deleted from A, which is common in many signal processing applications.

In some applications the cure to the high computational cost is to use less expensive rank-revealing decompositions. These decompositions do not provide as much information as the SVD or GSVD, but they often provide enough information to allow one to reliably solve rank-deficient problems. Some rank-revealing decompositions have the additional feature that they can be updated efficiently.

The most crucial problem in connection with rank-deficient problems is to reliably compute the numerical rank of the matrix A. These matters are discussed in detail in §3.1, and we shall merely summarize the definition here. Given a small threshold ϵ, we say that A has numerical ϵ-rank r_ϵ if there is a well-determined gap between the singular values σ_{r_ϵ} and $\sigma_{r_\epsilon+1}$, and if ϵ lies in this gap:

$$\sigma_{r_\epsilon} > \epsilon \geq \sigma_{r_\epsilon+1}.$$

Hence, the SVD of A is obviously a "rank-revealing" decomposition of A, since r_ϵ can immediately be found from inspection of the singular values of A. However, we are interested in alternative decompositions that can reveal the numerical rank of A at a smaller computational cost than that involved in

---

[5]Updating schemes for an "approximate SVD" are developed in [252], [253]. In these schemes the diagonal matrices Σ and M are replaced by triangular matrices whose elements decay rapidly in magnitude away from the diagonal.

computing the SVD.

Before turning our attention to the rank-revealing decompositions, it is instructive to consider the most basic problem in connection with rank-deficient matrices: is a given matrix A rank deficient? From the definition of numerical rank given above, this is equivalent to asking whether A has at least one small singular value, i.e., whether $\sigma_n \ll \sigma_1$. Hence, reliable numerical methods for estimating the smallest singular value of a matrix are very important tools in connection with rank-deficient problems. In fact, many of today's rank-revealing algorithms depend on such methods. Higham [207] has given an extensive survey of methods for estimating the smallest singular value of triangular matrices. In [175] some of these methods are compared in connection with Cholesky decompositions.

We emphasize that currently all algorithms for computing rank-revealing decompositions are efficient only if the numerical rank is either high, $r_\epsilon \approx n$, or low, $r_\epsilon \ll n$.

2.2.1. Rank-Revealing QR and LU Decompositions

A *rank-revealing QR decomposition* (RRQR decomposition) is a special QR factorization with column pivoting which is guaranteed to reveal the ϵ-rank of A by having a leading well-conditioned $r_\epsilon \times r_\epsilon$ upper triangular submatrix of R and a trailing triangular submatrix of R with elements of the order $\sigma_{r_\epsilon+1}$ or less. Thus, an RRQR decomposition has the form

$$A\Pi = QR = (Q_1, Q_2)\begin{pmatrix} R_{11} & R_{12} \\ 0 & R_{22} \end{pmatrix}, \qquad (2.21)$$

where Π is a permutation matrix, Q has orthonormal columns, R_{11} is an $r_\epsilon \times r_\epsilon$ triangular matrix with condition number approximately equal to $\sigma_1/\sigma_{r_\epsilon}$, and the norm $\|R_{22}\|_2$ is of the order $\sigma_{r_\epsilon+1}$.

In addition to the matrices Π, Q, and R in (2.21), many RRQR algorithms also provide a matrix $W_{r_\epsilon} \in \mathbb{R}^{n \times (n-r_\epsilon)}$ whose columns span an approximation to the numerical null space of R. Another approximate basis for $\mathcal{N}_{r_\epsilon}(A)$ can be computed as the columns of the matrix

$$Y_{r_\epsilon} = \Pi \begin{pmatrix} R_{11}^{-1}R_{12} \\ -I_{n-r_\epsilon} \end{pmatrix}. \qquad (2.22)$$

The existence of RRQR decompositions was proved in [213] and the necessity of the column permutations is treated in [131]. So-called "strong RRQR decompositions" are discussed in [163]; here there is an additional requirement that the norm of the submatrix $R_{11}^{-1}R_{12}$ in (2.22) stay small.

The key to computing a rank-revealing decomposition is to choose the appropriate column permutation matrix Π. We can illustrate this with a simple example. Assume that we are given an $m \times 2$ matrix $A = (a_1, a_2)$ with numerical ϵ-rank 1, and we wish to compute an RRQR decomposition of A. We can seek either a permutation that maximizes $|r_{11}|$ or a permutation that minimizes $|r_{22}|$. In this $m \times 2$ example it is easy to show that both goals are achieved by the same permutation matrix Π, namely, the ordinary column pivoting strategy [154, §5.4.1] which permutes the column of A with largest 2-norm up front.

For matrices with more than two columns, as pointed out in [56], the corresponding two goals of an RRQR algorithm become

- maximize the smallest singular value of R_{11}, and
- minimize the largest singular value of R_{22}.

RRQR decompositions are not unique, and different RRQR algorithms produce different decompositions. The algorithms typically start with a preprocessing stage in which an initial QR factorization is computed in $2n^2(m - \frac{1}{3}n)$ flops (plus an additional $4(m^2 n - mn^2 + \frac{1}{3}n^3)$ flops if Q is wanted). In the second stage, the permutation matrix Π is constructed such that the triangular matrix R has the desired features. This is done by means of a sequence of steps in which either the smallest or the largest singular values are estimated one at a time, each step involving $\mathcal{O}(n^2)$ flops.

The QR factorization algorithm with ordinary column pivoting—in which Π is constructed "on the fly" during the QR factorization—seeks to achieve the first goal by trying to move the most linearly independent columns up front in $A\Pi$. The RRQR algorithms by Foster [130] and Chan [50] seek to achieve the second goal, in that they try to move the most linearly dependent columns of A to the back in $A\Pi$. The "Hybrid III" algorithm presented in [56], on the other hand, seeks to achieve both goals.

The ith singular value of the matrix R always lies between $\sigma_{\min}(R(1\!:\!i,1\!:\!i))$ and $\|R(i\!:\!n,i\!:\!n)\|_2$, where $\sigma_{\min}(R(1\!:\!i,1\!:\!i))$ is the smallest singular value of the leading $i \times i$ submatrix of R, and $\|R(i\!:\!n,i\!:\!n)\|_2$ is the norm of the trailing $(n-i+1) \times (n-i+1)$ submatrix of R. For a particular RRQR algorithm there exist factors $f_i^{\mathrm{alg}} \geq 1$ such that

$$\sigma_i / f_i^{\mathrm{alg}} \leq \sigma_{\min}(R(1\!:\!i,1\!:\!i)) \leq \sigma_i \leq \|R(i\!:\!n,i\!:\!n)\|_2 \leq f_i^{\mathrm{alg}} \sigma_i . \qquad (2.23)$$

We can consider the factors f_i^{alg} as "quality measures" of the computed RRQR decomposition in that the smaller the f_i^{alg}, the tighter the bounds in (2.23) will be. As an example, consider the high-rank algorithms. For the QR algorithm with ordinary column pivoting, the following result is derived in [56]:

$$f_i^{\mathrm{ord}} = \sqrt{n-i+1}\, \|\bar{R}(1\!:\!i,1\!:\!i)^{-1}\|_2 , \quad i = 1, \ldots, n , \qquad (2.24)$$

where $R = \mathrm{diag}(r_{11}, \ldots, r_{nn})\,\bar{R}$. For the Chan–Foster algorithm we have

$$f_i^{\mathrm{C/F}} = \sqrt{n-i+1}\, \|W_{r_\epsilon}(i\!:\!n, i-r_\epsilon\!:\!n-r_\epsilon)^{-1}\|_2 , \quad i = r_\epsilon+1, \ldots, n , \qquad (2.25)$$

2.2. RANK-REVEALING DECOMPOSITIONS

where we have the worst-case a priori upper bound

$$\|W_{r_\epsilon}(i{:}n, i - r_\epsilon{:}n - r_\epsilon)^{-1}\|_2 < \sqrt{n}\, 2^{n-i+1}, \quad i = r_\epsilon + 1, \ldots, n. \qquad (2.26)$$

In practice, $\|W_{r_\epsilon}(i{:}n, i - r_\epsilon{:}n - r_\epsilon)^{-1}\|_2$ is usually much smaller than the upper bound. Similarly, for the algorithm "Hybrid III" from [56], as well as the "cyclic pivoting" algorithm from [276], we have

$$f_i^H = \sqrt{i\,(n - i + 1)}, \quad i = r_\epsilon + 1, \ldots, n. \qquad (2.27)$$

Similar bounds for some low-rank algorithms can be found in [53]. All these bounds are important because the decision about the numerical ϵ-rank is based on these bounds.

A difficulty with RRQR decompositions is that their null-space approximation, in the form of the columns of either Y_{r_ϵ} or $\Pi\, W_{r_\epsilon}$, is not easily amenable to updating when a row is added to A or deleted from A (see [34] for RRQR updating issues).

Much less work has been done for *rank-revealing LU decompositions*. A complete treatment of the case where $r_\epsilon = n - 1$ was given in [49], and this algorithm was extended to the general high-rank case $r_\epsilon \approx n$ in [215]. The corresponding algorithm in [215] potentially has high complexity.

2.2.2. Rank-Revealing URV and ULV Decompositions

As already mentioned above, the null-space information is not an explicit part of an RRQR decomposition. Therefore, RRQR decompositions are not well suited for updating problems where the null-space information is explicitly required (which is often the case in signal processing applications). It was for this reason that Stewart [322] introduced the rank-revealing URV decomposition

$$A = U R V^T = (U_1, U_2) \begin{pmatrix} R_{11} & R_{12} \\ 0 & R_{22} \end{pmatrix} (V_1, V_2)^T, \qquad (2.28)$$

where U and V have orthonormal columns, and where the upper triangular factor R satisfies the requirements that R_{11} be $r_\epsilon \times r_\epsilon$ and well conditioned and $\left(\|R_{12}\|_2^2 + \|R_{22}\|_2^2\right)^{1/2}$ be of the order σ_{r_ϵ}. Hence, in a rank-revealing URV decomposition the ϵ-rank of A is revealed in both submatrices R_{12} and R_{22} having small elements. Notice that an RRQR decomposition is a special URV decomposition where the matrix V is restricted to being a permutation matrix.

Stewart also introduced the "twin decomposition" of (2.28), namely, the ULV decomposition [323]

$$A = U L V^T = (U_1, U_2) \begin{pmatrix} L_{11} & 0 \\ L_{21} & L_{22} \end{pmatrix} (V_1, V_2)^T. \qquad (2.29)$$

The ULV decomposition of A is equivalent to the URV decomposition of A^T. Both decompositions are often referred to as *UTV decompositions*, and the name "complete orthogonal decomposition" is also used [154, §5.4.2]. See [322] and [323] for more details.

UTV algorithms consist of the same two main stages as RRQR algorithm, namely, an initial QR factorization followed by a rank-revealing stage in which the smallest—or the largest—singular values are estimated one at a time. Hence, as long as the numerical rank is either high or low, the computational cost is dominated by the QR factorization.

The last $n-r_\epsilon$ columns of V in both the URV and the ULV decompositions span an approximation to A's numerical null space $\mathcal{N}_{r_\epsilon}(A)$, while the first r_ϵ columns of U span an approximation to the numerical range $\mathcal{R}_{r_\epsilon}(A)$. The main difference between the URV and the ULV decompositions is the intrinsic accuracy of these approximations. In theory (see Theorem 3.3.3), as well as in practice, the approximate null space from the ULV decomposition is more accurate than that from the URV decomposition, while the approximate range from the URV decomposition is superior to that from the ULV decomposition. For this reason, the ULV decomposition is often the decomposition of choice in signal processing applications that rely on the numerical null space.

In analogy with the RRQR decomposition, the UTV decompositions yield approximations to the singular values of A. For example, for the URV decomposition (2.28), Mathias and Stewart [247] have derived the following bounds:

$$\sigma_i f^{\mathrm{URV}} \leq \sigma_i(R_{11}) \leq \sigma_i \,, \qquad i = 1, \ldots, r_\epsilon \,,$$

and

$$\sigma_i \leq \sigma_{i-k}(R_{22}) \leq \sigma_i / f^{\mathrm{URV}} \,, \qquad i = r_\epsilon + 1, \ldots, n \,,$$

where

$$f^{\mathrm{URV}} = \left(1 - \frac{\|R_{12}\|_2^2}{\sigma_{\min}(R_{11})^2 - \|R_{22}\|_2^2}\right)^{1/2}.$$

We see that the smaller the norm of the off-diagonal block R_{12}, the better the bounds will be. The corresponding bounds for the subspace approximations are discussed in §3.3.2. A perturbation analysis for the UTV decompositions is given in [119].

As already mentioned, the main advantage of the UTV decomposition to the RRQR decomposition is that the information about A's numerical null space is part of the decomposition: $\mathcal{N}_{r_\epsilon}(A)$ is spanned by the last $n-r_\epsilon$ columns of the right orthogonal matrix V. Consequently, this information is immediately updated when the UTV decomposition is updated.

Another difference between the RRQR and UTV decompositions is that the latter can be refined[6] by means of block QR iterations without shift, which

---

[6]The same refinement is not possible for the RRQR decomposition, where the permutation matrix Π only allows for a finite number of different factorizations. The RRQR algorithms in [56], [163], and [276] can be considered as "refinement algorithms" in that they, given an initial RRQR decomposition, seek to find the optimal Π that minimizes the factors f_i^{alg} in (2.23).

2.2. RANK-REVEALING DECOMPOSITIONS

diminish the norm of the off-diagonal block (either R_{12} or L_{21}), and as a result, the null-space approximation in $V(1\!:\!n, r_\epsilon + 1\!:\!n)$ improves. Each refinement step requires $\mathcal{O}((n-r_\epsilon)n^2)$ flops in the high-rank case; see [247] for details.

The RRQR decomposition has advantages over UTV decompositions in two important cases. In subset selection problems—where the goal is to determine the r_ϵ columns of A whose condition number is minimal—the RRQR decomposition is the only decomposition that provides this information explicitly, namely, in the permutation matrix Π. The solution consists of the first r_ϵ columns of $A\Pi$.

For large sparse matrices, where updating is rarely an issue, RRQR decompositions can be computed with much less fill-in than UTV decompositions, because the fill-in generated in the second stage is restricted to $n-r_\epsilon$ columns of R (in the high-rank case); see [31, p. 1334]. The UTV decompositions, on the other hand, are very likely to suffer from severe fill-in; see the discussion in [282]. As an alternative, Pierce [282] proposed a sparse URL decomposition $A = URL^{-T}$ in which L is a full-rank lower triangular matrix.

Neither the RRQR decomposition nor the UTV decompositions can exploit symmetry of the matrix A. A symmetric rank-revealing decomposition can take the following form:

$$A = V \begin{pmatrix} S_{11} & S_{12} \\ S_{12}^T & S_{22} \end{pmatrix} V^T,$$

where S is symmetric, S_{11} is $r_\epsilon \times r_\epsilon$ and well conditioned, and S_{12} and S_{22} are matrices with small norm. Such decompositions are discussed in [242].

2.2.3. What RRQR and URV Decompositions Look Like

To illustrate the rank-revealing QR and URV decompositions we use the model problem from §1.4.1 with a 122×7 Hankel matrix A derived from a signal of length $N = 128$, consisting of two sinusoids in white noise. The parameters used in this example are $a_1 = 1$, $a_2 = 0.4$, $\omega_1 = \pi/10$, $\omega_2 = \pi/3$, $\phi_1 = 0$, $\phi_2 = \pi/4$, and $\sigma_{\text{noise}} = 0.02$. The singular values of A are

$$17.33, \ 12.06, \ 6.28, \ 4.14, \ 0.25, \ 0.23, \ 0.21,$$

showing that the numerical rank is indeed 4, as expected. The size of the three smallest singular values reflects the random errors in the signal.

The upper triangular matrix R from an RRQR decomposition of A looks as follows:

$$R = \begin{pmatrix} 8.5585 & 0.7664 & 5.3574 & 7.5719 & 1.7952 & 3.1787 & 7.5944 \\ 0 & 8.4036 & 2.7098 & -1.5979 & 7.3611 & 5.0473 & 1.1335 \\ 0 & 0 & 5.9395 & -1.5196 & 2.5900 & 5.4658 & 3.4046 \\ 0 & 0 & 0 & 3.1850 & 2.5824 & 2.2700 & -1.3874 \\ 0 & 0 & 0 & 0 & 0.4768 & 0.3539 & -0.1225 \\ 0 & 0 & 0 & 0 & 0 & 0.3176 & -0.0305 \\ 0 & 0 & 0 & 0 & 0 & 0 & -0.3214 \end{pmatrix},$$

and the numerical rank is displayed in the bottom right 3×3 submatrix R_{22} having elements whose absolute values are of the same order as the three smallest singular values of A.

The upper triangular matrix $\hat{R}$ from a URV decomposition of A looks as follows:

$$\hat{R} = \begin{pmatrix} 10.5986 & -0.1280 & 7.1729 & 0.5098 & 0.0007 & 0.0004 & -0.0015 \\ 0 & 10.4035 & 5.1032 & 8.4933 & -0.0002 & -0.0017 & 0.0015 \\ 0 & 0 & 8.1933 & 5.3433 & -0.0003 & 0.0017 & 0.0021 \\ 0 & 0 & 0 & 6.0206 & 0.0007 & 0.0005 & -0.0040 \\ 0 & 0 & 0 & 0 & -0.2134 & 0.0041 & 0.0127 \\ 0 & 0 & 0 & 0 & 0 & -0.2409 & 0.0211 \\ 0 & 0 & 0 & 0 & 0 & 0 & 0.2368 \end{pmatrix},$$

and now the numerical rank is displayed in the last three columns of $\hat{R}$ whose norms are of the same order as the three smallest singular values of A. Notice that the norm $\|\hat{R}_{12}\|_2$ is quite small in contrast to the norm $\|R_{12}\|_2$, which is of the same order as $\|R_{11}\|_2$.

2.2.4. Generalized Rank-Revealing Decompositions

The rank-revealing decompositions discussed so far are designed as alternatives to the SVD. In applications where a generalized SVD of two matrices A and B is involved, we need a suitable generalization of a rank-revealing decomposition. Paige [270] has done preliminary work along these lines in connection with definitions of generalized product- and quotient-type generalized QR factorizations, but these factorizations are not specifically rank revealing.

More recently, Luk and Qiao [241] have defined a generalization of the ULV decomposition, called the *rank-revealing ULLV decomposition*, for two matrices $A \in \mathbb{R}^{m \times n}$ and $B \in \mathbb{R}^{p \times n}$ with the same number of columns, where A is a general matrix and B has full column rank n (i.e., $p \geq n$). The rank-revealing ULLV decomposition has the form

$$A = U_A L \hat{L} V^T, \qquad B = U_B \hat{L} V^T, \qquad (2.30)$$

where $U_A^T U_A = U_B^T U_B = V^T V = I_n$ and L and $\hat{L}$ are lower triangular $n \times n$ matrices. An algorithm for updating the ULLV decomposition is also described

2.2. RANK-REVEALING DECOMPOSITIONS

in [241]. The use of this ULLV decomposition in speech processing is described in [200].

If B, in addition to having full rank, is also square, then we immediately obtain
$$A B^{-1} = U_A L U_B^T,$$
showing that in this case the matrix L reveals the numerical rank of the matrix quotient $A B^{-1}$. The GSVD, in the same situation, also reveals the rank of $A B^{-1}$; cf. (2.10). Hence, in this respect the two decompositions are equivalent,[7] and therefore the ULLV decomposition might as well be called a quotient ULV decomposition.

It is also convenient to define a generalized rank-revealing decomposition for matrix pairs in which B is rectangular with full row rank, i.e., $\text{rank}(B) = p < n$. This case is important in connection with regularization problems as well as in signal processing applications where $B^T B$ represents a rank-deficient noise covariance matrix. In [190] and [243] the following rank-revealing extension of the ULLV decomposition in (2.30) is proposed:

$$A = U_A L \begin{pmatrix} \hat{L} & 0 \\ 0 & I_{n-p} \end{pmatrix} V^T, \qquad B = U_B (\hat{L}, 0) V^T, \qquad (2.31)$$

where $U_A^T U_A = V^T V = I_n$, $U_B^T U_B = I_p$, $L \in \mathbb{R}^{n \times n}$ and $\hat{L} \in \mathbb{R}^{p \times p}$ are lower triangular matrices, and $\hat{L}$ has full rank. Notice the block diagonal structure of the matrix $\text{diag}(\hat{L}, I_{n-p})$. The formulations in (2.30) and (2.31) are identical when $p = n$. An updating algorithm for the decomposition in (2.31) that preserves the block diagonal structure of $\text{diag}(\hat{L}, I_{n-p})$ is described in [243].

If we compare (2.31) with the GSVD of (A, B), and if we write B in the form $B = U_b (I_p, 0) \text{diag}(\hat{L}, I_{n-p}) V^T$, then we see that the three ULLV matrices U_A, U_B, and $\text{diag}(\hat{L}, I_{n-p}) V^T$ play the roles of the GSVD matrices U, V, and X^{-1}. Moreover, the ULLV matrices L and $(I_p, 0)$ correspond to the "middle" GSVD matrices $\text{diag}(\Sigma, I_{n-p})$ and $(M, 0)$. The null space of B is spanned by the last $n - p$ orthonormal columns of V.

2.2.5. Algorithms and Software

As already mentioned, rank-revealing QR and UTV algorithms consist of two main stages: an initial QR factorization (possibly combined with a column pivoting scheme), and a rank-revealing stage in which the triangular factor is modified in order to satisfy the necessary requirements. The second stage consists of a sequence of either $n - r_\epsilon + 1$ steps in the high-rank case or $r_\epsilon + 1$

---

[7] A different version of the ULLV decomposition for a general B matrix was defined in [42]. This decomposition is of the form $A = U_A L_A \hat{L} V^T$, $B = U_B L_B \hat{L} V^T$ with $L_A^T L_A + L_B^T L_B = I_n$. When B is invertible we obtain $A B^{-1} = U_A L_A L_B^{-1} U_B^T$, which is not practical for revealing the numerical rank of $A B^{-1}$. The same is true for the decomposition $A = U_A R_A V^T$, $B = U_B R_B V^T$ discussed in [289].

steps in the low-rank case, in which the smallest—or largest—singular values are approximated one at a time. The work in each step is $\mathcal{O}(n^2)$ flops, and therefore the algorithms are efficient only in the high- and low-rank cases when the work is dominated by the first stage.

A survey of RRQR algorithms is given in [56]. Various aspects of RRQR algorithms are discussed in [31], [130] (sparse matrices), [32] (block algorithm), [192] (Toeplitz matrices), [237] (systolic implementation), and [283] (multifrontal sparse algorithm). See [53] for more details about low-rank algorithms.

UTV algorithms (for the high-rank case) and updating issues are discussed in [322] and [323], while low-rank algorithms are discussed in [125].

Very little software is currently available for computation of rank-revealing decompositions. Fortran subroutines for implementation of the Chan–Foster RRQR algorithm have appeared in [292]. A complete RRQR algorithm, written in Fortran and based on the "Hybrid III" algorithm, is presented in [33]. A suite of MATLAB routines associated with the RRQR decomposition is available from the author.

2.3. Transformation to Standard Form

A regularization problem with a discrete smoothing norm $\Omega(x) = \|L\,(x-x^*)\|_2$ (1.14) is said to be in *standard form* if the matrix L is the identity matrix I_n. In many applications, regularization in standard form is not the best choice; i.e., one should use an $L \neq I_n$ (see §4.2). We assume that the $p \times n$ matrix L has full row rank; i.e., the rank of L is p.

From a numerical point of view it is much simpler to treat problems in standard form, basically because only one matrix, A, is involved instead of the two matrices A and L. Hence, it is convenient to be able to "absorb" the matrix L into the matrix A, i.e., to transform a given regularization problem with residual $\|A\,x - b\|_2$ and smoothing norm $\Omega(x) = \|L\,(x - x^*)\|_2$ into one in standard form with a new variable $\bar{x}$, a new residual $\|\bar{A}\,\bar{x} - \bar{b}\|_2$, and a new smoothing norm $\bar{\Omega}(\bar{x}) = \|\bar{x} - \bar{x}^*\|_2$. For example, for Tikhonov regularization we want to transform the general-form problem

$$\min\left\{\|A\,x - b\|_2^2 + \lambda^2\,\|L\,(x - x^*)\|_2^2\right\}$$

into the following standard-form problem:

$$\min\left\{\|\bar{A}\,\bar{x} - \bar{b}\|_2^2 + \lambda^2\,\|\bar{x} - \bar{x}^*\|_2^2\right\}.$$

In particular, we need numerically stable methods for computing the new matrix $\bar{A}$, the new right-hand side $\bar{b}$, and the new vector $\bar{x}^*$ from the original quantities A, L, b, and x^*. We also need a numerically stable scheme for transforming the standard-form solution $\bar{x}_{\text{reg}}$ back to the general-form setting, in such a way that the transformed solution solves the general-form problem.

2.3. Transformation to Standard Form

For the simple case where L is square and invertible, the transformation is obvious: $\bar{A} = A L^{-1}$, $\bar{b} = b$, $\bar{x}^* = L x^*$, and the back-transformation simply becomes $x_\lambda = L^{-1} \bar{x}_\lambda$.

In most applications, however, the matrix L is not square, and the transformation becomes somewhat more involved than just a matrix inversion. What is now needed is the *A-weighted generalized inverse of L*, denoted $L_A^\dagger$ and defined as

$$L_A^\dagger \equiv \left(I_n - \left(A \left(I_n - L^\dagger L \right) \right)^\dagger A \right) L^\dagger . \tag{2.32}$$

If $p \geq n$ then $L_A^\dagger = L^\dagger$, and we emphasize that $L_A^\dagger$ is generally different from the pseudoinverse $L^\dagger$ when $p < n$. In addition, we need the component x_0 of the regularized solution in $\mathcal{N}(L)$, given by

$$x_0 \equiv (A (I_n - L^\dagger L))^\dagger b . \tag{2.33}$$

See [100] for details. Given the GSVD of (A, L) in (2.6), $L_A^\dagger$ and x_0 can be expressed as

$$L_A^\dagger = X \begin{pmatrix} M^{-1} \\ 0 \end{pmatrix} V^T , \qquad x_0 = \sum_{i=p+1}^n u_i^T b\, x_i . \tag{2.34}$$

Then the standard-form quantities $\bar{A}$, $\bar{b}$, and $\bar{x}^*$ take the form

$$\bar{A} = A L_A^\dagger , \qquad \bar{b} = b - A x_0 , \qquad \bar{x}^* = L x^* , \tag{2.35}$$

while the transformation back to the general-form setting becomes

$$x = L_A^\dagger \bar{x} + x_0 . \tag{2.36}$$

The component x_0 is the unregularized component of x which is not affected by the regularization scheme.

This standard-form transformation was originally proposed by Eldén [100] and later advocated in [167], [168], [171], and [187]. The idea of "absorbing" L into A in the continuous setting of §1.3 was proposed by Hilgers [210].

One of the many advantages of the standard-transformation technique defined in (2.35) is that there is a simple relation between the GSVD of (A, L) and the SVD of $\bar{A}$. If the matrix U_p consists of the first p columns of U, i.e., $U_p = (u_1, \ldots, u_p)$, then

$$A L_A^\dagger = U_p \Sigma M^{-1} V^T , \tag{2.37}$$

showing that generalized singular values γ_i are the singular values of $A L_A^\dagger$, except for the reverse ordering. Moreover, the vectors u_i and v_i, $i = 1, \ldots, p$, are the left and right singular vectors of $A L_A^\dagger$, respectively. We note that we can also write $A L_A^\dagger$ in terms of the rank-revealing ULLV decomposition in (2.31) as

$$A L_A^\dagger = U_A \begin{pmatrix} L_{11} \\ 0 \end{pmatrix} U_B^T , \tag{2.38}$$

where L_{11} is the leading $p \times p$ submatrix of L.

Another advantage of the standard-form transformation in (2.35) is the simple relations $L\,x = \bar{x}$ (due to $L\,L_A^\dagger = I_p$ and $L\,x_0 = 0$) and $A\,x - b = \bar{A}\,\bar{x} - \bar{b}$ that immediately lead to the equations

$$\|L\,x\|_2 = \|\bar{x}\|_2 , \qquad \|A\,x - b\|_2 = \|\bar{A}\,\bar{x} - \bar{b}\|_2 . \tag{2.39}$$

These relations are important in connection with methods for choosing the regularization parameter; cf. Chapter 7.

When the standard-form transformation is implemented, then it is often a good idea to distinguish between direct and iterative regularization methods—cf. Chapters 5 and 6. For the direct methods we need to be able to compute the matrix $\bar{A}$ explicitly, preferably by means of orthogonal transformations, for stability reasons. For the iterative methods, we merely need to be able to compute the matrix-vector products $\bar{A}\,\bar{x}$ and $\bar{A}^T z$ efficiently. Below, we describe two methods for transformation to standard form which are suited for direct and iterative methods, respectively.

2.3.1. Explicit Transformation

The *explicit* standard-form transformation for direct methods was developed by Eldén [98], and it is based on two QR factorizations. Let subscripts p, o, and q denote matrices with p, $n - p$, and $m - (n - p)$ columns, respectively. First, compute a QR factorization of L^T,

$$L^T = K\,R = (\,K_p, \, K_o\,) \begin{pmatrix} R_p \\ 0 \end{pmatrix}. \tag{2.40}$$

If L is banded with bandwidth $n - p + 1$ (which is the case for matrices of the form (1.15) and (1.16)), then this QR factorization requires about $2n(n-p+1)^2$ flops if Givens rotations are used (general banded matrices are treated in [36, §6.2.3]). We remark that since L has full row rank, its pseudoinverse is simply $L^\dagger = K_p\,R_p^{-T}$. Moreover, the columns of K_o are an orthonormal basis for the null space of L. Next, form the "skinny" $m \times (n-p)$ matrix $A\,K_o$ and compute its QR factorization,

$$A\,K_o = H\,T = (\,H_o, \, H_q\,) \begin{pmatrix} T_o \\ 0 \end{pmatrix}. \tag{2.41}$$

This QR factorization requires about $2m(n - p)^2$ flops.

Given these two QR factorizations, we have $A\,(I_n - L^\dagger L) = A\,K_o\,K_o^T = H_o\,T_o\,K_o^T \Rightarrow (A\,(I_n - L^\dagger L))^\dagger = K_o\,T_o^{-1}\,H_o^T$, and therefore, according to (2.32) and (2.33),

$$L_A^\dagger = (I_n - K_o\,T_o^{-1}\,H_o^T A)\,L^\dagger , \qquad x_0 = K_o\,T_o^{-1}\,H_o^T b . \tag{2.42}$$

2.3. Transformation to Standard Form

Hence, since $A K_o T_o^{-1} H_o^T = H_o H_o^T$, we obtain

$$\bar{A} = A (I_p - K_o T_o^{-1} H_o^T A) L^\dagger = (I_m - H_o H_o^T) A L^\dagger = H_q H_q^T A L^\dagger ,$$
$$\bar{b} = b - A K_o T_o^{-1} H_o^T b = (I_m - H_o H_o^T) b = H_q H_q^T b .$$

When we insert these quantities into the residual norm $\|\bar{A}\bar{x} - \bar{b}\|_2$ we see that the leftmost factor H_q does not contribute to the 2-norms. Therefore, it is convenient to work with slightly redefined versions of $\bar{A}$ and $\bar{b}$ where this factor is omitted; i.e., the standard-form quantities become

$$\bar{A}' = H_q^T \bar{A} = H_q^T A L^\dagger = H_q^T A K_p R_p^{-T} , \qquad \bar{b}' = H_q^T \bar{b} = H_q^T b . \qquad (2.43)$$

This is the form used in [98]. The most efficient way to compute $\bar{A}'$ and $\bar{b}'$ is to apply the orthogonal transformations that make up K and H "on the fly" to A and b when the QR factorizations in (2.40) and (2.41) are computed. When the standard-form problem has been solved for $\bar{x}_{\text{reg}}$, the transformation back to the general-form setting takes the form

$$x_{\text{reg}} = L^\dagger \bar{x}_{\text{reg}} + K_o T_o^{-1} H_o^T (b - A L^\dagger \bar{x}_{\text{reg}}) . \qquad (2.44)$$

We also stress that the redefined quantities $\bar{A}'$ and $\bar{b}'$ still satisfy (2.39) [179, Theorem 6.3]; i.e.,

$$\|A x - b\|_2 = \|\bar{A}' \bar{x} - \bar{b}'\|_2 . \qquad (2.45)$$

2.3.2. Implicit Transformation

For iterative methods, where A is accessed only via matrix-vector products with A and A^T, it is not practical to form $\bar{A}$ or $\bar{A}'$ explicitly. Instead, one should exploit the matrix multiplications in (2.35) or (2.43) and use an *implicit* standard-form transformation. Both approaches are numerically very stable, and both are suited for use with the conjugate gradient algorithm.

If we use the approach based on (2.35)—which is the one used in [171] and, for pedagogical reasons, in REGULARIZATION TOOLS [187]—then we need to compute x_0 as well as the matrix-vector products $L_A^\dagger \bar{x}$ and $(L_A^\dagger)^T x$ efficiently. We shall now describe an approach based on the developments in [187] and [193]. Given a basis W for $\mathcal{N}(L)$, definition (2.33) leads to the expression

$$x_0 = W (A W)^\dagger b ; \qquad (2.46)$$

see also [171, §4.3]. This involves $\mathcal{O}(mn(n-p))$ operations. To compute $L_A^\dagger \bar{x}$ and $(L_A^\dagger)^T x$ efficiently, we need to compute the skinny $(n-p) \times n$ matrix

$$T = (A W)^\dagger A , \qquad (2.47)$$

where $(A W)^\dagger$ is from (2.46). We also need to partition L, T, and x as

$$L = (L_{11} , L_{12}) , \quad T = (T_{11} , T_{12}) , \quad x = \begin{pmatrix} x_1 \\ x_2 \end{pmatrix}$$

where $L_{11} \in \mathbb{R}^{p \times p}$, $T_{11} \in \mathbb{R}^{(n-p) \times p}$, and $x_1 \in \mathbb{R}^p$. Then $L_A^\dagger$ is given by

$$L_A^\dagger = (I_n - W\,T) \begin{pmatrix} L_{11}^{-1} \\ 0 \end{pmatrix} = \left(\begin{pmatrix} I_p \\ 0 \end{pmatrix} - W\,T_{11} \right) L_{11}^{-1} ,$$

which is proved by checking the four conditions in [100]. Hence, the two vectors $y = L_A^\dagger \bar{x}$ and $\bar{y} = (L_A^\dagger)^T x$ are given by the relations

$$y = L_A^\dagger \bar{x} = \left(\begin{pmatrix} I_p \\ 0 \end{pmatrix} - W\,T_{11} \right) L_{11}^{-1} \bar{x}$$

and

$$\bar{y} = (L_A^\dagger)^T x = (L_{11}^{-1})^T \left((I_p,\, 0) - T_{11}^T W^T \right) x ,$$

which lead to the following algorithms for computing y and $\bar{y}$:

$$\hat{x} \leftarrow L_{11}^{-1} \bar{x} , \qquad y \leftarrow \begin{pmatrix} \hat{x} \\ 0 \end{pmatrix} - W\,T_{11}\,\hat{x} , \qquad (2.48)$$

$$\hat{x} \leftarrow x_1 - T_{11}^T W^T x , \qquad \bar{y} \leftarrow L_{11}^{-T} \hat{x} . \qquad (2.49)$$

In the above formulas, W need not have orthonormal columns, although this is the best choice from a numerical point of view.

If L is banded and the bandwidth is $n - p + 1$ (like the matrices in (1.15) and (1.16)), then L_{11} is banded and upper triangular, and the operations $L_{11}^{-1}\bar{x}$ and $L_{11}^{-T}\hat{x}$ and (2.48) and (2.49) become particularly simple. Otherwise—e.g., if L represents the Laplacian—then an LU factorization of L_{11} is required. Variants of this algorithm are described in [187] and [193].

Consider now the approach based on (2.43), which is the one used in [35]. Then x_0 is computed by means of (2.42), and if the matrix $W\,T$ in (2.48) and (2.49) is replaced by $K_o K_o^T$, where K_o is an orthonormal basis for $\mathcal{N}(L)$, then the two algorithms compute $y = L^\dagger \bar{x}$ and $\bar{y} = (L^\dagger)^T x$.

2.3.3. GSVD Computations Based on Standard-Form Transformation

As already mentioned, the relation between the SVD of $\bar{A} = A\,L_A^\dagger$ and the GSVD of (A, L) is very simple, due to (2.37):

$$A\,L_A^\dagger = \sum_{i=i}^{p} u_i\,\gamma_i\,v_i^T .$$

Less simple SVD–GSVD relations for the redefined matrix $\bar{A}' = H_q^T \bar{A}$ in (2.43) are investigated in [179], [182].

These relations can be used to compute parts of the GSVD of (A, L) via the SVD of either $\bar{A}$ or $\bar{A}'$. For dense matrices, this possibility is explored in [179] and [182]. For sparse matrices, the relations are used in [193] to derive a Lanczos bidiagonalization algorithm for computing the largest generalized singular values and associated vectors. See these papers for details.

2.4. Computation of the SVE

The SVE mentioned in §1.2 is a powerful analysis tool, but unfortunately it is only known analytically in a limited number of cases; see, e.g., [27], [75], [114], [115], [116], [117]. Hence, we need a numerical technique for computing approximations to the SVE of a kernel K. The Galerkin method mentioned in §1.3 provides such a technique. The algorithm is described in detail in [178], and we summarize the main results here. The technique was applied in a specific analysis of first-kind Fredholm integral equations in potential theory in [191].

The algorithm takes the following form. Assume that we choose *orthonormal* basis functions $\phi_1, \ldots, \phi_n$ and $\psi_1, \ldots, \psi_n$, set up the matrix A with elements given by (1.12), and compute its SVD. Then the n singular values $\sigma_i^{(n)}$ of A are approximations to n singular values of K. Moreover, if we introduce the functions

$$\tilde{u}_j(s) = \sum_{i=1}^{n} u_{ij}\, \phi_i(s)\,, \qquad j=1,\ldots,n\,, \qquad (2.50)$$

$$\tilde{v}_j(t) = \sum_{i=1}^{n} v_{ij}\, \psi_i(t)\,, \qquad j=1,\ldots,n\,, \qquad (2.51)$$

where u_{ij} and v_{ij} are the elements of U and V, then these functions are approximations to n left and right singular functions of K, due to the following results.

Theorem 2.4.1. [178, Theorems 1–5]. *Let $\|K\|$ denote the norm of K,*

$$\|K\|^2 = \int_0^1 \int_0^1 |K(s,t)|^2\, ds\, dt = \sum_{i=1}^{\infty} \mu_i^2\,,$$

and define

$$\delta_n^2 \equiv \|K\|^2 - \|A\|_F\,. \qquad (2.52)$$

Then

$$\sum_{i=1}^{n} \left(\mu_i - \sigma_i^{(n)}\right)^2 \leq \delta_n^2\,, \qquad (2.53)$$

and for $i = 1, \ldots, n$ we have

$$0 \leq \mu_i - \sigma_i^{(n)} \leq \delta_n\,, \qquad (2.54)$$

$$\sigma_i^{(n)} \leq \sigma_i^{(n+1)} \leq \mu_i\,, \qquad (2.55)$$

$$\max\{\|u_i - \tilde{u}_i\|_2,\, \|v_i - \tilde{v}_i\|_2\} \leq \left(\frac{2\,\delta_n}{\mu_i - \mu_{i+1}}\right)^{1/2}\,. \qquad (2.56)$$

This theorem implies that if $\delta_n \to 0$ for $n \to \infty$, then the approximate singular values $\sigma_i^{(n)}$ converge uniformly in n to the true singular values μ_i, and the corresponding approximate singular functions $\tilde{u}_i$ and $\tilde{v}_i$ converge in the mean to the true singular functions. In case of multiple singular values $\mu_i = \cdots = \mu_{i+k}$ a similar result holds for the distances between $\tilde{u}_i$ and $\tilde{v}_i$ and the associated subspaces span$\{u_i, \ldots, u_{i+k}\}$ and span$\{u_i, \ldots, u_{i+k}\}$ (see [61, p. 326] for a discussion of these aspects for symmetric kernels).

Notice the square root in (2.56), which means that the singular value estimates $\sigma_i^{(n)}$ are usually much more accurate than the approximate singular functions. The same is true for the eigenvalues and eigenvectors of symmetric eigenvalue problems computed by means of Galerkin's method.

We stress that the norm $\|K\|$ of the kernel can often be evaluated or approximated rather precisely, and the quantity δ_n in (2.52) is therefore not only of theoretical interest but also of very practical importance. We refer to [178] for more details, for a discussion of the relations to regularization, and for numerical examples. Related results for the singular values can be found in [1].

We remark that the heart of the SVD computation involves computation of the stationary values of

$$\mathcal{H}[y, z] = \frac{y^T A z}{\|y\|_2 \|z\|_2},$$

which are exactly the singular values of A. If the basis functions $\phi_1, \ldots, \phi_n$ and $\psi_1, \ldots, \psi_n$ are not orthonormal, then the computational problem involved in the Galerkin approach to computing the SVE becomes computation of the stationary values of

$$\mathcal{H}_{S,T}[y, z] = \frac{y^T A z}{(y^T S y \, z^T T z)^{1/2}},$$

with symmetric matrices S and T whose elements are given by

$$s_{ij} = (\phi_i, \phi_j), \quad t_{ij} = (\psi_i, \psi_j), \quad i,j = 1, \ldots, n.$$

In this case, the stationary values of $\mathcal{H}_{S,T}$ are equal to the T^{-1}, S-singular values defined by Van Loan [350]. If S and T are well conditioned, then a change of variables, via computing the Cholesky factors of S and T, leads to the desired results. Otherwise, the RSVD (2.17) can be used.

3

Methods for Rank-Deficient Problems

In this chapter we discuss numerical methods that are suited for regularization of problems with a numerically rank-deficient coefficient matrix A, i.e., problems for which there is a well-determined gap between the large and small singular values of A.

Such problems come in two "flavors": those that involve the solution of a (possibly overdetermined) system of linear equations, and those that involve the computation of a rank-k matrix approximation to a given matrix. As we shall see, these two problems are strongly connected—although the applications in which the problems arise can be very different.

We start with a discussion of the numerical rank of a matrix, as defined via the singular value decomposition (SVD). Next we describe various solutions to rank-deficient problems, defined in terms of the SVD and generalized SVD (GSVD), and we give some perturbation bounds for these solutions. Finally, we focus on algorithms that use rank-revealing decompositions as computational alternatives to the SVD, and we relate the corresponding solutions to the SVD-based solutions.

While the difference between two vectors is naturally measured as the norm of their differences, it is perhaps not intuitively clear how to compare two subspaces. The subspace angle is a convenient tool for measuring the difference between two subspaces. Given the two subspaces S_1 and S_2 of the same dimension, spanned by the orthonormal columns of the matrices V_1 and V_2, the *subspace angle* Θ between S_1 and S_2 is defined via the relation

$$\sin \Theta = \|V_1 V_1^T - V_2 V_2^T\|_2 , \qquad (3.1)$$

where $V_i V_i^T$ is the orthogonal projection matrix associated with the subspace S_i. We note that $\sin \Theta$ is a distance function, and therefore $\sin \Theta = 0$ if and only if $S_1 = S_2$. See [154, §2.6] or [368] for more information about orthogonal projections and subspace angles.

3.1. Numerical Rank

The rank of a matrix A is defined as the number of linearly independent columns of A. The rank is equal to the number of strictly positive singular values of A. In the presence of errors (measurement errors, approximation and discretization errors, as well as rounding errors), this definition is not useful since columns of A that, from a mathematical point of view, are strictly linearly independent, may be considered almost linearly dependent from a practical point of view. Hence, a loose definition of *numerical rank* is the number of columns of A that, with respect to some error level, are practically linearly independent.

A strict and operational definition of numerical rank takes the following form. We define the *numerical ϵ-rank* r_ϵ of a matrix A, with respect to the tolerance ϵ, by

$$r_\epsilon = r_\epsilon(A, \epsilon) \equiv \min_{\|E\|_2 \leq \epsilon} \operatorname{rank}(A + E) . \qquad (3.2)$$

In other words, the ϵ-rank of A is equal to the number of columns of A that are guaranteed to be linearly independent for any perturbation of A with norm less than or equal to the tolerance ϵ. In terms of the singular values of A, the numerical ϵ-rank r_ϵ satisfies

$$\sigma_{r_\epsilon} > \epsilon \geq \sigma_{r_\epsilon+1} . \qquad (3.3)$$

This definition was already mentioned in §2.2 in connection with rank-revealing decompositions. It has been used by many authors; see, e.g., [93, §11.1], [154, §§2.5.5 and 5.5.8], [319]. Equivalent names for "numerical rank" are "effective rank" [129] and "pseudorank" [201].

We emphasize that the numerical ϵ-rank r_ϵ only makes sense when there is a *well-determined gap* between σ_{r_ϵ} and $\sigma_{r_\epsilon+1}$, as pointed out in [149]. The number r_ϵ should be robust to small perturbations of the threshold ϵ and the singular values. Otherwise, one should avoid the notion of numerical rank and instead use regularization methods designed for problems with no gap in the singular value spectrum—see Chapters 4–6.

There are two key relations that lead to more insight about the numerical ϵ-rank defined above. The first is the relation (2.2) between the singular values and vectors,

$$\|A \, v_i\|_2 = \sigma_i , \qquad i = 1, \ldots, n .$$

The second is a bound for the perturbation of the singular values when a matrix A^{exact} is perturbed (see, e.g., [154, Corollary 8.6.2]):

$$|\sigma_i(A^{\text{exact}}) - \sigma_i(A^{\text{exact}} + E)| \leq \|E\|_2 , \qquad i = 1, \ldots, n .$$

Here, $\sigma_i(M)$ denotes the ith singular value of the matrix M, and E is the perturbation of A^{exact}. Thus, in practice it is the matrix $A = A^{\text{exact}} + E$ which is the given matrix.

The first relation states that if σ_i is "small" compared to the norm $\sigma_1 = \|A\|_2$, then the corresponding right singular vector v_i is "almost" a null vector for A. The second relation quantifies the notion of "small": any singular value

3.1. NUMERICAL RANK

$\sigma_i(A^{\text{exact}} + E)$ of the perturbed matrix cannot be distinguished from a true zero singular value $\sigma_i(A^{\text{exact}})$ of the unperturbed matrix if $\sigma_i(A^{\text{exact}} + E) \leq \|E\|_2$.

In (3.2) one can substitute the Frobenius norm $\|\cdot\|_F$ for the 2-norm $\|\cdot\|_2$. With respect to this norm, the numerical ϵ-rank is the smallest integer k for which

$$\sigma_{k+1}^2 + \cdots + \sigma_n^2 \leq \epsilon^2 . \tag{3.4}$$

This definition can also be found in the literature; see, e.g., [319].

In both definitions of numerical rank, we see that the norm of the perturbation E plays a central role. Obviously, E is unknown, but we may have knowledge of the source of E, and from this information we can obtain an estimate of the norm of E.

Often E is a random matrix with elements from a certain statistical distribution. The 2-norm of random matrices $E_\mathcal{N}$ and $E_\mathcal{P}$ with elements from a normal distribution with zero mean and standard deviation σ, and from a Poisson distribution with mean value λ, respectively, are studied in [177]. The normal distribution arises in many applications. The Poisson distribution typically appears in applications where A has nonnegative elements, such as image processing. The results from [177] are that estimates $\mathcal{E}(\|E\|_2)$ of $\|E\|_2$ are given by

$$\mathcal{E}(\|E_\mathcal{N}\|_2) \approx \sigma \sqrt{\max(m,n)} \quad \text{and} \quad \mathcal{E}(\|E_\mathcal{P}\|_2) \approx \lambda \sqrt{mn} . \tag{3.5}$$

Estimes for the corresponding Frobenius norms are easy to derive due to the simple definition of this norm:

$$\mathcal{E}(\|E_\mathcal{N}\|_F) \approx \sigma \sqrt{mn} \quad \text{and} \quad \mathcal{E}(\|E_\mathcal{P}\|_F) \approx \lambda \sqrt{mn} . \tag{3.6}$$

The reason that $\mathcal{E}(\|E_\mathcal{P}\|_F) \approx \mathcal{E}(\|E_\mathcal{P}\|_2)$ is that the largest singular value of a matrix with Poisson distributed elements is much greater than the remaining singular values. See also [8] and [95] for related results.

When A is given exactly, then it is natural to let E represent the influence of rounding errors during the computation of the SVD. The singular values computed on a computer with roundoff unit $\mathbf{u}$ are the exact singular values of a slightly perturbed matrix $A + E$, where $\|E\|_2 \leq \phi \mathbf{u} \|A\|_2$ and ϕ is a slowly growing function of m and n (see, e.g., [154, §5.5.8]).

In order for definitions (3.3) and (3.4) of the numerical ϵ-rank to make sense, one must assume that the rows and columns of A are scaled such that the errors in all A's elements, on the average, are of the same size. Otherwise, (3.3) and (3.4) can lead to wrong conclusions—see [319, §6] for an illuminating example.

Even without errors in A, we are still faced with the influence of rounding errors. A good scaling strategy for full-rank problems is to equilibrate the rows or columns of A: if A is diagonally scaled from the left or right such that all rows or columns of the scaled matrix have equal 2-norm, then the condition number of the scaled matrix is no more than a factor $m^{1/2}$ or $n^{1/2}$ away from its minimum [344] (optimal diagonal scalings are related to the elements of the first and last columns of U and V in the SVD of A [155]).

We stress that one should always be careful with row scalings in (rank-deficient) least squares and regularization problems, because these scalings change the residual and thus the solution. Additional aspects of rank degeneracy, rank determination, and scaling are discussed in [319] and [325].

Associated with the numerical rank k are the *numerical null space* and the *numerical range* of A, defined as

$$\mathcal{N}_k(A) \equiv \mathrm{span}\{v_{k+1},\ldots,v_n\}, \tag{3.7}$$
$$\mathcal{R}_k(A) \equiv \mathrm{span}\{u_1,\ldots,u_k\}. \tag{3.8}$$

Obviously, $\mathcal{N}_k(A)$ is a subspace with dimension $n - k$, and for any nonzero vector $v \in \mathcal{N}_k(A)$, we have $\|Av\|_2/\|v\|_2 \leq \epsilon$; i.e., any unit vector in $\mathcal{N}_k(A)$ is mapped by A to a vector with small norm. Similarly, $\mathcal{R}_k(A)$ has dimension k, and for any nonzero vector $u \in \mathcal{R}_k(A)$, we have $\|u^T A\|_2/\|u\|_2 > \epsilon$; i.e., any unit vector in $\mathcal{R}_k(A)$ has a large component in the range of A.

We emphasize that in regularization problems that involve a regularization matrix $L \neq I_n$ and standard-form transformation, it is the numerical rank of the matrix $A L_A^\dagger$ that is important—and not that of A. Due to the SVD–GSVD relations the decision about the numerical rank should therefore ideally be based on the GSVD of the matrix pair (A, L).

Another case where the rank decision should not be based directly on A is the following. Assume that the covariance matrix for the errors in the right-hand side b is CC^T, where $C \in \mathbb{R}^{m \times q}$ has full column rank q. If $q = m$ and C is well conditioned, then one should solve the weighted least squares problem $\min \|C^{-1}(Ax - b)\|_2$ and base the rank decision on $C^{-1}A$. Otherwise, one should solve the general Gauss–Markov linear model

$$\min \|u\|_2 \quad \text{subject to} \quad Ax + Cu = b,$$

and the rank decision should be based on the matrix $((C^T)^\dagger_{A^T})^T A$, where $(C^T)^\dagger_{A^T}$ is the A^T-weighted pseudoinverse of C^T in the terminology of §2.3. Both these decisions about numerical rank involve the GSVD of the matrix pair (A^T, C^T). See §5.2 for more about regularization in the general Gauss–Markov linear model.

3.2. Truncated SVD and GSVD

Having defined the numerical rank in terms of the SVD or GSVD, we now turn to the use of these decompositions in the regularization of rank-deficient problems.

3.2. TRUNCATED SVD AND GSVD

3.2.1. Rank-Deficient Systems of Equations

In the ideal setting, without perturbations and rounding errors, the treatment of rank-deficient problems $Ax = b$ and $\min \|Ax - b\|_2$ is easy: simply ignore the SVD components associated with the zero singular values and compute the solution by means of (2.4):

$$x_{\mathrm{LS}} = \sum_{i=1}^{\mathrm{rank}(A)} \frac{u_i^T b}{\sigma_i} v_i \ .$$

In practice, A is never exactly rank deficient, but instead numerically rank deficient; i.e., it has one or more small but nonzero singular values such that $r_\epsilon < \mathrm{rank}(A) = n$. The small singular values inevitably give rise to difficulties. To see why, recall from §2.1.1 that the norm of x_{LS} is given by

$$\|x_{\mathrm{LS}}\|_2^2 = \sum_{i=1}^{n} \left(\frac{u_i^T b}{\sigma_i}\right)^2 .$$

Hence, $\|x_{\mathrm{LS}}\|_2$ is very large due to the small σ_i, unless b lies almost in the range of A, i.e., unless the last $n - r_\epsilon$ coefficients $u_i^T b$ satisfy

$$|u_i^T b| < \sigma_i , \qquad i = r_\epsilon + 1, \ldots, n \ . \tag{3.9}$$

Whenever errors are present in b, this requirement is very unlikely to be satisfied, and the solution x_{LS} is therefore dominated by the last $n - r_\epsilon$ SVD components in (2.4).

The most common approach to regularization of numerically rank-deficient problems is to consider the given matrix A as a noisy representation of a mathematically rank-deficient matrix, and to replace A by a matrix that is close to A and mathematically rank deficient. The standard choice is the rank-k matrix A_k defined as

$$A_k \equiv \sum_{i=1}^{k} u_i \sigma_i v_i^T \ ; \tag{3.10}$$

i.e., we replace the small nonzero singular values $\sigma_{k+1}, \ldots, \sigma_n$ with exact zeros. Among all rank-k matrices Z_k, the matrix A_k minimizes both the 2-norm and the Frobenius norm of the difference $A - Z_k$ (see, e.g., [154, Theorem 2.5.3]).

It is natural to choose the rank k of A_k as the numerical ϵ-rank of A, i.e., $k = r_\epsilon$, because $k < r_\epsilon$ leads to loss of information associated with large singular values, while $k > r_\epsilon$ leads to a solution with large norm and also a covariance matrix with large norm; see (4.26). Thus, the regularization (or stabilization) of the solution is achieved by projecting the ill-conditioned matrix A onto the set of rank-r_ϵ matrices.

When A is replaced by A_k, then we obtain a new least squares problem $\min \|A_k x - b\|_2$. The minimum-norm solution x_k to this problem, i.e., the unique solution with minimum 2-norm $\|x\|_2$, is given by

$$x_k = A_k^\dagger b = \sum_{i=1}^{k} \frac{u_i^T b}{\sigma_i} v_i . \tag{3.11}$$

The solution x_k is referred to as the *truncated SVD* solution, for obvious reasons. The complete method is referred to as *truncated SVD* (TSVD), and the matrix A_k in (3.10) is called the TSVD matrix. The use of this methods dates back to [201] and [352].

To summarize, the regularized (or stabilized) TSVD solution x_k is obtained by first replacing the ill-conditioned matrix A with the rank-k matrix A_k, followed by computing the minimum-norm least squares solution to

$$\min \|x\|_2 \quad \text{subject to} \quad \min \|A_k x - b\|_2 . \tag{3.12}$$

The norm of x_k is $\|x_k\|_2 = (\sum_{i=1}^{k}(u_i^T b)^2 \sigma_i^{-2})^{1/2}$, which can obviously be much smaller than the norm $\|x_{\text{LS}}\|_2$ of the least squares solution. Notice that—as in all regularization problems—we achieve this reduction in the solution's norm by allowing a larger residual norm.

A different SVD-based approach to treating a numerically rank-deficient matrix A, with numerical rank k, is to extract the k most linearly independent columns from A. To do this, we need the column permutation Π that minimizes the condition number of the submatrix $\hat{A}_k$ consisting of the first k columns of $A \Pi$. This process is called *subset selection*, [154, §12.2], and the corresponding *basic solution* x_k^{basic} is given by

$$x_k^{\text{basic}} = \Pi \begin{pmatrix} \hat{A}_k^\dagger b \\ 0 \end{pmatrix} ; \tag{3.13}$$

i.e., x_k^{basic} has zeros in those positions that correspond to the neglected columns of A.

The computation of Π involves the product of Π with the right singular matrix V of the SVD of A. If we partition this matrix as follows:

$$\Pi^T V = \begin{pmatrix} \hat{V}_{11} & \hat{V}_{12} \\ \hat{V}_{21} & \hat{V}_{22} \end{pmatrix} , \quad \hat{V}_{11} \in \mathbb{R}^{k \times k} ,$$

then minimization of the condition number of $\hat{A}_k$ is equivalent to minimizing the condition number of $\hat{V}_{11}$. The range of $\hat{A}_k$ approximates A's numerical range $\mathcal{R}_k(A)$, and the subspace angle $\hat{\Psi}_k$ between these two subspaces is bounded as follows [52, Theorem 4.1]:

$$\sin \hat{\Psi}_k \leq \|\hat{V}_{11}^{-1}\|_2 \frac{\sigma_{k+1}}{\sigma_k} . \tag{3.14}$$

3.2. Truncated SVD and GSVD

Hence, the larger the gap between σ_k and σ_{k+1}, the closer the two subspaces.

A heuristic approach to computing the permutation Π, developed in [149], is to compute a pivoted QR factorization, using ordinary column pivoting, of the submatrix $(\hat{V}_{11}^T, \hat{V}_{21}^T)$; see [154, §12.2] for more details. Alternatively, Π can be computed as part of an RRQR decomposition of A, and we return to this issue in §3.3.1.

Since the TSVD solution x_k is a regularized solution with minimum 2-norm, it is intimately connected with regularization in standard form, i.e., with side constraint $\Omega(x) = \|x\|_2$. However, as we mentioned in §2.1, it is common in regularization problems to use a more general side constraint $\Omega(x) = \|L\,x\|_2$.

To deal with such problems we can use a standard-form transformation (see §2.3) to compute the matrix $\bar{A}$ and the corresponding right-hand side $\bar{b}$, and then apply the TSVD method to $\bar{A}$ and $\bar{b}$. Hence, if k denotes the number of retained singular values of $\bar{A}$, then in terms of the GSVD of the matrix pair (A, L), we compute the intermediate TSVD matrix $Z_k = \sum_{i=p-k+1}^{p} u_i \gamma_i v_i^T$, which is the closest rank-k approximation to $A\,L_A^\dagger$. Then the solution is given by $x_{L,k} = L_A^\dagger Z_k^\dagger (b - A\,x_0) + x_0$ (cf. (2.36)), which leads to the expression

$$x_{L,k} = \sum_{i=p-k+1}^{p} \frac{u_i^T b}{\sigma_i} x_i + \sum_{i=p+1}^{n} u_i^T b\, x_i\;, \tag{3.15}$$

and $x_{L,k}$ is referred to as the *truncated GSVD* (TGSVD) solution. Notice that the last term in (3.15) is the component of $x_{L,k}$ in the null space of L. This method was defined in [179] and also analyzed in [183].

As an alternative to the TGSVD method, which is based on standard-form transformation, we can define a *modified TSVD* (MTSVD) solution $\hat{x}_{L,k}$ as the solution to the problem

$$\min \|L\,x\|_2 \quad \text{subject to} \quad \min \|A_k\,x - b\|_2\;, \tag{3.16}$$

where A_k is the TSVD matrix from (3.10). Notice the similarity to (3.12). This method was introduced in [199], and it is useful, e.g., when one has access to an SVD routine but not to a GSVD routine. If we define the matrix $V_k^o = (v_{k+1}, \ldots, v_n)$, then the MTSVD solution $\hat{x}_{L,k}$ has the form

$$\hat{x}_{L,k} = x_k - V_k^o (L\,V_k^o)^\dagger L\,x_k\;, \tag{3.17}$$

and we notice that $z = (L\,V_k^o)^\dagger L\,x_k$ is the solution to the following least squares problem:

$$\min \|(L\,V_k^o) z - L\,x_k\|_2\;. \tag{3.18}$$

For more details about the TSVD, TGSVD, and MTSVD methods we refer to [176], [179], [181], and [183].

Tikhonov regularization can also be applied successfully to rank-deficient problems—despite the fact that this method does not seem to involve the numerical rank of the matrix. In §5.3 we investigate the relationships between Tikhonov regularization and the TSVD and TGSVD methods.

3.2.2. Truncated Total Least Squares

The methods discussed so far assume that the errors in $Ax \approx b$ are confined to the right-hand side b. Although this is true in many applications, there are some problems in which A is also not precisely known. For example, A may be available only by measurements, or it may be an idealized approximation of the true operator. Discretization typically also adds errors to the matrix A. Hence, there is a need for developing methods that take into account the errors in A and their size relative to those in b. Total least squares (TLS) is one such method.

The development of the TLS technique [153], [348] was motivated by linear models $Ax \approx b$ in which both the coefficient matrix A and the right-hand side b are subject to errors. In the ordinary TLS method one allows a residual matrix as well as a residual vector, and the computational problem becomes

$$\min \|(A,b) - (\tilde{A}, \tilde{b})\|_F \qquad \text{subject to} \qquad \tilde{b} = \tilde{A}x \, . \tag{3.19}$$

(In contrast to this, the ordinary least squares method requires that $\tilde{A} = A$, and minimizes the 2-norm of the residual vector $b - \tilde{b}$.) The ordinary TLS solution can be written in terms of the SVD of (A,b): if

$$(A,b) = \bar{U}\,\bar{\Sigma}\,\bar{V}^T \, , \qquad \bar{\Sigma} = \text{diag}(\bar{\sigma}_1, \ldots, \bar{\sigma}_{n+1}) \, , \tag{3.20}$$

then the TLS solution is given by $x_{\text{TLS}} = -\bar{v}_{1:n,n+1}/\bar{v}_{n+1,n+1}$.

If A is rank deficient, then the rank-deficiency carries over to the matrix (A,b), and the same is true if A is numerically rank deficient, provided that the system $Ax \approx b$ is not highly incompatible (but then the TLS solution should be avoided). For numerically rank-deficient problems it is therefore common to neglect all the small but nonzero singular values of (A,b) and treat the matrix as exactly rank deficient—similar to the TSVD technique. This *truncated TLS* (T-TLS) technique was developed in [348, Chapter 3] and later analyzed in [120]. Both the TSVD and the T-TLS methods replace the ill-conditioned problem with a nearby one which is more well conditioned. The main difference between TSVD and T-TLS lies in the way that this is done: in TSVD the modification depends solely on A, while in T-TLS the modification depends on both A and b.

Let k denote the T-TLS truncation parameter, i.e., the number of retained singular values of (A,b), and partition the $(n+1) \times (n+1)$ matrix $\bar{V}$ such that

$$\bar{V} = \begin{pmatrix} \bar{V}_{11} & \bar{V}_{12} \\ \bar{V}_{21} & \bar{V}_{22} \end{pmatrix} \, , \qquad \text{where} \qquad \bar{V}_{11} \in \mathbb{R}^{n \times k} \, . \tag{3.21}$$

If $\bar{V}_{22} \neq 0$, then the T-TLS solution $\bar{x}_k$ exists and is given by

$$\bar{x}_k = -\bar{V}_{12}\,\bar{V}_{22}^\dagger = -\bar{V}_{12}\,\bar{V}_{22}^T\,\|\bar{V}_{22}\|_2^2 \, , \tag{3.22}$$

3.2. Truncated SVD and GSVD

and if $k = n$, then obviously $\bar{x}_n = x_{\text{TLS}}$. The norms of $\bar{x}_k$ and the corresponding TLS residual matrix are given by

$$\|\bar{x}_k\|_2 = \sqrt{\|\bar{V}_{22}\|_2^{-2} - 1} \qquad (3.23)$$

and

$$\|(A,\, b) - (\tilde{A},\, \tilde{b})\|_F = \sqrt{\bar{\sigma}_{k+1}^2 + \cdots + \bar{\sigma}_{n+1}^2} \,. \qquad (3.24)$$

We see that $\|\bar{x}_k\|_2$ increases with k, while the residual norm decreases with k.

When the T-TLS solution is computed, it is important to choose the truncation parameter k such that the resulting exactly rank-deficient TLS problem is not near-nongeneric, for then the T-TLS solution becomes unstable. This means that one should always choose k such that $\bar{\sigma}_k$ is isolated and $\|\bar{V}_{22}\|_2$ is not too small (this difficulty does not arise in TSVD).

The similarities and differences between the TSVD solution x_k and the T-TLS solution $\bar{x}_k$ are investigated in [120], [320], and [371], and perturbation theory for $\bar{x}_k$ is given in [123] and [371] (see also §3.2.4). The main conclusion is that if there is a distinct gap between σ_k and σ_{k+1}, and if the largest Fourier coefficients $|u_i^T b|$ correspond to the smallest singular values that are *retained* in $\bar{x}_k$, then the T-TLS method is superior to TSVD in suppressing the noise in A and b.

3.2.3. Matrix Approximations

As we have seen, matrix approximations play an important role in the regularization of rank-deficient systems of linear equations via the choice of rank-k approximation to A. In particular, according to the Eckart–Young–Mirsky theorem [154, Theorem 2.5.3], the TSVD matrix A_k (3.10) is the closest rank-k matrix to A in any unitarily invariant norm (including the 2-norm and the Frobenius norm), and from the SVD we immediately obtain

$$\|A - A_k\|_2 = \sigma_{k+1}\,, \qquad \|A - A_k\|_F = (\sigma_{k+1}^2 + \cdots + \sigma_n^2)^{1/2}\,. \qquad (3.25)$$

The same matrix approximation problem arises in disguise in the TGSVD method in the form of computing the closest rank-k approximation Z_k to the matrix $A L_A^\dagger$; cf. §3.2.1.

The matrix approximation view is also useful for the treatment of problems in which some of the columns of the coefficient matrix A are given exactly. Golub, Hoffman, and Stewart [147] have derived an extension of the Eckart–Young–Mirsky theorem for the closest rank-k matrix to A when ℓ columns of A are fixed, where $k \geq \ell$. Let $P = A A^\dagger$ and collect the nonfixed columns of A in the matrix $\hat{A}$. Then the modified columns are given by $P \hat{A} + \hat{B}_{k-\ell}$, where $\hat{B}_{k-\ell}$ is the TSVD approximation of rank $n - \ell$ to the matrix $(I_m - P)\,\hat{A}$.

Matrix approximations are important tools in many signal processing applications; see, e.g., the surveys in [307], [308] and the proceedings [92], [152],

[250], [343]. Here, the matrix A is derived from data constituting a noisy signal; for example, A may be a Hankel matrix as discussed in §1.4.1. If the signal were noise-free, then the matrix would be exactly rank deficient (this is part of the underlying mathematical model), but due to the noise, A has a cluster of small nonzero singular values. The numerical null space $\mathcal{N}_k(A)$ (called the noise subspace in signal processing) is then used as an approximation to the exact null space of A^{exact}, while the numerical range $\mathcal{R}_k(A)$ is used as an approximation to the range of A^{exact}.

To suppress—or filter out—the influence of the noise present in the signal, A is approximated by a rank-k matrix which, in turn, is taken to represent the new filtered signal. This process, which is identical to the TSVD method, is called rank reduction in signal processing and, quoting from [308]:

> Rank reduction is a general principle for finding the right tradeoff between model bias and model variance when reconstructing signals from noisy data.

Consider the situation where the given matrix $A = A^{\text{exact}} + E$ is a perturbation of an unknown, exactly given rank-k matrix A^{exact}, and assume that E satisfies $E^T A^{\text{exact}} = 0$ and $E^T E = \alpha^2 I_n$, where α is a constant. This E represents white noise. Then the singular values σ_i and $\bar{\sigma}_i$ of A and A^{exact}, respectively, are related as

$$\sigma_i^2 = \bar{\sigma}_i^2 + \alpha^2, \qquad i = 1, \ldots, n.$$

Hence, we can estimate A^{exact} in terms of the SVD of A as the rank-k matrix

$$A_k^{\text{LS}} = \sum_{i=1}^{k} u_i \left(\frac{\sigma_i^2 - \alpha^2}{\sigma_i^2} \right)^{1/2} \sigma_i v_i^T, \qquad (3.26)$$

which is a least squares estimate of A^{exact} [88]. Another estimate of A^{exact} is the *minimum variance estimate* A_k^{MV}, which is the rank-k matrix of the form AT that minimizes $\|AT - A^{\text{exact}}\|_2$; cf. [88]. With the same assumptions as before, this matrix can be written in terms of the SVD of A as

$$A_k^{\text{MV}} = \sum_{i=1}^{k} u_i \frac{\sigma_i^2 - \alpha^2}{\sigma_i^2} \sigma_i v_i^T. \qquad (3.27)$$

The corresponding least squares and minimum variance TSVD solutions x_k^{LS} and x_k^{MV} are given by

$$x_k^{\text{LS}} = \sum_{i=1}^{k} \left(\frac{\sigma_i^2}{\sigma_i^2 - \alpha^2} \right)^{1/2} \frac{u_i^T b}{\sigma_i} v_i \quad \text{and} \quad x_k^{\text{MV}} = \sum_{i=1}^{k} \frac{\sigma_i^2}{\sigma_i^2 - \alpha^2} \frac{u_i^T b}{\sigma_i} v_i.$$

3.2. Truncated SVD and GSVD

Both solutions x_k^{LS} and x_k^{MV} are deregularized versions of the TSVD solution x_k (3.11)—"deregularized" in the sense that the first k filter factors $\sigma_i^2/(\sigma_i^2 - \alpha^2)$ are greater than 1; see Chapter 5 for more on filter factors.

If the noise is not white, $E^T E \neq \alpha^2 I_n$, then we need algorithms that can take the actual noise into account. This is typically done by a process called "prewhitening." The central idea is that if the noisy signal is represented by a Hankel or Toeplitz matrix A, then the matrix $A^T A$ is a scaled approximation to the covariance matrix of the signal. Now, if R denotes the Cholesky factor of the covariance for the pure noise, and if R has full rank, then the matrix $(A R^{-1})^T A R^{-1}$ is a scaled approximation to the covariance matrix of a new signal whose noise component is white. Hence, we should apply TSVD to the matrix $A R^{-1}$, followed by a "dewhitening" by means of a right multiplication with R. The matrix R can be computed as the triangular QR-factor of a matrix E with the same number of columns as A, and derived from a pure noise signal.

As shown in [218], we can avoid the QR factorization of E. Let the GSVD of (A, E) be given by (2.6), i.e., $A = U \Sigma X^{-1}$ and $E = V M X^{-1}$. Then the GSVD of $(A, R) = (A, Q^T E)$ is simply

$$A = U \Sigma X^{-1}, \qquad R = (Q^T V) M X^{-1}.$$

Hence, the GSVDs of the two matrix pairs are identical except for the left singular matrix of E. Consequently, the rank-k matrix $A_{L,k}$ that we obtain after applying TGSVD to the matrix pair (A, E),

$$A_{L,k} = U \begin{pmatrix} \Sigma_k & 0 \\ 0 & I_{n-p} \end{pmatrix} X^{-1} \qquad (3.28)$$

with

$$\Sigma_k = \text{diag}(0, \ldots, 0, \sigma_{p-k+1}, \ldots, \sigma_p),$$

is identical to the matrix obtained when we "dewhiten" the rank-k TSVD matrix approximation derived from $A R^{-1}$.

In certain signal processing applications, such as signal separation and interference problems, the "noise" consists of a narrow-band signal whose covariance matrix is rank deficient. This case is discussed in [190], where it is shown that the appropriate prewhitening matrix is the A-weighted pseudoinverse of either the noise matrix E or its Cholesky factor R. This means that the rank-k matrix $A_{L,k}$ in (3.28) can also be used in connection with rank-deficient prewhitening.

In some applications, the TSVD and TGSVD matrices A_k and $A_{L,k}$—or the corresponding least squares or minimum variance estimates given by (3.26) and (3.27)—are the natural choices of matrix approximations. For example, this is the case in *direction of arrival* problems where the important information about the signal is associated with the range and the null space associated with A_k and $A_{L,k}$; see [233] for details.

In other applications, further processing is necessary to compute the desired results from the rank-k approximation. Consider noise reduction, where A is a Hankel matrix of the form (1.18) derived from a noisy signal, and where we approximate A by the TSVD matrix A_k. In order to recover a filtered signal from A_k we must convert this matrix into a nearby Hankel matrix that corresponds to the filtered signal.

A standard approach is to average along the antidiagonals (see, e.g., [44]). If the process of applying TSVD/TGSVD plus averaging is repeated iteratively, then the matrix converges to a rank-k matrix with the correct structure, but—as pointed out in [89]—this matrix may not be the structured rank-k matrix which is closest to A.

An application of the TGSVD technique to a reduction of broad-band non-white noise (i.e., $E^T E \neq I_n$) in speech processing can be found in [218]. In this application we found that repeating the TGSVD-plus-averaging process did not improve the signal-to-noise ratio in the filtered signal, compared to mere averaging.

In [194] it is shown that the filtering in the TSVD method (i.e., in the white-noise case) can be modeled by an array of parallelly connected analysis-synthesis finite-duration impulse response (FIR) filter pairs. The ith branch includes an amplification equal to σ_i, and the number of branches equals the truncation parameter k. In the ith branch, the FIR filter coefficients of the analysis filter are the elements of the right singular vectors v_i, and the synthesis filter coefficients are the same elements in reverse order. In the same paper, it is shown how the FIR filters that model the TGSVD method are related to the GSVD of the matrix pair (A, E).

In speech processing, we find experimentally [194] that the FIR filters have the shape of band-pass filters located at the speech signal's formants (i.e., maxima in the speech's power spectrum), and σ_i^2 is proportional to the power of each formant. Hence, using the TSVD/TGSVD in this application essentially corresponds to retaining the formants in the speech signal with highest energy.

3.2.4. Perturbation Bounds

We now give a few perturbation bounds related to the TSVD, TGSVD, and T-TLS methods. These and other bounds can be found in [176] and [179].[8] More general perturbation bounds can be found in [326].

Theorem 3.2.1. [176, Theorems 3.2–3.4]. *Let* $\tilde{A} = A + E$, $\tilde{b} = b + e$, *and let* A_k *and* $\tilde{A}_k$ *denote the rank-k TSVD matrices defined in* (3.10) *from* A *and* $\tilde{A}$, *respectively. Moreover, let* $x_k = A_k^\dagger b$ *and* $\tilde{x}_k = \tilde{A}_k^\dagger \tilde{b}$, *and let* Θ_k *denote the subspace angle between the null spaces* $\mathcal{N}_k(A)$ *and* $\mathcal{N}_k(\tilde{A})$. *If* $\|E\|_2 < \sigma_k - \sigma_{k+1}$, *then*

---

[8]Note the following misprints. In [176, Eqs. (26a) and (27a)], b should be replaced by $A x_k$. In [179, Eq. (4.1)], $\kappa(X)$ should be deleted.

3.2. TRUNCATED SVD AND GSVD

$$\frac{\|A_k^\dagger - \tilde{A}_k^\dagger\|_2}{\|A_k^\dagger\|_2} \leq 3 \frac{\sigma_1/\sigma_k}{\left(1 - \frac{\|E\|_2}{\sigma_k}\right)\left(1 - \frac{\|E\|_2}{\sigma_k} - \frac{\sigma_{k+1}}{\sigma_k}\right)} \frac{\|E\|_2}{\|A\|_2}, \qquad (3.29)$$

$$\frac{\|x_k - \tilde{x}_k\|_2}{\|x_k\|_2} \leq \frac{\sigma_1/\sigma_k}{1 - \frac{\|E\|_2}{\sigma_k}} \left(\frac{\|E\|_2}{\|A\|_2} + \frac{\|e\|_2}{\|A x_k\|_2} + \eta_k \frac{\|r_k\|_2}{\|A x_k\|_2} \right) + \eta_k, \qquad (3.30)$$

$$\sin \Theta_k \leq \eta_k, \qquad (3.31)$$

where $r_k = A x_k - b$ and η_k is given by

$$\eta_k = \frac{\sigma_1/\sigma_k}{1 - \frac{\|E\|_2}{\sigma_k} - \frac{\sigma_{k+1}}{\sigma_k}} \frac{\|E\|_2}{\|A\|_2}. \qquad (3.32)$$

The requirement $\|E\|_2 < \sigma_k - \sigma_{k+1}$ is necessary in order to ensure that the subspace angle Θ_k is acute; otherwise $\tilde{A}_k^\dagger$ can differ arbitrarily much from $A_k^\dagger$. Theorem 3.2.1 then states that a distinct gap between σ_k and σ_{k+1} ensures that the SVD can be successfully truncated at k, and the sensitivity of $\tilde{x}_k$ to the errors E and e is essentially governed by the ratio σ_1/σ_k—instead of $\text{cond}(A) = \sigma_1/\sigma_n$ for the ordinary solution x_{LS} in (2.4). The same is true for the numerical null space $\mathcal{N}_k(\tilde{A})$.

Theorem 3.2.2. [179, Theorem 4.1]. *Let $\tilde{b} = b + e$, and let $x_{L,k}$ and $\tilde{x}_{L,k}$ denote the TGSVD solutions (3.15) to the unperturbed and the perturbed problems, respectively. Then*

$$\frac{\|x_{L,k} - \tilde{x}_{L,k}\|_2}{\|x_{L,k}\|_2} \leq \frac{\|A\|_2 \|X\|_2}{\hat{\sigma}_{p-k+1}} \frac{\|e\|_2}{\|A x_{L,k}\|_2} \qquad (3.33)$$

$$\leq \frac{\sigma_1}{\sigma_k} \text{cond}(X) \frac{\|e\|_2}{\|A x_{L,k}\|_2}, \qquad (3.34)$$

where X and $\hat{\sigma}_{p-k+1}$ are from the GSVD of (A, L), while σ_1 and σ_k are from the SVD of A.

The inequality in (3.34) is easy to derive by means of Theorems 2.1.1 and 2.1.2. The condition number $\text{cond}(X) = \|X\|_2 \|X^{-1}\|_2$, in turn, is usually bounded approximately by $\text{cond}(L) = \|L\|_2 \|L^\dagger\|_2$, the condition number of L; see (2.15). Hence, the bound in (3.34) is roughly a factor $\text{cond}(L)$ greater than the corresponding bound in (3.30) with $E = 0$. This increase in sensitivity bound is the price that one pays for using general-form regularization with $L \neq I_n$.

Theorem 3.2.3. [371, Theorem 7.3]. *Let $\tilde{A} = A + E$ and $\tilde{b} = b + e$, and let $\bar{x}_k$ and $\tilde{\bar{x}}_k$ denote the T-TLS solutions corresponding to (A, b) and $(\tilde{A}, \tilde{b})$. Moreover, let (3.20) be the SVD of (A, b). If $\|(E, e)\|_2 < (\sigma_k - \bar{\sigma}_{k+1})/6$ then*

$$\|\bar{x}_k - \tilde{\bar{x}}_k\|_2 \leq 6 \frac{\sigma_1/\sigma_k}{1 - (\bar{\sigma}_{k+1}/\sigma_k)} \left(\frac{\|(E, e)\|_2}{\|(A, b)\|_2} + \frac{\bar{\sigma}_{k+1}}{\bar{\sigma}_1} \right) \|\bar{V}_{22}^\dagger\|_2. \qquad (3.35)$$

Theorem 3.2.3 states that if there is a distinct gap between σ_k and $\bar{\sigma}_{k+1}$—which requires a gap between σ_k and σ_{k+1} as well as b being not too inconsistent with A—then the sensitivity of $\bar{x}_k$ is governed by the ratio σ_1/σ_k. The norm $\|\bar{V}_{22}^\dagger\|_2$ in (3.35) controls the norm of $\bar{x}_k$ via Eq. (3.23). If it is large then $\|\bar{x}_k\|_2$ is also large; in this case the TLS problem is near-nongeneric and the truncation parameter k should be reduced [348, §3.4].

3.3. Truncated Rank-Revealing Decompositions

As already mentioned in §2.2, various rank-revealing decompositions have been designed as computationally advantageous alternatives to the SVD and GSVD when the numerical rank is either high or low. In the treatment of such numerically rank-deficient matrices—both linear systems of equations and matrix approximation problems with related subspace approximation problems—it is natural to make use of the rank-revealing decompositions to derive alternative methods to the SVD- and GSVD-based methods.

For example, in standard-form regularization problems, the key idea is simply to replace the rank-k matrix A_k (3.10) by another rank-k matrix which is easy to compute and update by means of a rank-revealing decomposition, and which is also close to A—although not the closest in the 2-norm or Frobenius norm.

So far, no efficient rank-revealing algorithms have been developed for matrices whose numerical rank is neither low nor high. Recently, it has been shown in [78] how approximate TSVD solutions to numerically rank-deficient problems can be computed efficiently by means of a bidiagonalization of A, followed by an iterative scheme whose steps require only $\mathcal{O}(n)$ flops. The numerical rank is "revealed" by computing all the singular values of the bidiagonal matrix, which typically costs $30n$ flops. Although no specific rank-revealing decomposition is involved, this approach is an alternative to the SVD.

3.3.1. The Use of Rank-Revealing QR Decompositions

One of the most obvious applications of the rank-revealing QR (RRQR) decomposition is in connection with subset selection problems and the computation of basic solutions. In terms of the RRQR decomposition $A\Pi = QR$, the first k columns of $A\Pi$ are given by

$$\hat{A}_k^{\mathrm{QR}} = Q \begin{pmatrix} R_{11} \\ 0 \end{pmatrix}, \qquad (3.36)$$

showing that the range of $\hat{A}_k^{\mathrm{QR}}$ is spanned by the first k columns of Q. Since the aim of the permutation matrix Π in the RRQR decomposition is to permute the most linearly independent columns of A to the front of $A\Pi$ (i.e., to maximize

3.3. Truncated Rank-Revealing Decompositions

the condition number of R_{11}), it is clear that the vector

$$x_k^{\text{basic,QR}} = \Pi \begin{pmatrix} (\hat{A}_k^{\text{QR}})^\dagger b \\ 0 \end{pmatrix} = \Pi \begin{pmatrix} R_{11}^{-1} & 0 \\ 0 & 0 \end{pmatrix}^\dagger Q^T b \qquad (3.37)$$

is a basic solution which, in general, is different from the basic solution x_k^{basic} (3.13) defined in terms of the SVD of A.

When the RRQR decomposition (2.21) is used to compute regularized solutions, then one should neglect the submatrix R_{22} with small norm, and base the regularized solution on the rank-k matrix

$$A_k^{\text{QR}} = Q \begin{pmatrix} R_{11} & R_{12} \\ 0 & 0 \end{pmatrix} \Pi^T . \qquad (3.38)$$

Obviously, $\|A - A_k^{\text{QR}}\|_2 = \|R_{22}\|_2 \leq f_{k+1}^{\text{alg}}$, where f_{k+1}^{alg} is defined in (2.23). If x_k^{QR} denotes the minimum-norm solution to $\min \|A_k^{\text{QR}} x - b\|_2$, given by

$$x_k^{\text{QR}} = (A_k^{\text{QR}})^\dagger b = \Pi \begin{pmatrix} R_{11} & R_{12} \\ 0 & 0 \end{pmatrix}^\dagger Q^T b , \qquad (3.39)$$

then x_k^{QR} is an approximation to the TSVD solution x_k, and we can derive the following approximation results (alternative results can be found in [126]).

Theorem 3.3.1. [52, Theorems 3.1 & 5.1], [126, Theorem 5.2]. *The matrix A_k^{QR} defined in (3.38) and the approximate TSVD solution x_k^{QR} satisfy*

$$\frac{\|A_k - A_k^{\text{QR}}\|_2}{\|A_k\|_2} \leq \left(1 + f_{k+1}^{\text{alg}}\right) \frac{\sigma_{k+1}}{\sigma_1} , \qquad (3.40)$$

$$\frac{\|A_k^\dagger - (A_k^{\text{QR}})^\dagger\|_2}{\|A_k^\dagger\|_2} \leq \mu \frac{\|A\|}{\|R_{11}\|_2} \frac{\|R_{22}\|_2}{\|R_{11}\|_2} \text{cond}(R_{11})^2 , \qquad (3.41)$$

$$\frac{\|x_k - x_k^{\text{QR}}\|_2}{\|x_k\|_2} \leq \text{cond}(R_{11}) \frac{\|R_{22}\|_2}{\|R_{11}\|_2}$$
$$\times \left(2 + \text{cond}(R_{11}) \frac{\|r_k\|_2}{\|R_{11}\|_2 \|x_k\|_2}\right), \qquad (3.42)$$

where $\mu = (1 + \sqrt{5})/2$ and $r_k = A x_k - b$.

RRQR decompositions also provide approximate bases for the numerical null space $\mathcal{N}_k(A) = \text{span}\{v_{k+1}, \ldots, v_n\}$ and the associated numerical range $\mathcal{R}_k(A) = \text{span}\{u_1, \ldots, u_k\}$. Two different approximate bases for $\mathcal{N}_k(A)$ are available, namely, the columns of the two matrices

$$Y_k = \Pi \begin{pmatrix} R_{11}^{-1} R_{12} \\ -I_{n-k} \end{pmatrix} \quad \text{and} \quad \Pi W_k ,$$

where the matrix W_k, as mentioned in §2.2.1, is a byproduct of the RRQR algorithm and whose columns span an approximation to the numerical null space of R. The approximate basis for $\mathcal{R}_k(A)$ consists of the first k columns of the matrix Q, and $\mathcal{R}_k(A)$ is identical to the SVD-based subset selection basis. The following approximation bounds have been derived for these subspaces.

Theorem 3.3.2. [51, Theorem 4.1], [126, Theorem 5.1]. *The subspace angle Φ_k^{QR} between $\mathcal{R}_k(A)$ and the range of the first k columns of Q satisfies*

$$\sin \Phi_k^{\mathrm{QR}} \leq \mathrm{cond}(R_{11}) \, \|R_{22}\|_2/\|R_{11}\|_2 \, , \tag{3.43}$$

and the subspace angle Ψ_k^{QR} between $\mathcal{N}_k(A)$ and $\mathcal{R}(Y_k)$ satisfies

$$\sin \Psi_k^{\mathrm{QR}} \leq \frac{\mathrm{cond}(R_{11})^2 \, \|R_{22}\|_2^2/\|R_{11}\|_2^2}{1 - \mathrm{cond}(R_{11})^2 \, \|R_{22}\|_2^2/\|R_{11}\|_2^2} \, . \tag{3.44}$$

The subspace angle $\tilde{\Psi}_k^{\mathrm{QR}}$ between $\mathcal{N}_k(A)$ and $\mathcal{R}(\Pi W_k)$ satisfies

$$\sin \tilde{\Psi}_k^{\mathrm{QR}} \leq \left(1 + f_{k+1}^{\mathrm{alg}}\right) \frac{\sigma_{k+1}}{\sigma_k} \, . \tag{3.45}$$

Since $\|R_{11}^{-1}\|_2 \approx \sigma_k^{-1}$ and $\|R_{22}\|_2 \approx \sigma_{k+1}$, we see that the "quality" of the solutions, as well as the subspaces, depends on the gap between σ_k and σ_{k+1}. We remark that the approximate TSVD solution x_k^{QR} can be improved by means of improvement of the approximate null space; see [51] and [126, §6] for more details.

The subspace angle Φ_k is important when comparing the RRQR-based basic solution $x_k^{\mathrm{basic,QR}}$ with the SVD-based basic solution x_k^{basic}. Although the two subset selection algorithms are likely to choose different sets of columns of A, the subspaces spanned by these two sets of columns may be close. If we compare (3.43) with (3.14) then we conclude that the subspaces are indeed close whenever σ_{k+1}/σ_k is small.

An important application of the matrices Y_k and ΠW_k is in connection with the T-TLS method from §3.2.2, where one can replace the matrix $\binom{\tilde{V}_{12}}{\tilde{V}_{22}}$ in (3.22) with the matrix obtained from orthonormalization of the columns of either Y_k or ΠW_k. See [52, §6] and [122] for more details.

3.3.2. The Use of UTV Decompositions

The rank-revealing UTV decompositions are used in the same manner as the RRQR decomposition to compute regularized solutions (but not basic solutions), by neglecting the submatrices R_{12} and R_{22} in the URV decomposition (2.28) and the submatrices L_{21} and L_{22} in the ULV decomposition (2.29). This leads to the two rank-k matrices

$$A_k^{\mathrm{ULV}} = U \begin{pmatrix} L_{11} & 0 \\ 0 & 0 \end{pmatrix} V^T \, , \quad A_k^{\mathrm{URV}} = U \begin{pmatrix} R_{11} & 0 \\ 0 & 0 \end{pmatrix} V^T \, , \tag{3.46}$$

3.3. Truncated Rank-Revealing Decompositions

defined in terms of (2.28) and (2.29), respectively, and the corresponding solutions

$$x_k^{\text{ULV}} = (A_k^{\text{ULV}})^\dagger b = V \begin{pmatrix} L_{11}^{-1} & 0 \\ 0 & 0 \end{pmatrix} U^T b \qquad (3.47)$$

and

$$x_k^{\text{URV}} = (A_k^{\text{URV}})^\dagger b = V \begin{pmatrix} R_{11}^{-1} & 0 \\ 0 & 0 \end{pmatrix} U^T b . \qquad (3.48)$$

The UTV decompositions also provide approximate basis vectors for the subspaces $\mathcal{N}_k(A)$ and $\mathcal{R}_k(A)$, given by the last $n - k$ columns of V and the first k columns of U. The approximate null-space basis can be used to compute an approximate T-TLS solution [349].

Approximation bounds for all these quantities are given below (the bound in (3.51) was also derived in [247]).

Theorem 3.3.3. [121, Corollaries 2.3 and 2.5]. *Let Ψ_k^{ULV} and Ψ_k^{URV} denote the subspace angles between $\mathcal{N}_k(A)$ and the spaces spanned by the last $n - k$ columns of the matrices V from the ULV and URV decompositions, respectively. Similarly, let Φ_k^{ULV} and Φ_k^{URV} denote the subspace angles between $\mathcal{R}_k(A)$ and the spaces spanned by the first k columns of the matrices U from the ULV and URV decompositions, respectively. If $\sigma_k(M)$ denotes the kth singular value of M, then*

$$\sin \Psi_k^{\text{ULV}} \leq \frac{\|L_{21}\|_2 \, \|L_{22}\|_2}{\sigma_k(L_{11})^2 - \|L_{22}\|_2^2}, \qquad (3.49)$$

$$\sin \Phi_k^{\text{ULV}} \leq \frac{\sigma_k(L_{11}) \, \|L_{21}\|_2}{\sigma_k(L_{11})^2 - \|L_{22}\|_2^2}, \qquad (3.50)$$

$$\sin \Psi_k^{\text{URV}} \leq \frac{\sigma_k(R_{11}) \, \|R_{12}\|_2}{\sigma_k(R_{11})^2 - \|R_{22}\|_2^2}, \qquad (3.51)$$

$$\sin \Phi_k^{\text{URV}} \leq \frac{\|R_{12}\|_2 \, \|R_{22}\|_2}{\sigma_k(R_{11})^2 - \|R_{22}\|_2^2} . \qquad (3.52)$$

Theorem 3.3.4. [126, Theorems 3.2–3]. *If $\|(L_{21}, L_{22})\|_2 \leq \sigma_k(L_{11})$ and $\|(R_{12}^T, R_{22}^T)\|_2 \leq \sigma_k(R_{11})$, then the quantities from (3.46)–(3.48) satisfy*

$$\frac{\|A_k^\dagger - (A_k^{\text{ULV}})^\dagger\|_2}{\|A_k^\dagger\|_2} \leq \mu \sin \Phi_k^{\text{ULV}} \|A\|_2 \|L_{11}^{-1}\|_2 , \qquad (3.53)$$

$$\frac{\|A_k^\dagger - (A_k^{\text{URV}})^\dagger\|_2}{\|A_k^\dagger\|_2} \leq \mu \sin \Psi_k^{\text{URV}} \|A\|_2 \|R_{11}^{-1}\|_2 , \qquad (3.54)$$

$$\frac{\|x_k - x_k^{\text{ULV}}\|_2}{\|x_k\|_2} \leq \text{cond}(L_{11}) \sin \Phi_k^{\text{ULV}} \frac{\|r_k\|_2}{\|L_{11}\|_2 \|x_k\|_2} + \sin \Psi_k^{\text{ULV}} , \qquad (3.55)$$

$$\frac{\|x_k - x_k^{\text{URV}}\|_2}{\|x_k\|_2} \leq \text{cond}(R_{11}) \left(\sin \Psi_k^{\text{URV}} \frac{\|R_{12}\|_2}{\|R_{11}\|_2} \right.$$

$$\left. + \sin \Phi_k^{\text{URV}} \frac{\|r_k\|_2}{\|R_{11}\|_2 \|x_k\|_2} \right) + \sin \Psi_k^{\text{URV}} , \qquad (3.56)$$

From the above two theorems we conclude that the quality of the approximate subspaces and solutions mainly depends on the norm of the off-diagonal submatrix, i.e., $\|L_{21}\|_2$ and $\|R_{12}\|_2$. It is for this reason that the block QR improvement scheme mentioned in §2.2.2 is important. The main differences between the bounds for the ULV and URV subspace angles in Theorem 3.3.3 are the factors $\|L_{22}\|_2$ and $\sigma_k(R_{11})$ (for the approximate null spaces) and the factors $\sigma_k(L_{11})$ and $\|R_{22}\|_2$ (for the approximate ranges). Since $\|L_{22}\|_2 \approx \sigma_{k+1}$ while $\sigma_k(R_{11}) \approx \sigma_k$, the ULV-based approximate null space is likely to be more accurate than the URV-based subspace. The reverse is true for the approximations to the numerical ranges. Numerical examples that illustrate the above bounds can be found in [126] as well as in §3.4.2.

We conclude this section with a brief discussion of the use of the rank-revealing ULLV decomposition defined in §2.2.4 for two matrices $A \in \mathbb{R}^{m \times n}$ and $B \in \mathbb{R}^{p \times n}$. As already mentioned in §3.2.3, the TGSVD method is the method of choice in signal processing applications where "prewhitening" is used to compensate for colored noise. In these applications, it is natural to replace the GSVD of the pair (A, B) with the ULLV decomposition. This is indeed possible due to the following theorem.

Theorem 3.3.5. [190]. *Let the ULLV decomposition of (A, B) be given by either (2.30) or (2.31), depending on p. If $p \geq n$ then*

$$B^\dagger = B_A^\dagger = V \hat{L}^{-1} U_B^T, \qquad A B^\dagger = U_A L U_B^T. \tag{3.57}$$

If $p < n$ then $B^\dagger$, $B_A^\dagger$, and $A B_A^\dagger$ are given by

$$B^\dagger = V \begin{pmatrix} \hat{L}^{-1} \\ 0 \end{pmatrix} U_B^T, \tag{3.58}$$

$$B_A^\dagger = V \begin{pmatrix} \hat{L}^{-1} \\ -L_{22}^{-1} L_{21} \end{pmatrix} U_B^T, \tag{3.59}$$

$$A B_A^\dagger = U_A \begin{pmatrix} L_{11} \\ 0 \end{pmatrix} U_B^T, \tag{3.60}$$

where L_{11} is the leading $p \times p$ submatrix of L.

We see that the ULLV decomposition of (A, B), defined in (2.30) and (2.31), immediately yields a ULV decomposition of the matrix $A B_A^\dagger$. Hence, the ULLV decomposition of (A, B) reveals the numerical rank of $A B_A^\dagger$, and it produces approximations to the numerical null space and range of this matrix. It can also be used to compute an approximation to the matrix $A_{L,k}$ defined in (3.28). The ULLV decomposition is therefore a computationally attractive alternative to the GSVD whenever a standard-form transformation is involved in treating the matrix pair (A, B).

3.4. Truncated Decompositions in Action

We conclude this chapter with numerical examples that illustrate the accuracy of some of the above-mentioned methods for treating numerically rank-deficient problems. We use the test matrix from §1.4.1, i.e., a Hankel matrix A derived from a signal consisting of two sinusoids plus additive noise. The signal has length $N = 128$, the number of columns in A is $n = 7$, and the signal parameters are $a_1 = 1$, $a_2 = 0.9$, $\omega_1 = \pi/10$, $\omega_2 = \pi/3$, $\phi_1 = 0$, and $\phi_2 = \pi/4$.

3.4.1. Subset Selection by SVD and RRQR

As mentioned in §3.2.1, the aim of subset selection is to determine the k most linearly independent columns of the matrix A. There are two algorithms for doing this: one is based on the SVD of A, and the other on the RRQR decomposition of A.

In this example, the numerical rank is $k = 4$, and we consider the subspaces $\mathcal{R}(\hat{A}_k)$ and $\mathcal{R}(\hat{A}_k^{\mathrm{QR}})$ spanned by the first four columns of the pivoted matrices $A\Pi$, where the column permutation Π is determined by either the SVD method or the RRQR decomposition. As a measure of the accuracy of these subspaces, we compute the subspace angle between these subspaces and the reference subspace[9] $\mathcal{R}_k(A^{\mathrm{exact}})$, i.e., the numerical range of the unperturbed matrix A^{exact} associated with the noise-free signal.

We add white (Gaussian) noise to the signal, and we use three noise levels (standard deviations) σ_{noise} equal to 10^{-1}, 10^{-2}, and 10^{-3}. For each noise level we generate 10 matrices, and the corresponding relative matrix perturbations $\|A - A^{\mathrm{exact}}\|_2/\|A^{\mathrm{exact}}\|_2$ are, on the average, $7 \cdot 10^{-2}$, $7 \cdot 10^{-3}$, and $7 \cdot 10^{-4}$.

Figure 3.1 shows the subspace angles between $\mathcal{R}_k(A^{\mathrm{exact}})$ and $\mathcal{R}(\hat{A}_k)$ for the SVD-based method (top part), and between $\mathcal{R}_k(A^{\mathrm{exact}})$ and $\mathcal{R}(\hat{A}_k^{\mathrm{QR}})$ for the RRQR-based method (bottom part). For each method, there are three clusters representing the three noise levels 10^{-1} (circles), 10^{-2} (plusses), and 10^{-3} (crosses). We see that the subspace angles are proportional to the noise level, and that the SVD-based subspace $\mathcal{R}(\hat{A}_k)$ is slightly more accurate than the RRQR-based subspace $\mathcal{R}(\hat{A}_k^{\mathrm{QR}})$, at the expense of the much larger computational effort involved in the SVD method.

3.4.2. Minimum-Norm Solutions and Null Spaces by SVD and UTV

In our second example, we use the SVD as well as the ULV and URV decompositions to compute minimum-norm solutions and approximate numerical null spaces. The unperturbed matrix A^{exact} is the same as above, and the corresponding right-hand side $b^{\mathrm{exact}} = A^{\mathrm{exact}} x^{\mathrm{exact}}$ is generated with

---
[9] Alternatively, one can apply the SVD-based subset selection algorithm to A^{exact} and use the associated subspace as reference subspace. This subspace is so close to $\mathcal{R}_k(A^{\mathrm{exact}})$ that the numerical results are practically the same.

64   3. METHODS FOR RANK-DEFICIENT PROBLEMS

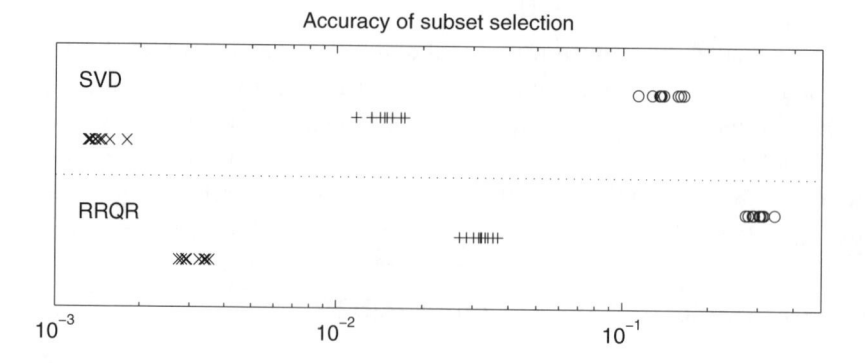

FIG. 3.1. *The subset selection example. Subspace angles between $\mathcal{R}_k(A^{\mathrm{exact}})$ and $\mathcal{R}(\hat{A}_k)$ (top part), and between $\mathcal{R}_k(A^{\mathrm{exact}})$ and $\mathcal{R}(A_k^{\mathrm{QR}})$ (bottom part), for three noise levels: 10^{-1} (circles), 10^{-2} (plusses), and 10^{-3} (crosses).*

TABLE 3.1. *Representative values of the relative perturbations in A and b, and representative singular value of A, for the three noise levels in the second example.*

σ_{noise}	0.4	0.1	0.04
$\|A - A^{\mathrm{exact}}\|_2 / \|A^{\mathrm{exact}}\|_2$	0.284	0.092	0.029
$\|b - b^{\mathrm{exact}}\|_2 / \|b^{\mathrm{exact}}\|_2$	0.092	0.031	0.009
σ_1	18.09	17.61	17.40
σ_2	15.85	15.74	15.51
σ_3	14.93	15.07	14.83
σ_4	8.54	7.80	7.63
σ_5	4.80	1.56	0.45
σ_6	4.71	1.38	0.39
σ_7	4.46	1.26	0.38

$x^{\mathrm{exact}} = (1, \ldots, 1)^T$. Then A and b are obtained by adding white noise with standard deviation σ_{noise} to the signal as well as to b^{exact}. Three noise levels σ_{noise} are used: 0.4, 0.1, and 0.04, and for each noise level we generate 25 problems. Representative values of the corresponding relative perturbations in A and b, as well as representative singular values of A, are shown in Table 3.1.

All our numerical results for this example are shown in Fig. 3.2 which, indeed, contains a lot of information.

First we consider the minimum-norm solutions, i.e., the TSVD solution x_k and the ULV- and URV-based approximations x_k^{ULV} and x_k^{URV}. All three solutions are compared to the TSVD solution x_k^{exact} to the unperturbed problem involving A^{exact} and b^{exact}. The top part of Fig. 3.2 shows the rela-

3.4. Truncated Decompositions in Action

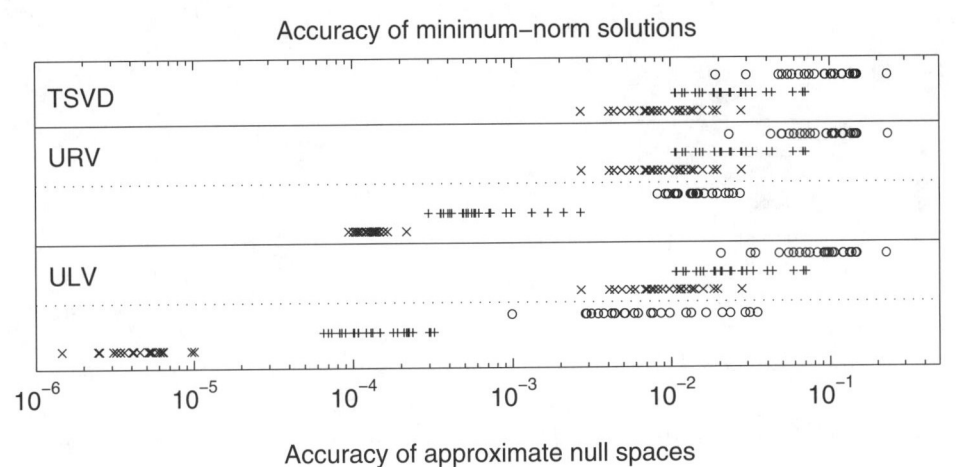

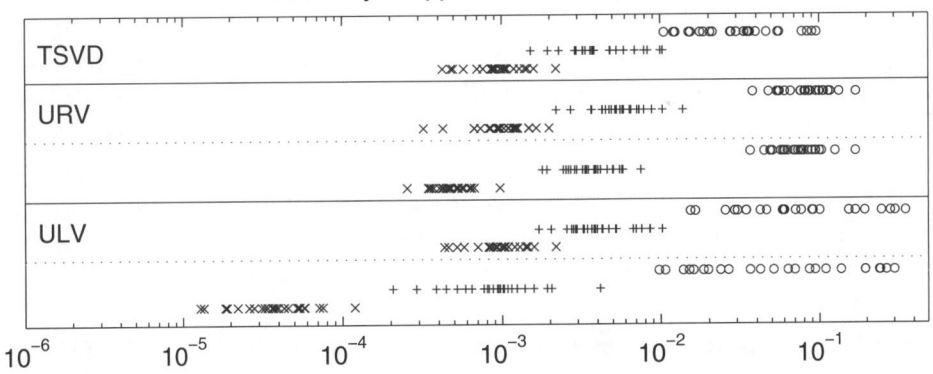

FIG. 3.2. *Numerical results for the second example, for three noise levels: 0.4 (circles), 0.1 (plusses), and 0.04 (crosses). The top figure shows the accuracy of three minimum-norm solutions, namely, the TSVD solution x_k and the UTV solutions x_k^{URV} and x_k^{ULV}. The bottom part shows the accuracy of the numerical null spaces computed by means of the SVD and the two UTV decompositions.*

tive errors $\|x_k - x_k^{\mathrm{exact}}\|_2 / \|x_k^{\mathrm{exact}}\|_2$, $\|x_k^{\mathrm{ULV}} - x_k^{\mathrm{exact}}\|_2 / \|x_k^{\mathrm{exact}}\|_2$, and $\|x_k^{\mathrm{URV}} - x_k^{\mathrm{exact}}\|_2 / \|x_k^{\mathrm{exact}}\|_2$ for all three noise levels; for ULV and URV the results are plotted above the dotted lines. All three methods produce minimum-norm solutions with almost the same accuracy (there is no superior method), and the accuracy is proportional to the noise level.

Figure 3.2 also shows the relative differences $\|x_k^{\mathrm{ULV}} - x_k\|_2 / \|x_k\|_2$ and $\|x_k^{\mathrm{URV}} - x_k\|_2 / \|x_k\|_2$ between the UTV solutions and the TSVD solutions; these results are plotted below the dotted lines. We see that the ULV solutions x_k^{ULV} are better approximations to the TSVD solutions than the URV solutions x_k^{URV}. However, we emphasize that both UTV solutions have the same accuracy when considered as approximations to x_k^{exact}.

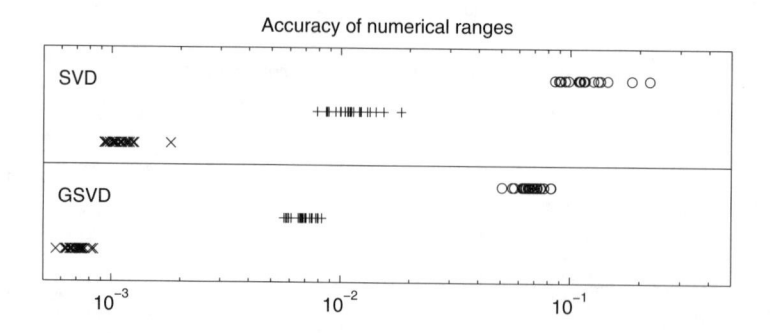

FIG. 3.3. *Numerical ranges computed by means of SVD and GSVD. Subspace angles between* $\mathcal{R}_k(A^{\mathrm{exact}})$ *and* $\mathcal{R}(A_k)$ *(top part), and between* $\mathcal{R}_k(A^{\mathrm{exact}})$ *and* $\mathcal{R}(A_{B,k})$ *(bottom part), for three noise levels:* 10^{-1} *(circles),* 10^{-2} *(plusses), and* 10^{-3} *(crosses).*

Next, we consider the accuracy of the approximate numerical null spaces as determined via the SVD and the UTV decompositions. The bottom part of Fig. 3.2 shows the subspace angles between these approximate null spaces and the reference null space $\mathcal{N}_k(A^{\mathrm{exact}})$, i.e., the numerical null space of A^{exact}. The ULV and URV results are plotted above the dotted lines. Similar to the minimum-norm solutions, we see that all three methods produce results with approximately the same accuracy, and the accuracy is again proportional to the noise level.

We also computed the subspace angles between the SVD-based null space and the two UTV-based null spaces, and these results are shown in Fig. 3.2 below the dotted lines. Considered as approximations to the SVD-based null spaces, the approximate null spaces from the ULV decomposition are clearly superior to those from the URV decomposition. But again we emphasize that, considered as approximations to "exact" null spaces associated with A^{exact}, both UTV-based subspaces have the same accuracy.

3.4.3. Numerical Ranges by SVD and GSVD

In our third example, we demonstrate that the use of the GSVD instead of the SVD can result in more accurate results. The test matrix is the same as before, except that we now add *colored* noise to the pure signal. The colored noise is generated by applying a recursive filter of the form

$$\eta_i \leftarrow \eta_i - 0.98\,\eta_{i-1}\,, \qquad i = 2, \ldots, N\,,$$

to the white noise η_i. The constant -0.98 leads to colored noise in which the high-frequency components are emphasized.

One approximation to the range $\mathcal{R}_k(A^{\mathrm{exact}})$ of the unperturbed matrix is the numerical range $\mathcal{R}_k(A)$ which, in turn, is identical to the range of the TSVD

3.4. TRUNCATED DECOMPOSITIONS IN ACTION

matrix A_k (3.10). Another approximation is obtained by instead considering the range of the TGSVD matrix $A_{E,k}$ (3.28), where E is a Hankel matrix of the same dimensions as A, derived from the pure noise signal. This corresponds to the use of "prewhitening" as discussed in §3.2.3, and the effect is that we compensate for the dominating high-frequency components of the noise. In terms of the GSVD of (A, E), the subspace $\mathcal{R}(A_{E,k})$ is spanned by the k left singular vectors u_i of A corresponding to the k largest generalized singular values.

Numerical results for three noise levels $5 \cdot 10^{-2}$, $5 \cdot 10^{-3}$, and $5 \cdot 10^{-4}$ are shown in Fig. 3.3. For each noise level we generate 25 test problems. The subspace angles between $\mathcal{R}_k(A^{\text{exact}})$ and $\mathcal{R}(A_k)$ are shown in the top part of the figure, while the subspace angles between $\mathcal{R}_k(A^{\text{exact}})$ and $\mathcal{R}(A_{E,k})$ are shown in the bottom part. The latter subspace is clearly a more accurate approximation to the exact range than the numerical range of A. This illustrates the advantage of "prewhitening" the signal via the standard-form transformation with the matrix E.

4

Problems with Ill-Determined Rank

The purpose of this chapter is to summarize important results about discrete ill-posed problems, i.e., systems of equations (either square or overdetermined) derived from discretization of ill-posed problems. The main feature of these problems is that all the singular values of the coefficient matrix decay gradually to zero, with no gap anywhere in the spectrum. Whatever threshold ϵ is used in Eq. (3.3), the numerical ϵ-rank is highly ill determined, and therefore the concept of "numerical rank" is not useful for these problems.

As a consequence, the regularization of discrete ill-posed problems is more complicated than merely filtering out a cluster of small singular values. For this reason, it is convenient to have a variety of mathematical tools at hand for obtaining more insight into the problem as well as the available regularization methods. Among these tools we find the filter factors, the resolution matrix, and the L-curve, all of which are described in detail below. Numerical examples that illustrate all these tools are presented in the last section of this chapter.

4.1. Characteristics of Discrete Ill-Posed Problems

From a strictly mathematical point of view, a finite-dimensional problem always satisfies the Picard condition (1.10), the minimum-norm solution is stable, and no regularization is required. Indeed, in a purely mathematical sense the transformation of a continuous problem to a discrete problem ("regularization by discretization") always has a regularizing effect; see, e.g., [160, Chapter 4], [226, Chapter 3], or [227, Chapter 17]. However, this point of view does not account for the disastrous effects of rounding errors when the system is solved, due to the huge condition number of the coefficient matrix; cf. [160, Eq. (4.10)].

In practical treatments of discrete ill-posed problems it is therefore necessary to incorporate some kind of regularization in the solution procedure for the discretized system $Ax = b$ or $\min \|Ax - b\|_2$, in order to compute a useful solution. It is also convenient to introduce the concept of a discrete Picard condition, equivalent to the Picard condition described in §4.5.

To summarize the results from Chapters 1 and 2, in the *absence of errors* a discrete ill-posed problem is characterized by a coefficient matrix A^{exact} whose singular values σ_i^{exact} all decay gradually to zero and whose singular vectors

u_i^{exact} and v_i^{exact} tend to have more sign changes in their elements as the index i increases, i.e., as the corresponding σ_i^{exact} decreases. Moreover, the coefficients $|(u_i^{\text{exact}})^T b^{\text{exact}}|$ decay, on the average, to zero at least as fast as the singular values σ_i^{exact}.

In practice, we are faced with various types of errors in A and b. One source of errors is the discretization process involved in setting up the linear system, and the approximation errors influence both A and b. Another common source of errors are measurement errors; these errors most often contribute to the right-hand side b. Finally, we cannot avoid the rounding errors involved in the numerical computations with A and b, and these rounding errors can be interpreted as perturbations of the input data A and b.

The effect of all these errors is that the singular values σ_i and Fourier coefficients $u_i^T b$ do not behave exactly as described above. The singular values σ_i decrease monotonically (by definition) until they tend to settle at a level τ_A determined by the errors in A. The coefficients $|u_i^T b|$ also decay, on the average, until they settle at a level τ_b determined by the errors in b.

For example, if the elements of the perturbation $e = b - b^{\text{exact}}$ are unbiased and uncorrelated with covariance matrix $\sigma_0^2 I_m$, then the expected value of $\|e\|_2$ satisfies $\mathcal{E}(\|e\|_2^2) = m\,\sigma_0^2$ and the expected value of the Fourier coefficients of e are given by

$$\mathcal{E}\big(|u_i^T e|\big) = \sigma_0 , \qquad i = 1, \ldots, n .$$

As a consequence, the Fourier coefficients $|u_i^T b|$ of the perturbed right-hand side level off at $\tau_b \approx \sigma_0$, because the Fourier coefficients are dominated by $|u_i^T e|$ for large indices i.

Similarly, if the elements of the matrix perturbation $E = A - A^{\text{exact}}$ are normally distributed with zero mean and standard deviation σ_0, then it follows from (3.5) that the expected value of $\|E\|_2$ is approximately $\sigma_0 \sqrt{m}$. Hence, if all the singular values of A^{exact} decay gradually to zero, then we expect the singular values of A to level off at $\tau_A \approx \sigma_0 \sqrt{m}$.

The two error levels τ_A and τ_b for the singular values and the Fourier coefficients, respectively, determine how much information about the underlying exact system (with A^{exact} and b^{exact}) can be extracted from the given system with A and b. Assume that the singular values σ_i level off at τ_A for $i \geq i_A$, and that the Fourier coefficients $|u_i^T b|$ level off at τ_b for $i \geq i_b$. Then we can only expect to recover those singular value decomposition (SVD) components of the solution for which the errors in σ_i and $u_i^T b$ do not dominate, i.e., the components $u_i^T b / \sigma_i$ for $i \leq \min(i_A, i_b)$.

The typical situation is that the measurement errors in b are larger than the other types of errors in A and b, and that the relative right-hand side errors $\|b^{\text{exact}} - b\|_2 / \|b^{\text{exact}}\|_2$ are therefore larger than the relative matrix errors $\|A^{\text{exact}} - A\|_2 / \|A^{\text{exact}}\|_2$. In this situation, the coefficients $|u_i^T b|$ level off at τ_b for $i \geq i_b$ *before* the σ_i level off at τ_A, and therefore we can—roughly—recover the first i_b SVD components of the solution. The remaining $n-i_b$ SVD components are dominated by the errors. These components dominate the undesired ordinary (least squares) solution (2.4), and therefore they should be filtered out in the regularized solution. This situation is similar to the difficulties that are encounted in connection with numerically rank-deficient problems; cf. §3.2.

Whenever the exact solution's Fourier coefficients $|v_i^T x^{\text{exact}}|$, on the average, decrease with i, it is informative to define the *effective resolution limit* η_{res} as the smallest coefficient $|v_i^T x^{\text{exact}}|$ that can be recovered from the given A and b.

If the right-hand side errors dominate, i.e., if $i_A > i_b$, then for all $i < i_A$ we have $\sigma_i \approx \sigma_i^{\text{exact}}$ while the solution's Fourier coefficients consist of two components, $v_i^T x = v_i^T x^{\text{exact}} + (u_i^T b)/\sigma_i$. At $i = i_b$, the two components are of approximately the same size, $|v_{i_b}^T x^{\text{exact}}| \approx |u_{i_b}^T b|/\sigma_{i_b} \approx \tau_b/\sigma_{i_b}$, and it follows that the effective resolution limit is

$$\eta_{\text{res}} = |v_{i_b}^T x^{\text{exact}}| \approx \frac{\tau_b}{\sigma_{i_b}} \quad \text{when} \quad i_A > i_b . \tag{4.1}$$

If the matrix errors dominate, i.e., if $i_A < i_b$, then for all $i < i_b$ we have $|v_i^T x| \approx |v_i^T x^{\text{exact}}|$. Moreover, for $i < i_A$ we have $\sigma_i \approx \sigma_i^{\text{exact}}$, while for $i > i_A$ the errors in the singular values dominate, $\sigma_i \approx \tau_A$, and therefore we have $|v_i^T x| \approx |u_i^T b|/\tau_A$. At $i = i_a$ we have $\sigma_{i_A} \approx \tau_A$, and the effective resolution limit is now

$$\eta_{\text{res}} = |v_{i_A}^T x^{\text{exact}}| \approx \frac{|u_{i_A}^T b|}{\tau_A} \quad \text{when} \quad i_A < i_b . \tag{4.2}$$

In both cases, η_{res} given by (4.1) and (4.2) is the size of the smallest SVD component that can be recovered in the solution, given the noisy A and b.

In the above analysis, we have assumed that the right singular vectors v_i are a suitable basis for the solution. If regularization in general form is needed to provide other basis vectors (namely, the generalized SVD (GSVD) vectors x_i), then a similar analysis can be carried out in terms of the GSVD quantities.

4.2. Filter Factors

In light of the above discussion, and speaking in general terms, we can say that regularization of a discrete ill-posed problem is a matter of finding out *which* erroneous SVD components to filter out, and *how* to filter them out. The *amount* of regularization or filtering is also extremely important, and we return to this issue in Chapter 7.

The choice of basis vectors for the regularized solution is another important aspect. In a sense, the singular vectors u_i and v_i are the optimal basis vectors for the column space and the row space of A, respectively, but the vectors v_i are not necessarily the best basis vectors for the desired regularized solution. It is for this reason that regularization in general form is so important. See, e.g., [58] and [354] for examples in helioseismology and inverse Laplace transformation, respectively.

In order to understand these aspects in more detail, it is convenient to introduce the concept of filter factors. If A has full rank then we can always write the regularized solution x_{reg} in the form

$$x_{\text{reg}} = V \Theta \Sigma^\dagger U^T b \quad \text{or} \quad x_{\text{reg}} = X \Theta \begin{pmatrix} \Sigma^\dagger & 0 \\ 0 & I_{n-p} \end{pmatrix} U^T b$$

using the SVD or GSVD, respectively, of A. If $\Theta \in \mathbb{R}^{n \times n}$ is a diagonal matrix, $\Theta = F = \text{diag}(f_i)$, then the diagonal elements f_i are called the *filter factors* for the regularization method.

Specifically, for many regularization methods in standard form with $L = I_n$, such as Tikhonov regularization

$$\min \{ \|A x - b\|_2^2 + \lambda^2 \|x\|_2^2 \} \, ,$$

we can write the regularized solution x_{reg} and the corresponding residual vector $b - A x_{\text{reg}}$ in terms of the SVD of A in the generic forms

$$x_{\text{reg}} = \sum_{i=1}^{n} f_i \frac{u_i^T b}{\sigma_i} v_i \tag{4.3}$$

and

$$b - A x_{\text{reg}} = \sum_{i=1}^{n} (1 - f_i) u_i^T b \, u_i + \sum_{i=n+1}^{m} u_i^T b \, u_i \, .$$

Similarly, for many regularization methods in general form with $L \neq I_n$, such as Tikhonov regularization in general form,[10]

$$\min \{ \|A x - b\|_2^2 + \lambda^2 \|L x\|_2^2 \} \, , \tag{4.4}$$

we can write the regularized solution x_{reg}, as well as the corresponding vectors $L x_{\text{reg}}$ and $b - A x_{\text{reg}}$, in terms of the GSVD of (A, L) in the generic forms

$$x_{\text{reg}} = \sum_{i=1}^{p} f_i \frac{u_i^T b}{\sigma_i} x_i + \sum_{i=p+1}^{n} u_i^T b \, x_i \, , \tag{4.5}$$

$$L x_{\text{reg}} = \sum_{i=1}^{p} f_i \frac{u_i^T b}{\gamma_i} v_i \, , \tag{4.6}$$

and

$$b - A x_{\text{reg}} = \sum_{i=1}^{p} (1 - f_i) u_i^T b \, u_i + (I_m - U U^T) b \, . \tag{4.7}$$

---

[10]We assume throughout that $\mathcal{N}(A) \cap \mathcal{N}(L) = \{0\}$, in which case the solution is unique.

4.2. Filter Factors

Recall that $x_0 = \sum_{i=p+1}^n u_i^T b\, x_i$ is the unregularized component of x_{reg}. The vector $(I_m - U U^T)\, b$ in (4.7) is the incompatible component of b which lies outside the range of A.

In the above equations,[11] the filter factors f_i for the particular regularization methods characterize the damping or filtering of the SVD/GSVD components. For some methods, there exist explicit formulas for the filter factors; for other methods, there are no known expressions for the filter factors. In the next two chapters we give several examples of filter factors, and filter factors for other methods can be found in [108, §4].

The formulations in (4.3) and (4.5) are also valid if the underlying regularization method is of the form

$$\min \|A\, x - b\|_2 \quad \text{subject to} \quad x \in \mathcal{S}_x, \qquad (4.8)$$

where $\mathcal{S}_x$ is a k-dimensional subspace and k is the regularization parameter. For example, regularizing conjugate gradient (CG) iterations (cf. §6.3) is a regularization method of the form (4.8) for which we can derive the corresponding filter factors. However, we emphasize that there are also regularization methods for which Θ is not diagonal (and the formulations (4.3) and (4.5) therefore do not hold); the mollifier methods and maximum entropy regularization described in §§5.5 and 5.6 are examples of such methods.

One important piece of insight that we obtain from Eqs. (4.3) and (4.5) is the way in which the SVD and GSVD components of the solution are filtered out by the particular regularization method, as described by the associated filter factors f_i. The filter factors are typically close to 1 for the large σ_i and much smaller than 1 for the small σ_i. In this way, the contributions to the regularized solution corresponding to the smaller σ_i are effectively filtered out. The difference between various regularization methods with the same L-matrix lies in the way that the filter factors f_i are defined.

As an example, the filter factors for Tikhonov regularization are

$$f_i = \frac{\sigma_i^2}{\sigma_i^2 + \lambda^2}, \quad L = I_n \quad \text{and} \quad f_i = \frac{\gamma_i^2}{\gamma_i^2 + \lambda^2}, \quad L \neq I_n,$$

and the filtering effectively sets in for those SVD/GSVD components for which $\sigma_i < \lambda$ and $\gamma_i < \lambda$, respectively. In particular, this shows that discrete ill-posed problems are essentially unregularized by Tikhonov's method for $\lambda > \sigma_1$ and $\lambda > \gamma_p$, respectively, because all the filter factors are approximately 1.

Another important insight we get from Eqs. (4.3) and (4.5) is that, for general-form regularization, we can control the choice of GSVD basis vectors x_i for the regularized solution x_{reg} by means of the regularization matrix L. This observation goes back to Varah [353], [354]; see also [236] for a recent reference. Only if $L = I_n$ do we obtain the SVD basis.

---

[11] If A is exactly rank deficient, then simply exclude the terms with $\sigma_i = 0$ in the relevant summations.

Finally, we shall use the filter factors to study the influence of the a priori estimate x^* in (1.17) on the regularized solutions. In the standard-form case where $\Omega(x) = \|x - x^*\|_2$, the expressions for x_{reg} and $x_{\text{reg}} - x^*$ become

$$x_{\text{reg}} = \sum_{i=1}^{n} \left(f_i \frac{u_i^T b}{\sigma_i} + (1 - f_i) v_i^T x^* \right) v_i , \qquad (4.9)$$

$$x_{\text{reg}} - x^* = \sum_{i=1}^{n} f_i \left(\frac{u_i^T b}{\sigma_i} - v_i^T x^* \right) v_i . \qquad (4.10)$$

Similarly, in the general-form case where $\Omega(x) = \|L(x - x^*)\|_2$, we obtain the following relations:

$$x_{\text{reg}} = \sum_{i=1}^{p} \left(f_i \frac{u_i^T b}{\sigma_i} + (1 - f_i) \theta_i^T x^* \right) x_i + \sum_{i=p+1}^{n} u_i^T b \, x_i , \qquad (4.11)$$

$$L(x_{\text{reg}} - x^*) = \sum_{i=1}^{p} f_i \left(\frac{u_i^T b}{\gamma_i} - v_i^T \bar{x}^* \right) v_i , \qquad (4.12)$$

where θ_i is the ith row of X^{-1}, $\bar{x} = L x^*$, and $v_i^T \bar{x}^* = \mu_i \theta_i^T x^*$. Notice the two factors f_i and $1 - f_i$ multiplying the two contributions to x_{reg} from b and x^*, respectively.

Equations (4.9) and (4.11) show that the SVD/GSVD components of the regularized solution x_{reg} corresponding to filter factors near one, and thus to large (generalized) singular values, are dominated by contributions from the right-hand side b. On the other hand, the SVD/GSVD components of x_{reg} corresponding to filter factors near zero, and thus to smaller (generalized) singular values, are dominated by contributions from the a priori estimate x^*. As the regularization parameter varies, the filter factors also vary, and the balance between the two contributions changes accordingly. The less regularization introduced, the more b dominates. The more regularization introduced, the more x^* dominates.

The situation is, of course, different for the quantities in (4.10) and (4.12). Here, the balance between the two contributions is fixed, and both contributions are weighted by the filter factors f_i.

4.3. Working with Seminorms

In many applications, the 2-norm $\|\cdot\|_2$ of the solution is not the optimal choice. For one thing, this norm may not always be affected as much by the errors as the norm of a derivative of the solution. Second, the SVD basis vectors v_i associated with the 2-norm may not be well suited for the desired solution, whereas choosing an $L \neq I_n$—which essentially corresponds to a change of basis vectors for the solution—can lead to much better basis vectors. Hence,

4.3. Working with Seminorms

the GSVD analysis combined with inspection of the GSVD basis vectors x_i is a powerful aid in choosing a suitable L.

One example of such an analysis can be found in [58], where the SVD and GSVD basis vectors are compared for the case where L approximates the second derivative operator. Both the right singular vectors v_i of the SVD and the right vectors x_i of the GSVD have the oscillation properties mentioned in §2.1. But the ordinary singular vectors v_i have undesired localized oscillations which are not found in the GSVD vectors x_i, showing that $L = I_n$ is a bad choice for computing smooth solutions.

Another example where $L \neq I_n$ is preferable to $L = I_n$ is data approximation by bivariate splines [64]. In regions without data, $L = I_n$ gives a bivariate spline that is not geometrically invariant (which is unsatisfactory); but if L is a discrete approximation to the Laplace operator then the computed solution becomes geometrically invariant and thus independent of the origin of the axis system.

We emphasize that the change of basis involved in choosing $L \neq I_n$ usually maintains the intrinsic properties of ill-posed problems, cf. §2.1.2. That is, the generalized singular values γ_i decay gradually to zero, and the generalized singular vectors u_i, v_i, and x_i have an increasing number of sign changes in their elements as the corresponding γ_i decreases.

When working with regularization matrices L in practice, it is important to realize that L is usually a sparse matrix, and very often a banded matrix. Whenever this is the case, the overhead involved in using general-form regularization, compared to standard-form regularization, is marginal.

If the matrix L approximates a derivative operator, then it can usually be constructed explicitly; see, e.g., Eqs. (1.15) and (1.16). On other occasions, L is computed as the Cholesky factor of a given covariance matrix for the solution, and symmetric pivoting may be necessary if this matrix is rank deficient [208]; see also the discussion in [171, §4.2].

On yet other occasions, L must be computed from a given matrix M satisfying $M^T M = L^T L$, in such a way that L has full row rank which is required by all the algorithms. An important example of this situation is in connection with the use of the *Sobolev norm*

$$\omega(f)^2 = \sum_{i=0}^{q} \alpha_i^2 \, \|f^{(i)}\|_2^2 \, , \tag{4.13}$$

where $f^{(i)}$ denotes the ith derivative of f and α_i is the associated weight. The discrete version of this Sobolev norm takes the following form when the a priori estimate x^* is incorporated:

$$\Omega(x)^2 = \sum_{i=0}^{q} \alpha_i^2 \, \|L_i \, (x - x^*)\|_2^2 \, , \tag{4.14}$$

where L_i approximates the ith derivative operator. The corresponding matrix L is the Cholesky factor of the matrix $\sum_{i=0}^{q} \alpha_i^2 L_i^T L_i$. Hence, L is computed as the triangular factor in the QR factorization of the "stacked" matrix

$$M = \begin{pmatrix} \alpha_q L_q \\ \vdots \\ \alpha_0 L_0 \end{pmatrix}. \tag{4.15}$$

This triangular QR factor can be computed very efficiently as long as the L_i are banded matrices.

For example, if $q = 1$ and both $\alpha_0 \neq 0$ and $\alpha_1 \neq 0$ then the sequence of Givens rotations proposed by Eldén [98] for annihilating a diagonal matrix below a bidiagonal matrix can be used; see p. 102 for details.

For other combinations of q and $\alpha_0, \ldots, \alpha_q$, a general row ordering scheme from [137] can be used: if ℓ_i denotes the column index of the last nonzero element of row i, then the rows are processed in order of increasing ℓ_i. This algorithm does not introduce any unnecessary fill-in [102], and the work involved in computing the triangular factor is only $\mathcal{O}(qn)$ flops.

Here we illustrate the algorithm for the case $q = 2$, when rows with identical ℓ_i are processed in order of increasing row index i. The notation is as follows: "0" denotes a produced zero, "*" denotes an element that is changed, "x" denotes an untouched element, and the arrow "$\rightarrow$" indicates an active row. Consider first the case $\alpha_0 \neq 0$, $\alpha_1 \neq 0$, $\alpha_2 \neq 0$. The first step, which annihilates the first row of the diagonal matrix, is identical to Eldén's algorithm [98]:

The next three steps annihilate the first row of the bidiagonal matrix:

Continue this process until a triangular 2×2 matrix is left in the bottom right corner. Next, we consider the case $\alpha_0 \neq 0$, $\alpha_1 = 0$, $\alpha_2 \neq 0$. The first three

4.3. Working with Seminorms

steps annihilate the first row of the bottom submatrix:

$$
\rightarrow \begin{pmatrix} x & x & x \\ & x & x & x \\ & & x & x & x \\ \rightarrow & x & & & \\ & & x & & \\ & & & x & \\ & & & & x \end{pmatrix} \rightarrow \begin{pmatrix} * & * & * \\ & x & x & x \\ & & x & x & x \\ 0 & * & * & \\ & & x & & \\ & & & x & \\ & & & & x \end{pmatrix} \rightarrow \begin{pmatrix} x & x & x \\ & x & x & x \\ & & x & x & x \\ 0 & * & & \\ * & * & & \\ & & x & & \\ & & & x \end{pmatrix} \begin{pmatrix} x & x & x \\ & x & x & x \\ & & x & x & x \\ & & 0 & & \\ & x & x & & \\ & & * & & \\ & & & x \end{pmatrix}.
$$

The next three similar steps annihilate the second row of the bottom submatrix. Continue this process until a triangular 2×2 matrix is left in the bottom. Finally, we consider the case $\alpha_0 = 0$, $\alpha_1 \neq 0$, $\alpha_2 \neq 0$. First reduce the bidiagonal matrix to a diagonal matrix (except for the last row):

$$
\rightarrow \begin{pmatrix} x & x & x \\ & x & x & x \\ & & x & x & x \\ \rightarrow & x & x & & \\ & & x & x & \\ & & x & x & \\ & & & x & x \end{pmatrix} \rightarrow \begin{pmatrix} * & * & * \\ & x & x & x \\ & & x & x & x \\ 0 & * & * & \\ & x & x & & \\ & & x & x & \\ & & & x & x \end{pmatrix} \rightarrow \begin{pmatrix} x & x & x \\ * & * & * \\ & x & x & x \\ * & 0 & & \\ * & * & & \\ & & x & x & \\ & & & x & x \end{pmatrix} \rightarrow \cdots \rightarrow \begin{pmatrix} x & x & x \\ & x & x & x \\ & & x & x & x \\ & x & & \\ & & x & \\ & & & x \\ & & & x & x \end{pmatrix}
$$

and then apply the scheme above to annihilate the diagonal submatrix.

Once a particular L has been chosen, it is sometimes possible to modify it in such a way that the regularized solution will possess certain features. Specifically, given an ℓ-dimensional subspace $\mathcal{Q}$, we can enforce the condition that the component of the regularized solution in the subspace $\mathcal{Q}$ is unaffected by the regularization process [310]. To do this, we need ℓ orthonormal basis vectors $q_1, \ldots, q_\ell$ of $\mathcal{Q}$. Then the matrix $\sum_{i=1}^{\ell} q_i q_i^T$ is the orthogonal projection matrix onto the subspace $\mathcal{Q}$, and if we define the modified regularization matrix

$$L_\mathcal{Q} = L \left(I_n - \sum_{i=1}^{\ell} q_i q_i^T \right), \tag{4.16}$$

then $\mathcal{Q} \subseteq \mathcal{N}(L_\mathcal{Q})$; $\mathcal{Q} = \mathcal{N}(L_\mathcal{Q})$ if L has full rank. Hence, general-form regularization with $L_\mathcal{Q}$ as a regularization matrix will leave any component of the solution in $\mathcal{Q}$ unaffected. We illustrate the use of $L_\mathcal{Q}$ in §4.8.2.

Unfortunately, there is very little literature on general criteria for choosing the matrix L. In [268] it is argued—somewhat heuristically—that an L-matrix whose null space includes the vector $(1, \ldots, 1)^T$ is superior to the choice $L = I_n$.

An analytical investigation of convolution operators K was given in [67]. The main result is that the order q of the Sobolev norm (4.13) critically influences the convergence rate of the solution, and that the optimal order q of $\omega(f)$ depends on both the decay rate of the singular values of K and the smoothness of the desired solution. These quantities are often unknown in a given application.

One slight disadvantage of choosing $L \neq I_n$ is that a smooth regularized solution x_{reg}, as measured by $\|L x_{\text{reg}}\|_2$, is not guaranteed to be close to the exact solution x^{exact} in the sense that $\|x^{\text{exact}} - x_{\text{reg}}\|_2$ is close to its minimum. For example, MacLeod [245] showed that the existence of a minimum of the expected value of $\|x^{\text{exact}} - x_{L,\lambda}\|_2$, where $x_{L,\lambda}$ is the Tikhonov solution, can only be guaranteed when $L = I_n$. Whether this is an important issue in practice depends on the particular application.

4.4. The Resolution Matrix, Bias, and Variance

Another useful diagnostic tool in connection with linear regularization problems is the resolution matrix. To define this matrix we note that for any linear regularization method there always exists a matrix $A^\#$, which we call the *regularized inverse*, such that the regularized solution x_reg can be written as

$$x_\text{reg} = A^\# b . \tag{4.17}$$

In particular, when x_reg can be expressed in terms of the SVD of A and a set of filter factors $f_1, \ldots, f_n$ (cf. (4.3)), then the matrix $A^\#$ can be written as

$$A^\# = V F \Sigma^\dagger U^T , \qquad F = \text{diag}(f_1, \ldots, f_n) . \tag{4.18}$$

Similarly, if x_reg can be expressed in terms of the GSVD of (A, L) and filter factors $f_1, \ldots, f_p$ (cf. (4.5)), then

$$A^\# = X \begin{pmatrix} F \Sigma^\dagger & 0 \\ 0 & I_{n-p} \end{pmatrix} U^T , \qquad F = \text{diag}(f_1, \ldots, f_p) . \tag{4.19}$$

We now define the $n \times n$ *resolution matrix* Ξ as

$$\Xi \equiv A^\# A , \tag{4.20}$$

and if $A^\#$ is of the form (4.18) or (4.19) then

$$\Xi = V F V^T = \sum_{i=1}^n f_i v_i v_i^T \qquad \text{if} \qquad L = I_n , \tag{4.21}$$

in which case Ξ is symmetric, and

$$\Xi = X \begin{pmatrix} F & 0 \\ 0 & I_{n-p} \end{pmatrix} X^{-1} \qquad \text{if} \qquad L \neq I_n . \tag{4.22}$$

In the special case of the truncated SVD (TSVD), we have $A^\# = A_k^\dagger$ (3.10) and $\Xi = V_k V_k^T$, where $V_k = (v_1, \ldots, v_k)$, and we see that Ξ is simply the projection matrix onto the k-dimensional row space of A_k (i.e., the range of A_k^T).

The importance of the resolution matrix lies in the fact that Ξ quantifies the smoothing of the exact solution by the particular regularization method. Specifically, if $b = b^\text{exact} + e = A\, x^\text{exact} + e$, then

$$x_\text{reg} = A^\# b = \Xi\, x^\text{exact} + A^\# e .$$

4.4. THE RESOLUTION MATRIX, BIAS, AND VARIANCE

The first component $\Xi\, x^{\text{exact}}$ is a regularized or "smoothed" version of the exact solution x^{exact}, while the second component $A^\# e$ is the contribution to x_{reg} from the noise in b. The pure regularization error in x_{reg} is therefore $(I_n - \Xi)\, x^{\text{exact}}$, and the deviation of Ξ from the identity matrix characterizes the regularization error. In other words, Ξ describes how well the exact solution x^{exact}, in the noise-free case, is approximated by the regularized solution $A^\# b^{\text{exact}}$.

In applications where Ξ is too large to display, it is customary to monitor only the diagonal elements of Ξ. Notice, however, that although Ξ can deviate considerably from I_n, the vector $\Xi\, x^{\text{exact}}$ is still close to x^{exact} if those spectral components of x^{exact} which are "damped" by the multiplication with Ξ are small.

There is more information in Ξ than just its deviation from the identity matrix. Again consider $\Xi\, x^{\text{exact}}$, the regularized version of the exact solution. If ξ_i^T denotes the ith row of Ξ, then clearly the ith component of the vector $\Xi\, x^{\text{exact}}$ is given by

$$(\Xi\, x^{\text{exact}})_i = \xi_i^T x^{\text{exact}}, \qquad i = 1, \ldots, n. \tag{4.23}$$

Thus, the ith *row* ξ_i^T of Ξ expresses the ith component $(\Xi\, x^{\text{exact}})_i$ as a weighted average of all the elements in x^{exact}. For this reason, the rows ξ_i^T—or the equivalent functions in the continuous setting—are sometimes referred to as the *averaging kernels* [59]. The vectors ξ_i, $i = 1, \ldots, n$, can be computed from the relations

$$\xi_i = V F \begin{pmatrix} v_{i1} \\ \vdots \\ v_{in} \end{pmatrix} = \sum_{j=1}^{n} f_j\, v_{ij}\, v_j, \qquad L = I_n, \tag{4.24}$$

and

$$\xi_i = X^{-T} \begin{pmatrix} F & 0 \\ 0 & I_{n-p} \end{pmatrix} \begin{pmatrix} x_{i1} \\ \vdots \\ x_{in} \end{pmatrix}, \qquad L \neq I_n. \tag{4.25}$$

The ith row ξ_i^T of Ξ is usually distinctly peaked at its ith component, but the width of the peak and the behavior of the "sidelobes" depend on the particular regularization method. Examples of Ξ for various regularization methods in helioseismology are given in [58] and [188]. See also the examples in §4.8.1. A thorough discussion of averaging kernels and the resolution of regularization methods is given in [278, Chapter 4].

As $\Xi = A^\# A$ is the resolution matrix for the solution, the matrix $A A^\#$ is the resolution matrix for the predicted right-hand side; i.e., it describes how well the vector $A x_{\text{reg}}$ predicts the given right-hand side b. The matrix $A A^\#$ is often referred to as the *influence matrix* [366, p. 13].

In most applications the actual Ξ is a consequence of the choice of regularization method and regularization parameter. However, we note that in the mollifier methods discussed in §5.5, the resolution matrix is the starting point from which $A^{\#}$ is constructed.

In the statistical framework (see, e.g., [246]), the vector $\Xi\, x^{\text{exact}}$ is the expectation of the estimator x_{reg}, i.e., $\mathcal{E}(x_{\text{reg}}) = \Xi\, x^{\text{exact}}$. Since $\mathcal{E}(x_{\text{reg}}) \neq x^{\text{exact}}$ except when x_{reg} is the least squares solution x_{LS}, it follows that x_{reg} is always a biased estimator. In the same framework, if CC^T is the covariance matrix for e, then the covariance matrix for x_{reg} is given by $\text{Cov}(x_{\text{reg}}) = A^{\#} C C^T (A^{\#})^T$. If $C = \sigma_0 I_m$, then this relation simplifies to

$$\text{Cov}(x_{\text{reg}}) = \begin{cases} \sigma_0^2 \, V F^2 (\Sigma^{\dagger})^2 V^T , & L = I_n , \\ \sigma_0^2 \, X \begin{pmatrix} F^2(\Sigma^{\dagger})^2 & 0 \\ 0 & I_{n-p} \end{pmatrix} X^T, & L \neq I_n , \end{cases} \qquad (4.26)$$

and this expression illustrates why $\|\text{Cov}(x_{\text{reg}})\|_2$ is generally much smaller than $\|\text{Cov}(x_{\text{LS}})\|_2$ for the least squares solution, because the influence from the large σ_i^{-1} is damped by the corresponding small filter factors f_i.

Until now, we have used the GSVD of (A, L) to examine the role that the regularization matrix L plays in the computation of the regularized solution. We now briefly examine the "opposite" situation where we are explicitly given the matrix $A^{\#}$ in (4.17)—this is the case, e.g., in mollifier methods; cf. §5.5—and we shall derive the corresponding filter factors f_i and regularization matrix L.

If an L exists such that $x_{\text{reg}} = A^{\#} b$ can be written in the form (4.17), then we have

$$A A^{\#} = U \begin{pmatrix} F & 0 \\ 0 & I_{n-p} \end{pmatrix} U^T , \qquad A^{\#} A = X \begin{pmatrix} F & 0 \\ 0 & I_{n-p} \end{pmatrix} X^{-1} . \qquad (4.27)$$

If both $A A^{\#}$ and $A^{\#} A$ are symmetric, we conclude that $L = I_n$ and the filter factors are the eigenvalues of $A^{\#} A$. On the other hand, if neither $A A^{\#}$ nor $A^{\#} A$ is symmetric, one regularization matrix is not enough to characterize $A^{\#}$ and a restricted SVD (RSVD) analysis (see §2.1.4) is necessary.

Assume now that $A A^{\#}$ is symmetric and that $A^{\#} A$ is nonsymmetric. Then the eigendecompositions in (4.27) yield unique F and U (since $U^T U = I_n$), up to the nonuniqueness in U associated with multiple eigenvalues. The matrix X is only determined within an unknown column scaling $D = \text{diag}(d_1, \ldots, d_n)$; i.e., we compute $\tilde{X} = X D$. Then Σ is determined within the same scaling,

$$\tilde{\Sigma} = \Sigma D_p = U(:, 1{:}p)^T A \, \tilde{X}(:, 1{:}p) , \qquad D_p = \text{diag}(d_1, \ldots, d_p) ,$$

and without loss of generality we can assume that D_p is chosen such that $0 \leq \tilde{\Sigma} \leq I_p$. Then L is given by

$$L = V \left((I_n - \Sigma^2)^{1/2}, 0 \right) X^{-1} = V \left((D_p^2 - \tilde{\Sigma}^2)^{1/2}, 0 \right) \tilde{X}^{-1} , \qquad (4.28)$$

where V (from the GSVD) and D_p are unknown. Since $\|L \cdot\|_2$ is independent of V we can always choose $V = I_p$, but there are still p free parameters in D_p.

Any choice of D_p that satisfies $D_p > \tilde{\Sigma}$ yields a valid L such that x_{reg} can be written in the form (4.17).

The parameters in D_p can be fixed if we assume that $A^\#$ corresponds to a particular regularization method. For example, if we assume that $A^\#$ corresponds to Tikhonov regularization in general form, then the matrix F of filter factors is fixed and therefore D_p must satisfy $F = \Sigma^2(\Sigma^2 + \lambda^2 M^2)^{-1} = \tilde{\Sigma}^2(\tilde{\Sigma}^2 + \lambda^2(D_p^2 - \tilde{\Sigma}^2)^{1/2})^{-1}$, from which we obtain

$$d_i^2 = \frac{\tilde{\sigma}_i^2(1-(1-\lambda^2)f_i)}{\lambda^2 f_i}, \qquad f_i \neq 0,$$

where $\tilde{\sigma}_i$ are the diagonal elements of $\tilde{\Sigma}$.

4.5. The Discrete Picard Condition

Due to the appearance of the ratios $(u_i^T b)/\sigma_i$ in the two equations (4.3) and (4.5) for the regularized solutions, it is clear that the decay of the coefficients $|u_i^T b|$ relative to the decay of the σ_i plays an important role in the treatment of discrete ill-posed problems.

Intuitively, we expect that if the coefficients $|u_i^T b|$ decay more slowly than the σ_i, then Tikhonov regularization and other methods that filter out the small singular values cannot produce a useful regularized solution. We shall now investigate the importance of the relative decay between the coefficients $u_i^T b$ and the σ_i in the important case of Tikhonov regularization. Let $A_\lambda^\# = (A^T A + \lambda^2 L^T L)^{-1} A^T$, let $A_\lambda^\# b^{\text{exact}}$ denote the solution obtained by applying Tikhonov regularization in general form ($L \neq I_n$) to the unperturbed right-hand side b^{exact}, and consider the regularization error

$$L\left(x^{\text{exact}} - A_\lambda^\# b^{\text{exact}}\right) = \sum_{i=1}^{p}(1-f_i)\frac{u_i^T b^{\text{exact}}}{\gamma_i} v_i, \qquad (4.29)$$

where the exact solution is $x^{\text{exact}} = A^\dagger b^{\text{exact}}$. Moreover, notice that for all the small generalized singular values, (2.7) yields $\gamma_i = \sigma_i(1-\sigma_i^2)^{-1/2} \approx \sigma_i$. Then we have the following result.

Theorem 4.5.1. [183, Theorem 2]. *Assume that the Fourier coefficients $u_i^T b$ and the generalized singular values γ_i are related by the following model:*

$$u_i^T b = \begin{cases} \gamma_i^{\alpha+1}, & i = 1,\ldots,p, \\ \gamma_p^{\alpha+1}, & i = p+1,\ldots,n, \end{cases} \qquad (4.30)$$

where α is a real number. Then the regularization error for Tikhonov's method satisfies

$$\frac{\|L(x^{\text{exact}} - A_\lambda^\# b^{\text{exact}})\|_2}{\|L x^{\text{exact}}\|_2} \leq \begin{cases} p^{1/2}, & \alpha \leq 0, \\ p^{1/2}(\lambda/\gamma_p)^\alpha, & 0 \leq \alpha < 2, \\ p^{1/2}(\lambda/\gamma_p)^2, & 2 \leq \alpha. \end{cases} \qquad (4.31)$$

We see that in order to guarantee small regularization errors in the model problem (4.30), the *relative decay rate* α must be somewhat greater than zero; i.e., the absolute value of the Fourier coefficients must decay *faster* than the generalized singular values γ_i. Moreover, the faster the decay, the better $L A_\lambda^\# b^{\text{exact}}$ approximates $L x^{\text{exact}}$. We can only guarantee that Tikhonov regularization is able to produce a useful regularized solution that approximates x^{exact} if α is somewhat greater than zero.

The decay of the Fourier coefficients and the generalized singular values is also important when comparing different regularization methods with different filter factors. Let f_i and ϕ_i denote the filter factors for Tikhonov regularization and some other regularization method, and let f_i and ϕ_i be related by

$$\begin{aligned} |f_i - \phi_i| &\leq c f_i, & \gamma_i &\leq C\lambda, \\ f_i &= \phi_i, & \gamma_i &> C\lambda, \end{aligned} \quad (4.32)$$

where c is a small positive constant and C is a positive constant satisfying $1 \leq C < \gamma_p/\lambda$ (thus ensuring that $C\lambda < \gamma_p$). Moreover, let $x_{L,\lambda}$ and x_{reg} denote the regularized solutions corresponding to f_i and ϕ_i, respectively.

Theorem 4.5.2. [184, Theorem 4]. *Assume that the Fourier coefficients satisfy (4.30) and that the filter factors for the Tikhonov solution $x_{L,\lambda}$ and the other regularized solution x_{reg} satisfy (4.32). Then*

$$\frac{\|L(x_{L,\lambda} - x_{\text{reg}})\|_2}{\|L x_{L,\lambda}\|_2} \leq \begin{cases} 2c, & \alpha \leq 0, \\ 2c \left(\frac{C\lambda}{\gamma_p}\right)^\alpha, & 0 \leq \alpha. \end{cases} \quad (4.33)$$

We see that if the Fourier coefficients $|u_i^T b|$ decay more slowly than the γ_i then we get a large upper bound in (4.33) even if $\phi_i \approx f_i$, and $L x_{\text{reg}}$ can therefore differ considerably from $L x_{L,\lambda}$. Only if the Fourier coefficients decay somewhat faster than the γ_i can we ensure a small upper bound in (4.33), provided that $C\lambda$ is somewhat smaller than γ_p. Considerations like these have led to the following definition.

The Discrete Picard Condition. Let b^{exact} denote the unperturbed right-hand side, and let τ_A denote the level at which the singular values of A level off. Then the discrete Picard condition is satisfied if, for all generalized singular values $\gamma_i > \tau_A \|L^\dagger\|_2$, the corresponding coefficients $|u_i^T b^{\text{exact}}|$, on the average, decay to zero faster than the γ_i.

This definition is from [183], where more details can be found. The importance of the discrete Picard condition and the decay rates of $|u_i^T b|$ and σ_i was also studied in [181], [184], and [378].

When the discrete Picard condition is tested numerically, we note that it is the ratios of nearby coefficients $|u_i^T b|$ and γ_i—rather than their individual values—that are important. Therefore, we propose to base a numerical check for satisfaction of the discrete Picard condition on the *moving geometric mean*

$$\omega_i = \gamma_i^{-1} \left(\prod_{j=i-q}^{i+q} |u_i^T b| \right)^{1/(2q+1)} , \qquad i = 1+q, \ldots, n-q , \qquad (4.34)$$

where q is a small integer, thus ensuring the locality of ω_i. Note that ρ_i should only be computed for numerically nonzero γ_i, i.e., for $\gamma_i > \tau_A \|L^\dagger\|_2$.

The relative decay rate α in the model problem (4.30) plays an important role in connection with the effective resolution limit η_{res} defined in (4.1). Assume that the right-hand side errors dominate. Then for $i = i_b$ we have that $|u_i^T b| \approx |u_i^T e|$, and when we insert the model (4.30) we get $\sigma_{i_b}^{\alpha+1} \approx \tau_b \Rightarrow \sigma_{i_b} \approx \tau_b^{1/(1+\alpha)}$. When we insert this relation into Eq. (4.1), we obtain

$$\eta_{\text{res}} \approx \tau_b^{\alpha/(1+\alpha)} = \left(m^{-1/2} \|b - b^{\text{exact}}\|_2 \right)^{\alpha/(1+\alpha)} . \qquad (4.35)$$

From this relation we see that the larger the α, i.e., the faster the relative decay of the Fourier coefficients, the smaller the η_{res}, and therefore the more accurately we can compute a regularized solution.

The parameter α in (4.30) has a counterpart in the study of regularization methods for ill-posed problems $Kf = g$ in the Hilbert space setting. Here, the equivalent formulation of (4.30) takes the form $f \in \mathcal{R}((K^*K)^\nu)$, where K^* is the adjoint of the operator K, and $\nu > 0$ is a real parameter that expresses the "smoothness" of f; see, e.g., [160, §1.2]. We see that α is related to ν by $\alpha = 2\nu$.

4.6. L-Curve Analysis

In the previous sections we mentioned several important analytical analysis tools, such as the filter factors and the decay of the Fourier coefficients and the (generalized) singular values.

Perhaps the most convenient graphical tool for analysis of discrete ill-posed problems is the so-called *L-curve* which is a plot for all valid regularization parameters of the discrete smoothing norm $\Omega(x_{\text{reg}})$—e.g., the (semi)norm $\|L x_{\text{reg}}\|_2$—of the regularized solution versus the corresponding residual norm $\|A x_{\text{reg}} - b\|_2$. The L-curve clearly displays the compromise between minimization of these two quantities, which is the heart of any regularization method.

The L-curve is a continuous curve when the regularization parameter is continuous, as in Tikhonov regularization. For regularization methods with a discrete regularization parameter, such as TSVD, the L-curve consists of a discrete set of points. The use of such plots in connection with ill-conditioned least squares problems goes back to Miller [249] and Lawson and Hanson [231, Chapter 26].

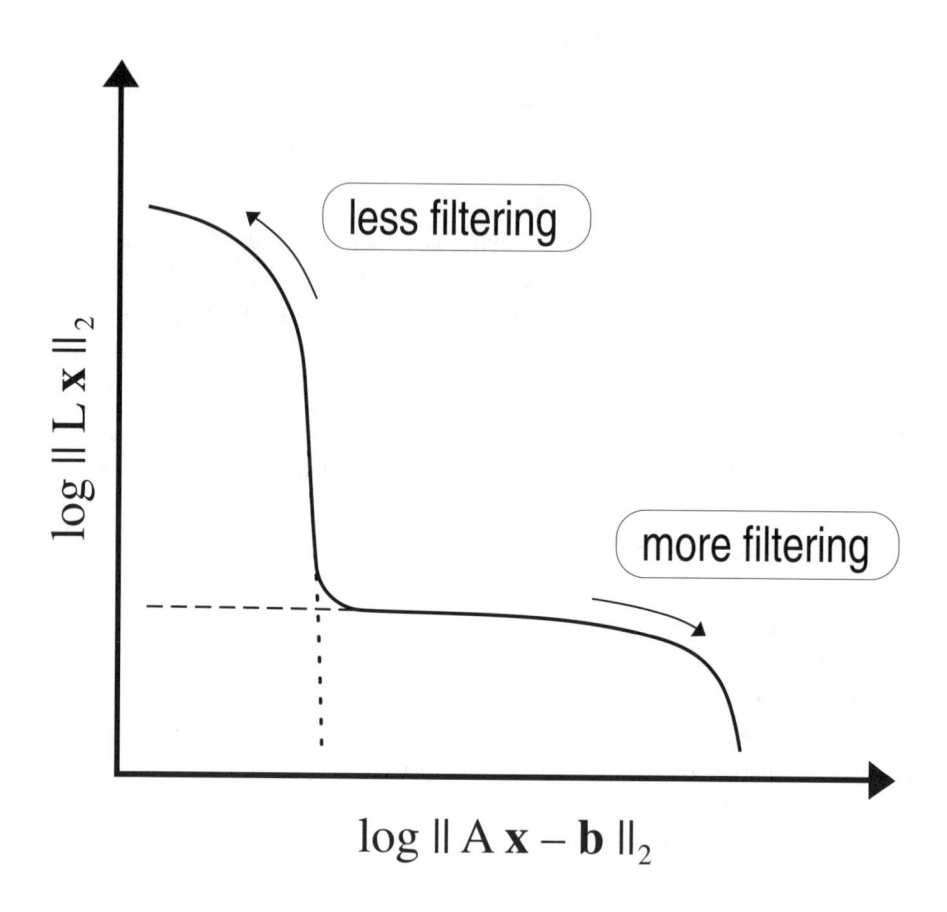

FIG. 4.1. *The generic form of the L-curve. Notice the log-log scale.*

Our present discussion is limited to the important case $\Omega(x_{\text{reg}}) = \|L\, x_{\text{reg}}\|_2$. However, as emphasized in [196] (see also §5.7), we note that different norms, seminorms, and other measures of "size" of the regularized solution define different regularization algorithms, and it can sometimes be advantageous to plot the L-curve in these norms instead of $\|L\, x_{\text{reg}}\|_2$.

For discrete ill-posed problems it turns out that the L-curve, when plotted in *log-log scale*, very often has a characteristic L-shaped appearance (hence its name) with a distinct corner separating the vertical and the horizontal parts of the curve; see Fig 4.1.

The L-curve for Tikhonov regularization plays a central role in connection with regularization methods for discrete ill-posed problems because it divides the first quadrant into two regions. It is impossible to construct any solution that corresponds to a point below the Tikhonov L-curve; any regularized solution must lie on or above this curve. The Tikhonov L-curve has the following properties (see also [231, Theorem (25.49)] and [291, Lemma 2]).

4.6. L-CURVE ANALYSIS

Theorem 4.6.1. [184, Theorem 1]. *Let $x_{L,\lambda}$ denote the Tikhonov regularized solution; cf. (4.4). Then $\|L\,x_{L,\lambda}\|_2$ is a monotonically decreasing convex function of $\|A\,x_{L,\lambda} - b\|_2$, and if we introduce the extreme residual norms corresponding to zero and infinite regularization,*

$$\delta_0 = \|(I_m - U\,U^T)\,b\|_2\,, \qquad \delta_\infty = \left(\delta_0^2 + \|U_p\,U_p^T b\|_2^2\right)^{1/2}, \qquad (4.36)$$

where $U_p = (u_1, \ldots, u_p)$, then

$$\delta_0 \le \|A\,x_{L,\lambda} - b\|_2 \le \delta_\infty\,, \qquad 0 \le \|L\,x_{L,\lambda}\|_2 \le \|L\,x_{\mathrm{LS}}\|_2\,. \qquad (4.37)$$

Moreover, any point (δ, η) on the L-curve is a solution to the following two inequality-constrained least squares problems:

$$\delta = \min\,\|A\,x - b\|_2 \quad \text{subject to} \quad \|L\,x\|_2 \le \eta\,, \quad 0 \le \eta \le \|L\,x_{\mathrm{LS}}\|_2\,, \qquad (4.38)$$

$$\eta = \min\,\|L\,x\|_2 \quad \text{subject to} \quad \|A\,x - b\|_2 \le \delta\,, \quad \delta_0 \le \delta \le \delta_\infty\,. \qquad (4.39)$$

Note that if $m = n$ then $\delta_0 = 0$, and if $n = p$ then $\delta_\infty = \|b\|_2$. Writing once again the right-hand side as $b = b^{\mathrm{exact}} + e$, where e is the perturbation and $b^{\mathrm{exact}} = A\,x^{\mathrm{exact}}$, the regularized solution takes the form $x_{\mathrm{reg}} = A^\# b^{\mathrm{exact}} + A^\# e$. Hence, in the general case ($L \ne I_n$), the error is given by

$$\begin{aligned}
x^{\mathrm{exact}} - x_{\mathrm{reg}} &= (x^{\mathrm{exact}} - A^\# b^{\mathrm{exact}}) - A^\# e \\
&= \sum_{i=1}^p (1 - f_i)\,\frac{u_i^T b^{\mathrm{exact}}}{\sigma_i}\,x_i \\
&\quad - \left(\sum_{i=1}^p f_i\,\frac{u_i^T e}{\sigma_i}\,x_i + \sum_{i=p+1}^n u_i^T e\,x_i\right). \qquad (4.40)
\end{aligned}$$

In [11] this equation is studied in detail for Tikhonov's method. The error consists of two components, namely, the perturbation error $A^\# e$ from the error e in the given right-hand side, and the regularization error $x^{\mathrm{exact}} - A^\# b^{\mathrm{exact}}$ due to the regularization of the error-free component b^{exact} in the right-hand side. When very little regularization is introduced, most of the filter factors f_i are approximately 1, and the error $x^{\mathrm{exact}} - x_{\mathrm{reg}}$ is dominated by the perturbation error $A^\# e$. This situation is called *undersmoothing*, and it corresponds to the uppermost part of the L-curve above the middle "corner." When a large amount of regularization is introduced, then most filter factors are small, $f_i \ll 1$, and the error $x^{\mathrm{exact}} - x_{\mathrm{reg}}$ is dominated by the regularization error $x^{\mathrm{exact}} - A^\# b^{\mathrm{exact}}$. This situation is called *oversmoothing*, and it corresponds to the rightmost part of the L-curve to the right of the "corner."

To explain the characteristic L-shape of the L-curve more precisely, we take a closer look at the quantities $\|L\,x_{\mathrm{reg}}\|_2$ and $\|b - A\,x_{\mathrm{reg}}\|_2$. From the general expressions in Eqs. (4.6) and (4.7) we get

$$\|L x_{\text{reg}}\|_2^2 = \sum_{i=1}^{p} \left(f_i \frac{u_i^T b}{\gamma_i} \right)^2 , \qquad (4.41)$$

$$\|A x_{\text{reg}} - b\|_2^2 = \sum_{i=1}^{p} \left((1 - f_i) u_i^T b \right)^2 + \delta_0^2 . \qquad (4.42)$$

These relations were used in [181] and [184][12] to analyze the relationship between the error in $x_{\text{reg}} = x_{L,\lambda}$ and the behavior of the L-curve, for Tikhonov regularization with filter factors $f_i = \gamma_i^2/(\gamma_i^2 + \lambda^2)$, and we summarize the analysis here. Throughout, we assume that the discrete Picard condition is satisfied, and we write $x_{L,\lambda} = A_\lambda^\# b$ with $A_\lambda^\# = (A^T A + \lambda^2 L^T L)^{-1} A^T$. The analysis carries over to other regularization methods for which the amount of regularization varies monotonically with the regularization parameter (continuous or discrete).

Let us first consider the behavior of the L^{exact}-curve for an unperturbed problem with $e = 0$ and $b = b^{\text{exact}}$. If $\lambda \ll \gamma_1$ then for all i we have $\gamma_i^2 + \lambda^2 \approx \gamma_i^2$ such that $f_i \approx 1$ and $f_i - 1 \approx \lambda^2/\gamma_i^2$. Then $A_\lambda^\# b^{\text{exact}} \approx x^{\text{exact}}$ and

$$\|A A_\lambda^\# b^{\text{exact}} - b\|_2^2 \approx \sum_{i=1}^{p} \left(\frac{\lambda^2}{\gamma_i^2} u_i^T b^{\text{exact}} \right)^2 + \delta_0^2 .$$

Hence, for very small λ the L^{exact}-curve is approximately a horizontal line at $\|L A_\lambda^\# b^{\text{exact}}\|_2 = \|L x^{\text{exact}}\|_2$. As λ increases, it follows from (4.41) that $\|L A_\lambda^\# b^{\text{exact}}\|_2$ slowly starts to decrease, while $\|A A_\lambda^\# b^{\text{exact}} - b\|_2$ steadily grows towards δ_∞. Eventually the L^{exact}-curve starts to bend down towards the abscissa axis, which happens when λ is comparable with the largest γ_i. For those values of λ, the residual norm is still somewhat smaller than δ_∞ because some of the filter factors f_i are less than 1.

Consider now the L^{pert}-curve associated with the mere perturbation e of the right-hand side, and assume that the covariance matrix for e is $\sigma_0^2 I_m$. Then the corresponding vector $A^\# e$ satisfies

$$\|L A_\lambda^\# e\|_2 = \sum_{i=1}^{p} \left(\frac{\gamma_i}{\gamma_i^2 + \lambda^2} u_i^T e \right)^2 \approx \sigma_0^2 \left(\sum_{i=1}^{p} \frac{\gamma_i}{\gamma_i^2 + \lambda^2} \right)^2 ,$$

$$\|A A_\lambda^\# e - e\|_2 = \sum_{i=1}^{p} \left(f_i u_i^T e \right)^2 + \|(I_m - U U^T) e\|_2^2$$

$$\approx \sigma_0^2 \left(\sum_{i=1}^{p} f_i^2 + (m - n) \right) .$$

---

[12] Note the misprint in [184, §3]: below Characterization 2, γ_n should be replaced by γ_1.

4.6. L-Curve Analysis

For very small $\lambda \ll \gamma_1$ this L^{pert}-curve is approximately a horizontal line at $\|L A_\lambda^\# e\|_2 \approx \sqrt{p}\,\sigma_0/\gamma_1$. The L^{pert}-curve starts to bend down towards the abscissa axis for much smaller values of λ than the L^{exact}-curve, namely, when $\lambda \approx \gamma_1$. As λ increases, $\|A A_\lambda^\# e - b\|_2$ becomes almost independent of λ, while $\|L A_\lambda^\# e\|_2$ is dominated by a few terms (ℓ, say) for which $\gamma_i/(\gamma_i^2 + \lambda^2) \approx 1/(2\lambda)$ such that $\|L A_\lambda^\# e\|_2 \approx \ell\,\sigma_0/(s\lambda) \approx \sigma_0/\lambda$. Hence, the L^{pert}-curve soon becomes almost a vertical line at $\|A A_\lambda^\# e - b\|_2 \approx \sigma_0 \sqrt{m - n + p}$ as $\lambda \to \infty$.

The two L-curves characterized above are indicated in Fig. 4.1 by the dashed lines. The actual L-curve for a given problem with $b = b^{\text{exact}} + e$ is a combination of these two special L-curves. For small λ the behavior of the L-curve is entirely dominated by the contributions from e. This corresponds to the uppermost part[13] and the vertical part where $\|L x_{L,\lambda}\|_2$ is very sensitive to changes in the regularization parameter. For large λ the L-curve is completely dominated by contributions from b^{exact}. This corresponds to the horizontal part and the rightmost part where it is the residual norm $\|A x_{L,\lambda} - b\|_2$ that is most sensitive to the regularization parameter. In this way, the L-curve clearly displays the tradeoff between minimizing the residual norm and the side constraint.

In between, there is a region where both b^{exact} and e contribute, and this region defines the L-shaped corner of the L-curve. If we assume that e has zero mean and covariance matrix $\sigma_0^2 I_m$, and that $\|e\|_2 < \|b^{\text{exact}}\|_2$, then the corner of the L-curve appears approximately at

$$\left(\|A x_{L,\lambda} - b\|_2,\, \|L x_{L,\lambda}\|_2\right) \approx \left(\sqrt{\sigma_0^2(m - n + p) + \delta_0^2},\, \|L x^{\text{exact}}\|_2\right) ;$$

cf. [184, Characterization 2]. The faster the Fourier coefficients $u_i^T b^{\text{exact}}$ decay to zero, the smaller this cross-over region and, thus, the sharper the L-shaped corner. This explains our choice of the name "L-curve." It is shown in [196, §4.2] that a log-log scale emphasizes the different appearances of the vertical and the horizontal parts of the L-curve.

For a given fixed right-hand side $b = b^{\text{exact}} + e$, there is obviously an optimal regularization parameter that balances the perturbation error and the regularization error in x_{reg}. An essential feature of the L-curve is that this optimal regularization parameter—defined in the above sense—is not far from the regularization parameter that corresponds to the L-curve's corner [184]. In other words, by locating the corner of the L-curve one can compute an approximation to the optimal regularization parameter and thus, in turn, compute a regularized solution with a good balance between the two types of errors. We return to these aspects in §7.5.

---

[13] For operators K with infinite rank, where $\mu_i \to 0$ as $i \to \infty$, there is no "shoulder" in the uppermost part of the L-curve as shown in Fig. 4.1. Instead, the vertical part extends to infinity.

For more details about the L-curve and its use in the analysis of discrete ill-posed problems we refer to [181], [184], and [196]. Apart from its role as an analysis tool, the L-curve has many other applications, most notably as a method for choosing the regularization parameter, and we return to this aspect in §7.5.

4.7. Random Test Matrices for Regularization Methods

Test matrices, either with fixed or random elements, play an important role in numerical analysis; see the discussion in [209]. For example, they are used for testing numerical software, and they are used for getting insight and understanding into new algorithms. Both fixed and random test matrices are important, because they tend to reveal different properties of the algorithm under consideration.

In connection with regularization methods for discrete ill-posed problems, we need to generate matrices for which all the singular values decay gradually to zero and the number of sign changes in the left and right singular vectors increases with the index.

One means of generating fixed test matrices with these properties is to discretize a given ill-posed problem. This strategy is used in the MATLAB package REGULARIZATION TOOLS [187] where 12 test problems are available. All these test problems are discretizations of Fredholm integral equations of the first kind, including the three inverse problems from §1.4 as well as the inverse Laplace transformation and Phillips' notorious test problem [280]—see the manual [186] for more detailed information.

Other fixed test matrices with the desired properties are available in the collection of test matrices in MATLAB compiled by Higham [209]. Among the many test matrices in this collection, six of them satisfy the above-mentioned requirements, namely, cauchy, hilbert, lotkin, pascal, prolate, and vand.

Although fixed test matrices are very useful in many circumstances, there is also a need for constructing random matrices to test regularization methods. It is easy to generate a random matrix A with a given singular value spectrum Σ: generate random U and V matrices (e.g., by Stewart's method [318]) and then set $A \leftarrow U \Sigma V^T$. However, such random matrices do not have singular vectors with the desired behavior of the sign changes.

An algorithm for computing random test matrices with the desired properties is described in [189]. The key idea in this algorithm is to compute the U and V matrices via the SVD of a random bidiagonal matrix B with positive elements on the diagonal and the upper bidiagonal. For such matrices, it is shown in [189] that the ith column of both U and V has exactly $i-1$ sign changes. This result is based on the fact that both $B^T B$ and $B B^T$ are so-called "oscillating matrices."

A further advantage of this strategy is that many routines for computing

4.7. RANDOM TEST MATRICES FOR REGULARIZATION METHODS

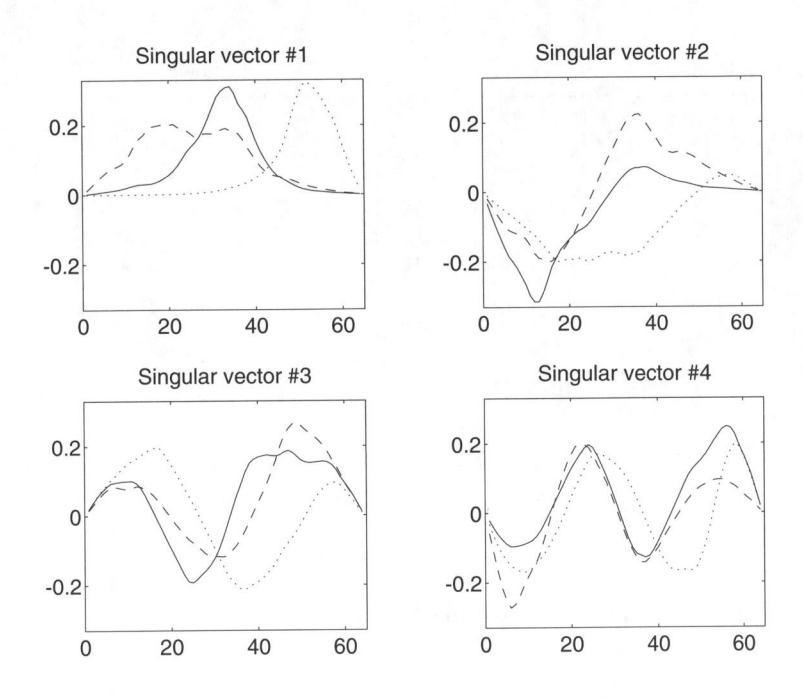

FIG. 4.2. *The first four left singular vectors of three* 64×64 *upper bidiagonal matrices with random elements generated by our heuristic formula.*

the SVD of bidiagonal matrices are know to have extremely good rounding error properties. Such routines are usually part of general SVD programs.

The appearance of the singular vectors depends on the distribution of the nonzero elements in B. For example, if these elements are from a uniform distribution in the range $(0,1]$, then most of the elements in the left and right singular vectors of B are very close to zero, and the oscillations are difficult to recognize when the vectors are plotted. In order to produce left and right singular vectors for which all the maxima and minima of the oscillations are large, we found it convenient to use a normal distribution with unit standard deviation and a large mean value μ given by the following heuristic formula:

$$\mu = \begin{cases} .222\, n + .0278\, n^2 & \text{for } n < 100\ , \\ 3\, n & \text{for } n \geq 100\ . \end{cases}$$

Figure 4.2 shows the first four left singular vectors of three 64×64 bidiagonal matrices generated with the above μ.

The condition number of the matrix B is of interest, because the accuracy produced by some bidiagonal SVD routines depends on this condition number. In [189] we found experimentally that the distribution of the singular values is almost linear with $\sigma_i \approx 7\, i$. Hence, B is always well conditioned, and we do

90   4. PROBLEMS WITH ILL-DETERMINED RANK

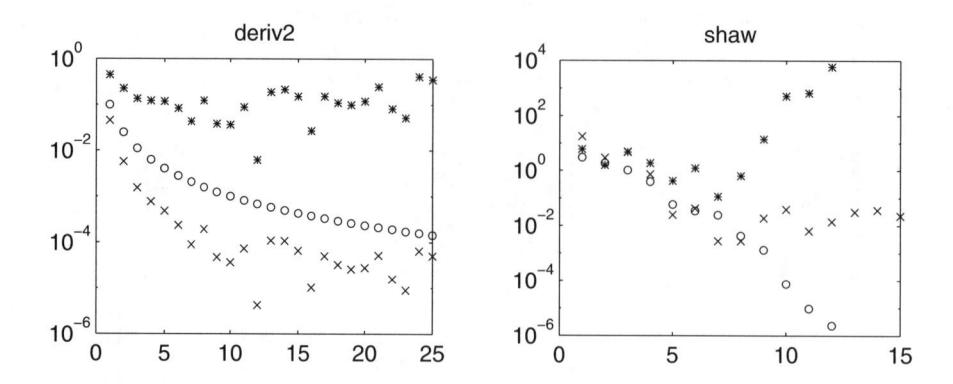

FIG. 4.3. *The first 25 singular values σ_i (circles), Fourier coefficients $|u_i^T b|$ (crosses), and coefficients $|u_i^T b|/\sigma_i$ (asterisks) for the two test problems* deriv2 *and* shaw.

not expect a bidiagonal SVD routine to have any difficulties with computing accurate singular vectors of the B-matrices.

4.8. The Analysis Tools in Action

We finish this chapter with a collection of numerical examples that illustrate the analytical and numerical tools described in the previous sections. Throughout, we use test problems $A x = b$ with a square unperturbed coefficient matrix (except for rounding errors) and a right-hand side $b = b^{\text{exact}} + e$ with a perturbation e whose elements are normally distributed with zero mean and standard deviation chosen such that the noise-to-signal ratio is $\|e\|_2/\|b^{\text{exact}}\|_2 = 0.01$.

4.8.1. Standard-Form Regularization of deriv2 and shaw

In the first example, we apply standard-form Tikhonov regularization to the two test problems deriv2 and shaw from §§1.4.2 and 1.4.3 with dimensions $m = n = 64$. The singular values of the first problem decay fairly slowly (the problem is moderately ill posed), and the condition number is $\text{cond}(A) \approx 5 \cdot 10^3$. The singular values of the second problem decay much faster (this problem is severely ill conditioned) until they hit the machine precision times σ_1, and the condition number of this matrix is therefore approximately the reciprocal of the machine precision.

   Figure 4.3 shows plots of the first 25 singular values, Fourier coefficients $|u_i^T b|$, and coefficients $|u_i^T b|/\sigma_i$ of the solution. In the first problem, the Fourier coefficients level off at $\tau_b \approx 10^{-4}$, which happens at $i_b = 7$. We cannot recover the SVD components of the solution for $i > 7$, because the coefficients $|u_i^T b|/\sigma_i$ are dominated by the noise. The singular value at $i = i_b$ is $\sigma_7 = 2.0 \cdot 10^{-3}$, and

4.8. THE ANALYSIS TOOLS IN ACTION

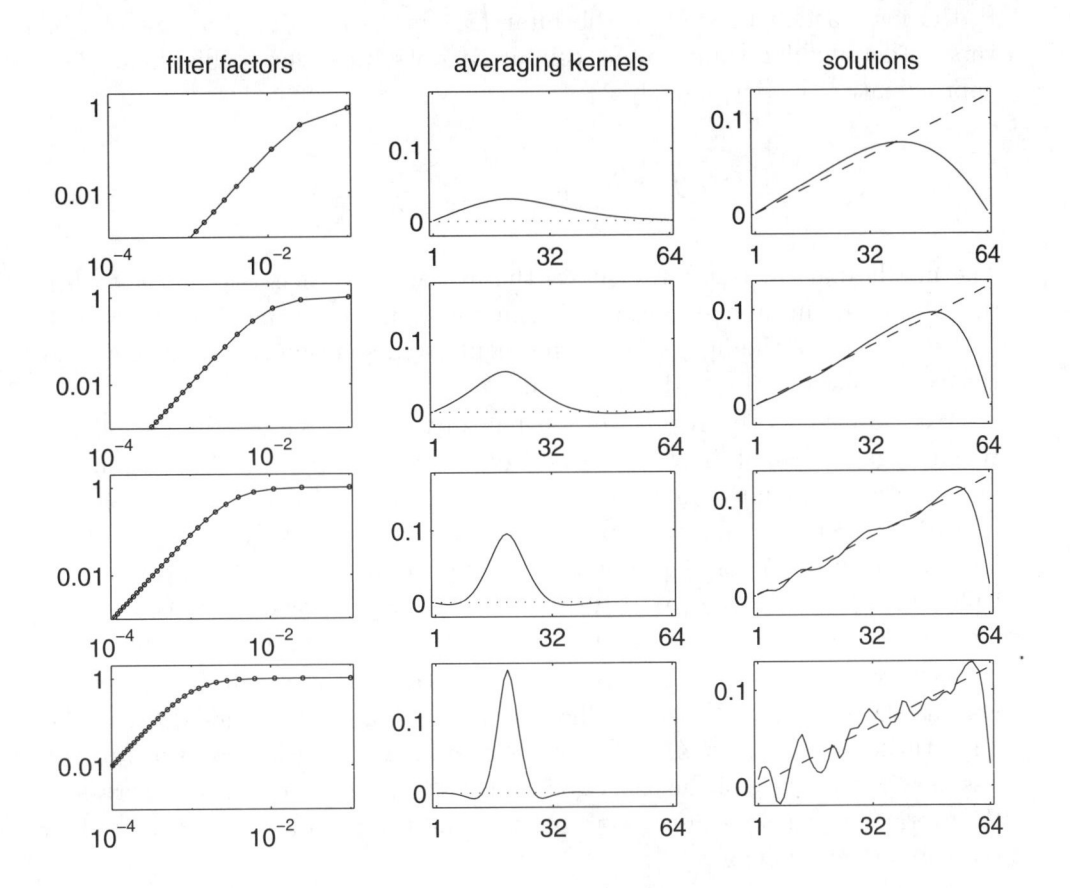

FIG. 4.4. *Test problem* deriv2 *with standard-form Tikhonov regularization. The left part shows the filter factors f_i versus the singular values σ_i for $i = 1, \ldots, n$. The middle part shows the averaging kernels ξ_{20} corresponding to the abscissa $t = 0.3$. The right part shows the regularized solution x_λ (solid line) and the exact solution x^{exact} (dashed line). The regularization parameters λ are, from top to bottom, $3.3 \cdot 10^{-2}$, 10^{-2}, $3.3 \cdot 10^{-3}$, and 10^{-3}.*

the effective resolution limit (4.1) is therefore $\eta_{\text{res}} \approx 5 \cdot 10^{-2}$, which is in fairly good agreement with the behavior of the coefficients $|u_i^T b|/\sigma_i$. In the second problem, the Fourier coefficients level off for $i > i_b = 5$ at the level $\tau_b \approx 10^{-2}$. Since $\sigma_5 = 5.9 \cdot 10^{-2}$, the effective resolution limit is here $\eta_{\text{res}} \approx 0.2$, and again this value is in fairly good agreement with the actual coefficients $|u_i^T b|/\sigma_i$.

In Fig. 4.4 we show corresponding filter factors, averaging kernels, and regularized solutions $x_\lambda = x_{I_n, \lambda}$ for the first test problem and the following four values of the regularization parameter λ:

$$3.3 \cdot 10^{-2}, \quad 10^{-2}, \quad 3.3 \cdot 10^{-3}, \quad 10^{-3}.$$

In the left part of the figure, the filter factors f_i, $i = 1,\ldots,n$, are plotted versus the singular values σ_i as connected dots, in order to highlight their approximate behavior, namely,

$$f_i \approx \begin{cases} 1, & \sigma_i \gg \lambda, \\ \sigma_i^2/\lambda^2, & \sigma_i \ll \lambda. \end{cases}$$

The middle part of the figure shows the averaging kernels ξ_{20}, i.e., the 20th row of the resolution matrix Ξ in (4.21), corresponding to the abscissa $t = 0.3$. The right part of the figure shows the regularized solution x_λ (solid line) and the exact solution x^{exact} (dashed line).

We see that as the regularization parameter λ decreases, the cut-off between large and small filter factors also decreases, and more SVD components are effectively included in the regularized solution. As a consequence, the perturbation error increases as more noise is introduced in the regularized solution. Since $i_b = 7$ for this problem, we expect that the perturbation errors will dominate x_λ for $\lambda \gtrsim \sigma_7 = 2 \cdot 10^{-3}$. Small oscillations are indeed visible for $\lambda = 3.3 \cdot 10^{-3}$, and for $\lambda = 10^{-3}$ the oscillations have a large amplitude.

The averaging kernels ξ_{20} give an impression of the regularization error. Due to the bell-shaped form of the averaging kernels, we conclude that the unperturbed component $A^\# b^{\text{exact}}$ of the regularized solution x_λ is a more or less localized average of the exact solution x^{exact} near $t = 0.3$. As λ decreases, the averaging kernel becomes more peaked, and the narrower the peak the less the regularization error.

The fact that the regularized solution always bends down towards the abscissa axis near $i = n$ is due to the choice $L = I_n$ which, for this positive exact solution, seeks to minimize the area below the regularized solution. We return to this test problem with a different choice of L below.

Similar plots are shown in Fig. 4.5 for the second test problem and the following four values of λ:

$$1, \quad 10^{-1}, \quad 10^{-2}, \quad 10^{-3}.$$

The general behavior of the filter factors, the averaging kernels, and the solutions is the same as above: as λ decreases and more SVD components are included in x_λ, the perturbation errors increase, and they become visible for $\lambda = 10^{-1}$. This agrees with the value $i_b = 5$ which we found above, for which $\sigma_{i_b} = \sigma_5 = 5.9 \cdot 10^{-2}$. In this test problem the averaging kernels have larger side lobes, and it is interesting that the location of the peak varies slightly with λ. This illustrates that not all averaging kernels are as nicely peaked as those for the first test problem.

The Tikhonov L-curves for the two test problems are shown in Fig. 4.6. Both curves have the typical "L"-shaped appearance described in §4.6. The

4.8. THE ANALYSIS TOOLS IN ACTION

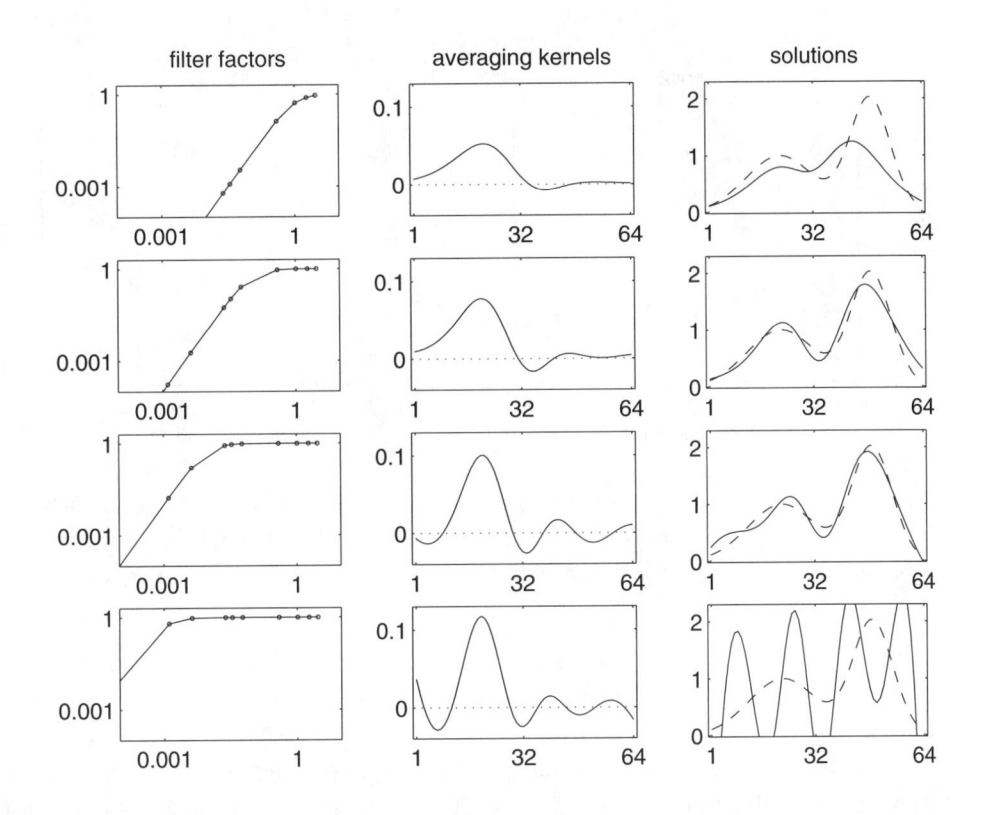

FIG. 4.5. *Test problem* shaw *with standard-form Tikhonov regularization. The left part shows the filter factors f_i versus the singular values σ_i for $i = 1, \ldots, n$. The middle part shows the averaging kernels ξ_{20} corresponding to the abscissa $t = 0.3$. The right part shows the regularized solution x_λ (solid line) and the exact solution x^{exact} (dashed line). The regularization parameters λ are, from top to bottom, 1, 10^{-1}, 10^{-2}, and 10^{-3}.*

horizontal parts of the curves correspond to smooth regularized solutions dominated by the regularization error, and the vertical parts correspond to solutions dominated by the perturbation error. According to §4.6, the corner of the L-curve for these test problems should appear at approximately $(\|e\|_2, \|x^{\text{exact}}\|_2)$, which takes the values $(4.5 \cdot 10^{-4}, 0.58)$ and $(0.19, 8.0)$ for deriv2 and shaw, respectively. These values agree fairly well with the actual L-curves in Fig. 4.6.

The four circles on each L-curve represent the regularized solutions x_λ corresponding to the four regularization parameters used in Figs. 4.4 and 4.5. In the second problem, the large oscillations for $\lambda = 10^{-3}$ cause the norm $\|x_\lambda\|_2$ of the corresponding regularized solution to be much greater than the norm $\|x^{\text{exact}}\|_2$ of the exact solution, and therefore the associated point on the L-curve lies on the vertical part above the corner. The regularized solution

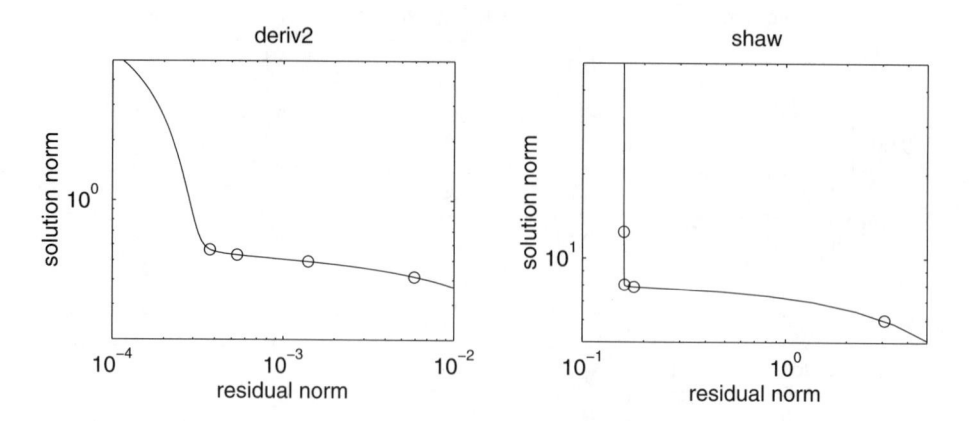

FIG. 4.6. *Tikhonov L-curves for the two test problems* deriv2 *and* shaw, *i.e., log-log plots of* $\|x_\lambda\|_2$ *versus* $\|A x_\lambda - b\|_2$. *The four circles on each curve represent the λ-values used in the previous two figures.*

corresponding to $\lambda = 1$ is too smooth (i.e., the regularization error dominates), and the associated point on the L-curve lies in the far right part of the L-curve's horizontal leg.

Similar observations are made for the first test problem, except that the unwanted oscillations in x_λ for $\lambda = 10^{-3}$ contribute very little to the norm $\|x_\lambda\|_2$ of the regularized solution. Consequently, the associated point does not lie on the L-curve's vertical part. As we shall see below, a different choice of L can change this situation.

4.8.2. General-Form Regularization of deriv2 and wing

To illustrate the influence that the matrix L has on the regularized solution, we also apply Tikhonov regularization in general form to the test problem deriv2, with $L = L_1$ (1.15) approximating the first derivative operator. The dimensions are still $m = n = 64$.

We computed regularized solutions $x_{L,\lambda}$ for three values of the regularization parameter λ:

$$10^{-1}, \quad 3 \cdot 10^{-2}, \quad 5 \cdot 10^{-3}.$$

These solutions are shown in Fig. 4.7 as solid lines, together with the exact solution shown as the dashed line. Figure 4.7 also shows the associated Tikhonov L-curve, i.e., a log-log plot of $\|L x_{L,\lambda}\|_2$ versus $\|A x_{L,\lambda} - b\|_2$, with the three circles representing the above three choices of λ.

With this regularization matrix $L = L_1$, we are able to produce better reconstructions of the exact solution, in the sense that the regularized solutions do not bend down towards the abscissa axis near $i = n$ (as is the case when

4.8. THE ANALYSIS TOOLS IN ACTION

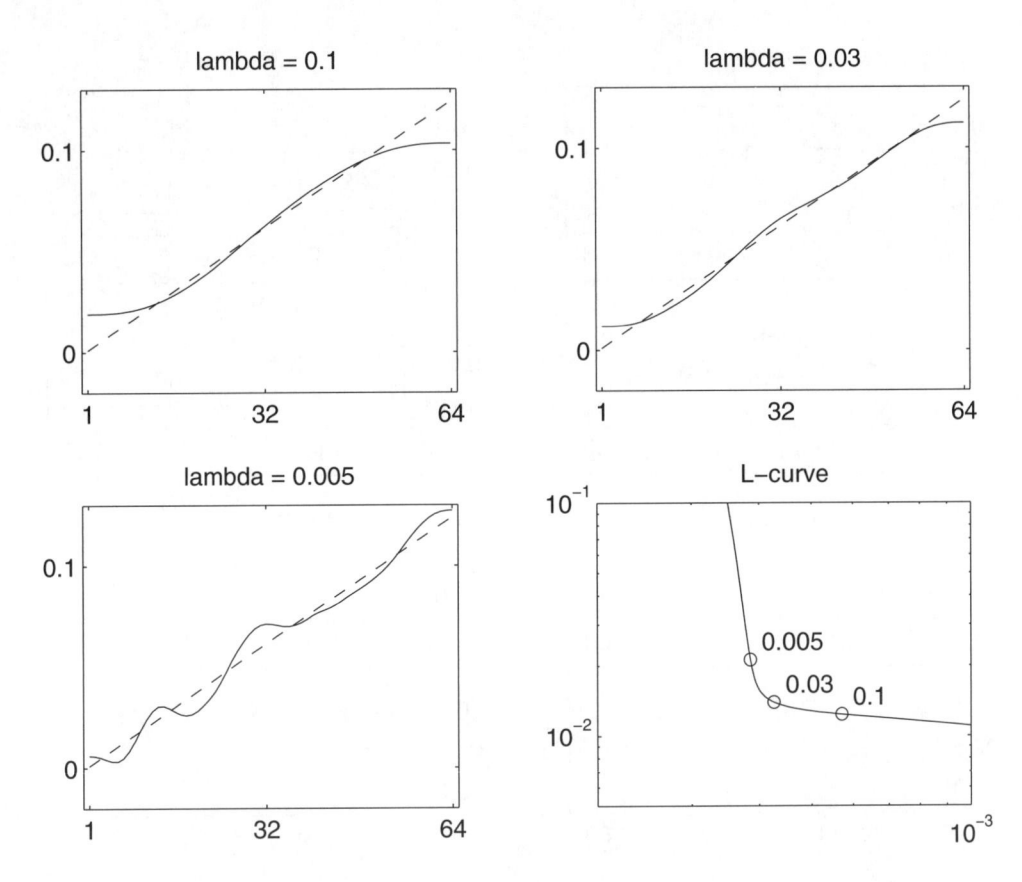

FIG. 4.7. *Tikhonov regularization in general form, with $L = L_1$ approximating the first derivative operator, applied to the test problem* **deriv2**. *We compare the regularized solution $x_{L,\lambda}$ (solid line) and the exact solution x^{exact} (dashed line) for three values of λ. In addition, we show the associated L-curve, i.e., a log-log plot of $\|L x_{L,\lambda}\|_2$ versus $\|A x_{L,\lambda} - b\|_2$. The three regularized solutions are represented by the three circles on the L-curve.*

$L = I_n$). Moreover, the unwanted oscillations in $x_{L,\lambda}$ for $\lambda = 10^{-1}$ now contribute so much to the seminorm $\|L x_{L,\lambda}\|_2$ that the associated point on the L-curve lies somewhat above the corner.

Next, we illustrate how the modified regularization matrix L_Q (4.16) can be used to enforce discontinuities in the regularized solution. This approach is useful when the solution is known to possess a given number of discontinuities at a priori known positions. We use the test problem **wing** from §1.4.4 with dimensions $m = n = 64$, and again $L = L_1$ approximates the first derivative operator. The exact solution to the underlying integral equation has two discontinuities at the abscissas $t = 1/3$ and $t = 2/3$, and the exact solution vector x^{exact} has two large gaps between elements 21 and 22, and between elements 43 and 44.

96   4. Problems with Ill-Determined Rank

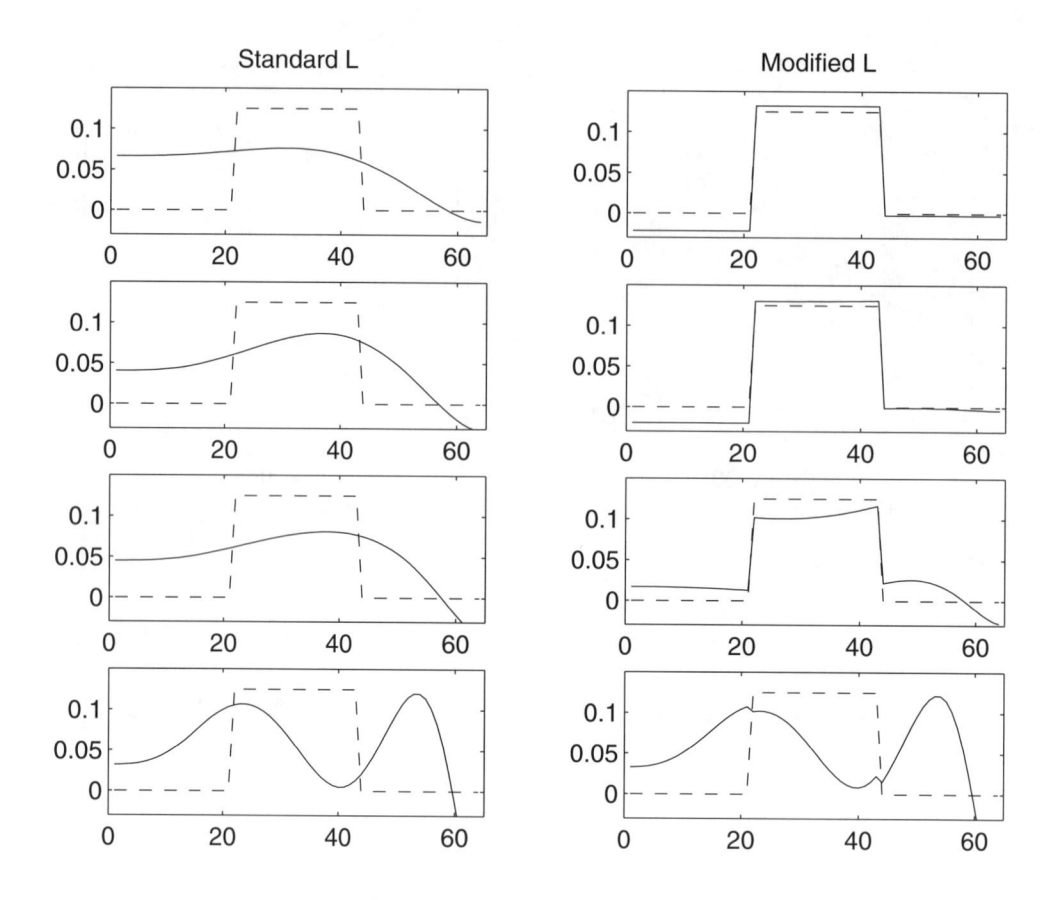

FIG. 4.8. *Test problem* wing *with a discontinuous solution. Exact solutions* x^{exact} *(dashed lines) and regularized solutions* $x_{L,\lambda}$ *(solid lines) computed for four values of* λ *and using either the original* $L = L_1$ *or the modified* $L = L_\mathcal{Q} = L_1 \left(I_n - \sum_{i=1}^{3} q_i q_i^T \right)$ *as the regularization matrix. The* λ*-values are, from top to bottom,* 10^{-2}, $2 \cdot 10^{-3}$, $5 \cdot 10^{-4}$, *and* 10^{-4}.

The left part of Fig. 4.8 shows the Tikhonov regularized solutions (solid lines), as well as the exact solution (dashed line), for the following four values of λ:

$$10^{-2}, \quad 2 \cdot 10^{-3}, \quad 5 \cdot 10^{-4}, \quad 10^{-4}.$$

Clearly, with this choice of L it is not possible to reconstruct the discontinuous solution in any way.

Since we know the positions of the "discontinuities" in x^{exact}, we can modify L to take this a priori knowledge into account. If we define the following three

4.8. THE ANALYSIS TOOLS IN ACTION

vectors (using MATLAB notation):

$$q_1 = \begin{pmatrix} \text{ones}(21,1) \\ \text{zeros}(22,1) \\ \text{zeros}(21,1) \end{pmatrix}, \quad q_2 = \begin{pmatrix} \text{zeros}(21,1) \\ \text{ones}(22,1) \\ \text{zeros}(21,1) \end{pmatrix}, \quad q_3 = \begin{pmatrix} \text{zeros}(21,1) \\ \text{zeros}(22,1) \\ \text{ones}(21,1) \end{pmatrix},$$

then any solution vector representing a piecewise constant solution with break points at $t = 1/3$ and $t = 2/3$ lies in the null space of $L_Q = L_1\left(I_n - \sum_{i=1}^{3} q_i q_i^T\right)$. The right part of Fig. 4.8 shows the Tikhonov regularized solutions computed with this modified regularization matrix, for the same four values of λ as before. Now we achieve a much better approximation to the exact solution, except when λ is so small that the perturbation error dominates the solution.

The residual norms for all the eight regularized solutions shown in Fig. 4.8 are approximately $1.4 \cdot 10^{-3}$, illustrating that residual norms are not suited for judging the quality of the regularized solutions.

4.8.3. The Importance of the Relative Decay Rate

We conclude this collection of numerical examples with an illustration of the importance of the relative decay rate α that appears in the model problem (4.30) and in the related results in §4.5. We generate severely ill posed model problems with $m = n = 32$ whose singular values σ_i and Fourier coefficient $u_i^T b$ are given by

$$\sigma_i = \rho^{i-1}, \quad u_i^T b = \sigma_i^{\alpha+1}, \quad i = 1, \ldots, n,$$

and then noise is added as before. We use the following two values of ρ and α:

$$\rho = 0.6, 0.8, \quad \alpha = 0.4, 0.8.$$

The results are shown in Fig. 4.9, where the dotted line represents the noise level for the Fourier coefficients, i.e., $\tau_b = m^{-1/2}\|b - b^{\text{exact}}\|_2$, and the dashed line represents the effective resolution limit $\eta_{\text{res}} = \tau_b^{\alpha/(1+\alpha)}$; cf. (4.35). This value of η_{res} agrees quite well with the actual behavior of the solution coefficients $u_i^T b/\sigma_i$. The figure illustrates the fact that the resolution limit decreases with the relative decay rate α, while it is independent of the actual decay of the singular values, modeled by the parameter ρ.

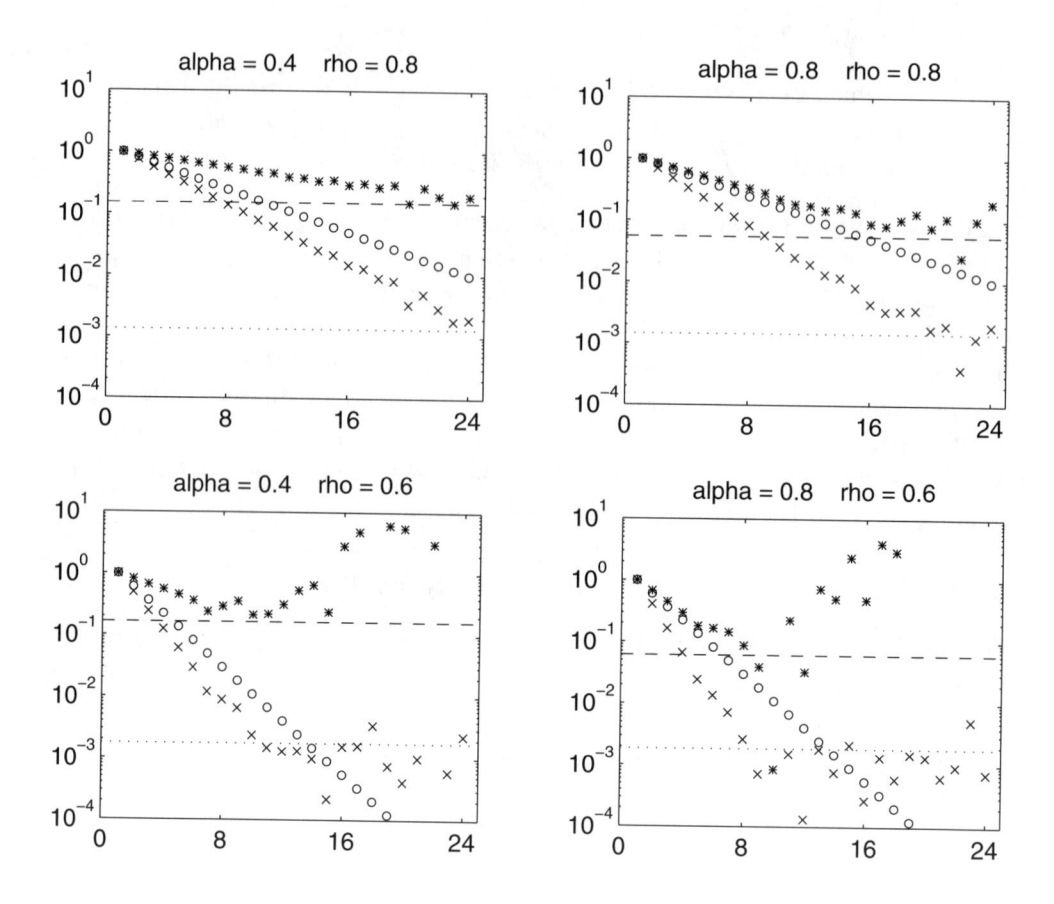

FIG. 4.9. *Singular values σ_i (circles), Fourier coefficients $u_i^T b$ (crosses), and solution coefficients $u_i^T b / \sigma_i$ (asterisks) for the 32×32 model problem. The dotted and dashed lines represent the noise level and the effective resolution limit η_{res}, respectively.*

5

Direct Regularization Methods

From a general perspective, we can essentially choose between two classes of regularization methods: those that are based on some kind of "canonical decomposition" such as the QR factorization or the singular value decomposition (SVD), and those that avoid such decompositions. In the first class of methods we find the direct methods treated in this chapter, while the iterative methods (treated in the next chapter) belong to the second class of algorithms.

In this chapter we also briefly consider a third class of regularization problems, namely, those which lead to nonlinear optimization problems.

Since all direct regularization methods are based on standard operations and decompositions in numerical linear algebra (such as matrix multiplications, backsubstitutions, orthogonal transformations, QR factorizations, bidiagonal reductions, SVDs, etc.), the computational effort involved in a direct method can be estimated a priori.

The standard linear algebra operations and decompositions are almost always available on high-performance computers in highly efficient implementations, often based on the basic linear algebra subprograms (BLAS) and/or the LAPACK Library [4]. Efficient parallel numerical linear algebra routines [25], [83], [165] are available on most parallel computers. Hence, it is not too difficult to develop an efficient implementation of a direct regularization method.

We wish to emphasize again that there are no computational difficulties involved with the use of rectangular regularization matrices L in connection with general-form regularization, when the transformation algorithms from §2.3 are used. Moreover, as long as L is a banded matrix, the computational overhead is negligible compared to standard-form regularization.

5.1. Tikhonov Regularization

The numerical treatment of Tikhonov regularization has been described by many authors, and we include it here mainly for completeness. The method was developed independently by Phillips [280] (see also Twomey's paper [342]) and Tikhonov [336]. Some recent presentations of the method can be found in [107], [111, Chapter 5], [160], [161, §5.1], and [171].

5.1.1. Formulations and Algorithms

The key idea in Tikhonov's method is to incorporate a priori assumptions about the size and smoothness of the desired solution, in the form of the smoothing function $\omega(f)$ in the continuous case, or the (semi)norm $\|L\,x\|_2$ in the discrete case. For discrete ill-posed problems, Tikhonov regularization in general form leads to the minimization problem

$$\min\left\{\|A\,x - b\|_2^2 + \lambda^2\|L\,x\|_2^2\right\}, \tag{5.1}$$

where the regularization parameter λ controls the weight given to minimization of the regularization term, relative to the minimization of the residual norm. The influence of the matrix L is discussed in Chapter 4.

Underlying the formulation in (5.1) is the assumption that the errors in the right-hand side are uncorrelated and with covariance matrix $\sigma_0^2 I_m$. If the covariance matrix is of the more general form $C\,C^T$, where C has full rank m, then one should scale the least squares residual with C^{-1} and solve the scaled problem

$$\min\left\{\|C^{-1}(A\,x - b)\|_2^2 + \lambda^2\|L\,x\|_2^2\right\}. \tag{5.2}$$

The amount of computational work involved in this rescaling depends on the structure of C, but as long as the covariance matrix is either diagonal or banded (which is very often the case in practice), then the cost of the rescaling will only contribute marginally to the total cost of the algorithm.

The Tikhonov problem (5.1) has two important alternative formulations:

$$(A^T A + \lambda^2 L^T L)\,x = A^T b \quad\text{and}\quad \min\left\|\begin{pmatrix} A \\ \lambda L \end{pmatrix} x - \begin{pmatrix} b \\ 0 \end{pmatrix}\right\|_2.$$

From these formulations we see that if $\mathcal{N}(A) \cap \mathcal{N}(L) = \{0\}$, i.e., if the null spaces of A and L intersect trivially such that the coefficient matrix has full rank—which we assume throughout the book—then the Tikhonov solution $x_{L,\lambda}$ is unique, and it is formally given by

$$x_{L,\lambda} = A_\lambda^\# b \quad\text{with}\quad A_\lambda^\# = (A^T A + \lambda^2 L^T L)^{-1} A^T, \tag{5.3}$$

where $A_\lambda^\#$ is the Tikhonov regularized inverse. For $\lambda = 0$ and $\lambda = \infty$, we have $x_{L,0} = x_{\text{LS}}$ and $x_{L,\infty} = x_0 = \sum_{i=p+1}^n u_i^T b\, x_i$, respectively, and $x_0 = 0$ if $p = n$. If $L = I_n$ then we skip the subscript "L" in the Tikhonov solution; i.e., $x_\lambda \equiv x_{I_n,\lambda}$.

If we insert the generalized SVD (GSVD) of (A, L) into this equation, then it is easy to show that the filter factors for Tikhonov regularization in standard and general form are given by

$$f_i = \frac{\sigma_i^2}{\sigma_i^2 + \lambda^2},\ L = I_n \quad\text{and}\quad f_i = \frac{\gamma_i^2}{\gamma_i^2 + \lambda^2},\ L \neq I_n \tag{5.4}$$

5.1. TIKHONOV REGULARIZATION

for $i = 1, \ldots, n$ and $i = 1, \ldots, p$, respectively. For the case $L = I_n$, we see that if σ_i is somewhat greater than λ, then $f_i \approx 1$, while $f_i \approx \sigma_i^2/\lambda^2$ when σ_i is somewhat smaller than λ; and similarly for the case $L \neq I_n$ with σ_i replaced by γ_i. Thus, the Tikhonov filters effectively set in at (generalized) singular values equal to λ.

We note in passing that for a square invertible L the alternative formulation $x_{L,\lambda} = (L^T L)^{-1} A^T (A (L^T L)^{-1} A^T + \lambda^2 I_m)^{-1} b$ occasionally appears in the literature.

In the statistical literature, Tikhonov's method is known as ridge regression; see [246] for a classical paper. The method also appears when the least squares problem is augmented with statistical a priori information about the solution, in the form of a covariance matrix $C_x C_x^T$ for the desired solution (considered as a stochastic variable); cf. [132], [217], and [331, Chapter 1]. In this setting—including the covariance matrix $C C^T$ for the right-hand side errors—the estimator with minimum variance [217, p. 143] and, simultaneously, maximum probability [331, §1.7.2] is the solution to

$$\left(A^T (C C^T)^{-1} A + (C_x C_x^T)^{-1} \right) x = A^T (C C^T)^{-1} b \,.$$

Rewriting this as $((C^{-1}A)^T (C^{-1}A) + (C_x^{-1})^T (C_x^{-1})) x = (C^{-1}A)^T (C^{-1}b)$, we see that the estimator is the scaled Tikhonov solution to (5.2) when λL is replaced by C_x^{-1}.

To solve (5.1) numerically we could in principle form the normal equation matrix $A^T A + \lambda^2 L^T L$ and compute its Cholesky factorization. In practice, this should be avoided: forming $A^T A$ explicitly can lead to loss of information in finite-precision arithmetic, and a new Cholesky factorization is required for each regularization parameter λ.

Eldén's *bidiagonalization algorithm* [98] is the most efficient and numerically stable way to compute the solution to the Tikhonov problem in (5.1). If the problem is given in general form $(L \neq I_n)$, then one should first transform it into standard form with $\bar{A}$ and $\bar{b}$ by means of the algorithm from §2.3.1. Then it should be treated as a least squares problem of the form

$$\min \left\| \begin{pmatrix} \bar{A} \\ \lambda I_p \end{pmatrix} \bar{x} - \begin{pmatrix} \bar{b} \\ 0 \end{pmatrix} \right\|_2 .$$

This problem can be reduced to an equivalent sparse and highly structured problem. The key idea is to transform $\bar{A}$ into a $p \times p$ upper bidiagonal matrix $\bar{B}$ by means of alternating left and right orthogonal transformations,

$$\bar{A} = \bar{U} \bar{B} \bar{V}^T \,, \tag{5.5}$$

as described in [154, §5.4.3]. Software for performing the bidiagonal reduction is available in most of the currently available mathematical libraries, and some of them are listed in Table 5.1. Of these libraries, only LAPACK and NAG

TABLE 5.1. *Software for performing the bidiagonal reduction.*

Software package	Subroutines
EISPACK [136]	Part of SVD
IMSL [216]	Part of LSVRR
LAPACK [4]	_GEBRD
LINPACK [93]	Part of _SVDC
NAG [257]	F01QCF + F02SWF
Numerical Recipes [287]	Part of SVDCMP

contain stand-alone bidiagonalization subroutines; in the other libraries this operation is part of the SVD subroutine.

Once $\bar{A}$ has been reduced to a bidiagonal matrix $\bar{B}$, we make the substitution $\bar{x} = \bar{V}\bar{\xi}$ and obtain the problem

$$\min \left\| \begin{pmatrix} \bar{B} \\ \lambda I_p \end{pmatrix} \bar{\xi} - \begin{pmatrix} \bar{U}^T b \\ 0 \end{pmatrix} \right\|_2, \tag{5.6}$$

which can be solved for $\bar{\xi}_\lambda = \bar{V}^T \bar{x}_\lambda$ in only $\mathcal{O}(p)$ operations. Strategic Givens rotations are used to annihilate the diagonal submatrix λI_p (same notation as in §4.3),

followed by backsubstitution with the resulting upper bidiagonal matrix and the corresponding components of the resulting right-hand side.[14] Then $\bar{x}_\lambda$ is given by $\bar{x}_\lambda = \bar{V}\bar{\xi}_\lambda$. This algorithm is particularly attractive when $\bar{\xi}_\lambda$ must be computed for many values of λ. Notice that only numerically stable orthogonal transformations are involved in this algorithm for computing λ.

We emphasize here that the solution's (semi)norm and the residual's norm satisfy the relations

$$\|L\, x_{L,\lambda}\|_2 = \|\bar{\xi}_\lambda\|_2, \qquad \|A\, x_{L,\lambda} - b\|_2 = \|\bar{B}\bar{\xi}_\lambda - \bar{U}^T b\|_2.$$

This means that it is not necessary to perform the transformations back to $\bar{x}_\lambda = \bar{V}\bar{\xi}_\lambda$ or $x_{L,\lambda} = L_A^\dagger \bar{x}_\lambda + x_0$ in order to compute these norms. This is

---

[14]Voevodin [355] also used bidiagonalization of $\bar{A}$, but then he explicitly computed the matrix $\bar{B}^T \bar{B} + \lambda^2 I_p$ (and he suggested using extended precision when dealing with this matrix).

5.1. Tikhonov Regularization

important for the efficiency in connection with the parameter-choice methods described in Chapter 7, which require the computation of $\|L\,x_{L,\lambda}\|_2$ and/or $\|A\,x_{L,\lambda} - b\|_2$ for many values of λ.

As mentioned in §1.3, we can incorporate an a priori estimated x^* into the smoothing norm $\Omega(x)$ and thus "bias" the regularized solution towards this estimate. Although such an estimate $x^* \neq 0$ can have a dramatic effect on the regularized solution, compared to the choice $x^* = 0$, the computational overhead involved in working with an $x^* \neq 0$ is minor. The least squares formulation now takes the form

$$\min \left\| \begin{pmatrix} A \\ \lambda L \end{pmatrix} x - \begin{pmatrix} b \\ \lambda L x^* \end{pmatrix} \right\|_2,$$

and after standard-form transformation, we obtain

$$\min \left\| \begin{pmatrix} \bar{A} \\ \lambda I_p \end{pmatrix} \bar{x} - \begin{pmatrix} \bar{b} \\ \lambda \bar{x}^* \end{pmatrix} \right\|_2.$$

After the bidiagonalization phase the resulting sparse least squares problem becomes

$$\min \left\| \begin{pmatrix} \bar{B} \\ \lambda I_p \end{pmatrix} \bar{\xi} - \begin{pmatrix} \bar{U}^T b \\ \lambda \bar{V}^T \bar{x}^* \end{pmatrix} \right\|_2, \tag{5.7}$$

which is solved by the same algorithm as (5.6), using the same sequence of strategic Givens rotations, followed by backsubstitution. The (semi)norms of the solutions are now related as $\|L\,(x_{L,\lambda} - x^*)\|_2 = \|\bar{\xi}_\lambda - \bar{V}^T \bar{x}^*\|_2$.

It should be noted that if A and L have special structures, then it may be possible to use these structures in a highly efficient algorithm. A variant of the bidiagonalization algorithm for banded A and L is described in [102], and an $\mathcal{O}(n^2)$ algorithm for the case when A and L are upper triangular Toeplitz matrices is described in [101] (such problems arise in connection with certain Volterra integral equations).

Alternatively, the Tikhonov solution can be computed by means of the SVD of A or the GSVD of (A, L) and the corresponding filter factors in (5.4). This technique is computationally more expensive than the above approach using $\bar{A}$ and $\bar{B}$, but it also provides much more insight into the regularization problem (cf. Chapter 4).

5.1.2. Perturbation Bounds

Perturbation bounds for Tikhonov regularization in standard and general form were derived in [179], [180], and [181], and they illuminate the tradeoff between adding too much regularization (such that the solution becomes too "smooth") and too little regularization (such that the solution is too sensitive to the perturbations). The results are summarized in the following theorem.

5. Direct Regularization Methods

Theorem 5.1.1. [179, Theorem 4.1], [180]. *Let $x_{L,\lambda}$ and $\tilde{x}_{L,\lambda}$ denote the solutions to*

$$\min\{\|A\,x - b\|_2^2 + \lambda^2 \|L\,x\|_2^2\} \quad \text{and} \quad \min\{\|\tilde{A}\,x - \tilde{b}\|_2^2 + \lambda^2 \|L\,x\|_2^2\}$$

computed with the same λ, and let $E = A - \tilde{A}$ and $e = b - \tilde{b}$. If $\epsilon = \|E\|_2/\|A\|_2$ and $\kappa_\lambda = \|A\|_2 \|X\|_2/\lambda$, where X is from the GSVD of (A, L), and if $0 < \lambda \leq 1$, then

$$\frac{\|x_{L,\lambda} - \tilde{x}_{L,\lambda}\|_2}{\|x_{L,\lambda}\|_2} \leq \frac{\kappa_\lambda}{1 - \epsilon\,\kappa_\lambda} \left(\left(1 + \mathrm{cond}(X)\right)\epsilon + \frac{\|e\|_2}{\|b_\lambda\|_2} + \epsilon\,\kappa_\lambda \frac{\|r_\lambda\|_2}{\|b_\lambda\|_2} \right), \quad (5.8)$$

where $b_\lambda = A\,x_{L,\lambda}$, $r_\lambda = b - b_\lambda$, and $\mathrm{cond}(X)$ is the condition number of X. If $E = 0$ and $\lambda < 1/\sqrt{2}$, then we obtain the tighter bound

$$\frac{\|x_{L,\lambda} - \tilde{x}_{L,\lambda}\|_2}{\|x_{L,\lambda}\|_2} \leq \frac{\kappa_\lambda}{2\,(1 - \lambda^2)^{1/2}} \frac{\|e\|_2}{\|b_\lambda\|_2}. \quad (5.9)$$

In particular, if L is square ($p = n$) and invertible, and if we define $\bar{\kappa}_\lambda = \|A\|_2 \|L^{-1}\|_2/\lambda$, then (5.8) and (5.9) can be sharpened to

$$\frac{\|x_{L,\lambda} - \tilde{x}_{L,\lambda}\|_2}{\|x_{L,\lambda}\|_2} \leq \frac{\bar{\kappa}_\lambda}{1 - \epsilon\,\bar{\kappa}_\lambda} \left(\left(1 + \mathrm{cond}(L)\right)\epsilon + \frac{\|e\|_2}{\|b_\lambda\|_2} + \epsilon\,\bar{\kappa}_\lambda \frac{\|r_\lambda\|_2}{\|b_\lambda\|_2} \right) \quad (5.10)$$

and

$$\frac{\|x_{L,\lambda} - \tilde{x}_{L,\lambda}\|_2}{\|x_{L,\lambda}\|_2} \leq \frac{1}{2} \bar{\kappa}_\lambda \frac{\|e\|_2}{\|b_\lambda\|_2}. \quad (5.11)$$

If $L = I_n$, then (5.10) and (5.11) hold with $\bar{\kappa}_\lambda = \|A\|_2/\lambda$ and $\mathrm{cond}(L) = 1$.

We see that the quantities κ_λ and $\bar{\kappa}_\lambda$ play the role of a condition number for the Tikhonov regularized solution; see Theorem 2.1.2 concerning $\mathrm{cond}(X)$. The larger the λ, the smaller the condition numbers, and thus the less sensitive the regularized solutions are to perturbations—but increasing λ also increases the regularization error; cf. Theorems 4.5.1 and 4.5.2.

An alternative perturbation bound that involves the shape of the L-curve from §4.6 was derived in [184].

Theorem 5.1.2. [184, Corollary 7]. *Let $\mathcal{F}$ denote the function defined by*

$$\|L\,x_{L,\lambda}\|_2 = \mathcal{F}(\|A\,x_{L,\lambda} - b\|_2).$$

For a given λ, let $\tilde{\lambda}$ be chosen such that

$$\|A\,x_{L,\lambda} - b\|_2 = \|\tilde{A}\,\tilde{x}_{L,\tilde{\lambda}} - \tilde{b}\|_2.$$

Then

$$\frac{\|L\,(x_{L,\lambda} - \tilde{x}_{L,\tilde{\lambda}})\|_2}{\|\tilde{x}_{L,\tilde{\lambda}}\|_2} < 4\sqrt{|\mathcal{F}'(\delta)|\,\|A\|_2\,\epsilon_{A,b}} \left(1 + \mathcal{O}(\epsilon_{A,b})\right), \quad (5.12)$$

5.1. Tikhonov Regularization

where
$$\epsilon_{A,b} = \frac{\|E\|_2}{\|A\|_2} + \frac{\|e\|_2}{\|b_\lambda\|_2} \tag{5.13}$$

and where $\mathcal{F}'(\delta)$ denotes the derivative of $\mathcal{F}$ at $\delta = \|A\,x_{L,\lambda} - b\|_2$.

We see that in order to guarantee a small perturbation bound in (5.12), we should choose the regularization parameter λ such that $|\mathcal{F}'(\|A\,x_{L,\lambda} - b\|_2)|$ is small; i.e., $x_{L,\lambda}$ should preferably correspond to a point on the L-curve immediately to the right of the corner.

5.1.3. Least Squares with a Quadratic Constraint

There are two other regularization methods which are almost equivalent to Tikhonov's method, and which can be treated numerically by essentially the same techniques as mentioned above, i.e., standard-form transformation plus bidiagonalization or GSVD. These two methods are formulated as the following least squares problems with a quadratic constraint:

$$\min \|A\,x - b\|_2 \quad \text{subject to} \quad \|L\,(x - x^*)\|_2 \leq \alpha, \tag{5.14}$$
$$\min \|L\,(x - x^*)\|_2 \quad \text{subject to} \quad \|A\,x - b\|_2 \leq \delta, \tag{5.15}$$

where α and δ are nonzero parameters each playing the role of the regularization parameter in (5.14) and (5.15), respectively. For $\alpha \geq 0$ and $\delta \geq \delta_0$ (where δ_0 is from (4.36)), the solution to both problems is identical to $x_{L,\lambda}$ from Tikhonov's method for suitably chosen values of λ that depend in a nonlinear way on α and δ. This follows from the characterization of the Tikhonov L-curve in Theorem 4.6.1. The theory of least squares problems with a quadratic constraint is discused by Gander [135].

Graphically, the solution to (5.14) lies on the intersection of the Tikhonov L-curve and the horizontal line $\|L\,(x - x^*)\|_2 = \alpha$. Similarly, the solution to (5.15) lies on the intersection of the L-curve and the vertical line $\|A\,x - b\|_2 = \delta$.

The solutions to both (5.14) and (5.15) can be computed by iterative schemes such as Newton iteration (cf. [255, §26]), possibly augmented with a Hebden rational model (cf. [55]). The latter reference also discusses a good initial value of λ for (5.14), namely,

$$\gamma_1 \left(\sqrt{\|L\,x_{\text{LS}}\|_2/\alpha} - 1 \right)^{1/2}.$$

For (5.15) we found in [186, discrep] that a good initial value for λ is that generalized singular value γ_k of (A, L) for which the corresponding truncated GSVD (TGSVD) residual norm $\|A\,x_{L,k} - b\|_2$ is closest to δ. An efficient algorithm for solving (5.14) when A is large and sparse, based on Gauss quadrature and Lanczos bidiagonalization, is described in [156].

5.1.4. Inequality Constraints

It is sometimes convenient to add certain constraints to the Tikhonov solution, such as nonnegativity, monotonicity, or convexity. All three constraints can be formulated as inequality constraints of the form $G x \geq 0$, taking the special forms

$$x \geq 0 \quad \text{(nonnegativity)}, \tag{5.16}$$
$$L_1 x \geq 0 \quad \text{(monotonicity)}, \tag{5.17}$$
$$L_2 x \geq 0 \quad \text{(convexity)}, \tag{5.18}$$

where L_1 and L_2 approximate the first and second derivative operators, respectively. The constraints can, of course, also be combined in the matrix G. Thus, the inequality-constrained Tikhonov solution solves the problem

$$\min \left\{ \|A x - b\|_2^2 + \lambda^2 \|L x\|_2^2 \right\} \quad \text{subject to} \quad G x \geq 0 \,. \tag{5.19}$$

The constraints can also be incorporated in other direct regularization methods discussed in this chapter. For example, the TSVD method with inequality constraints takes the form

$$\min \|A_k x - b\|_2 \quad \text{subject to} \quad G x \geq 0 \,, \tag{5.20}$$

and if the solution is nonunique, then a unique solution can be determined by also minimizing the (semi)norm $\|L x\|_2$. Algorithms for solving (5.19) and (5.20), as well as problems with more general inequality constraints of the form $c_{\min} \leq G x \leq c_{\max}$, are surveyed in [36, §5.2] and [37, §6].

5.1.5. Related Methods

The Tikhonov solution can be modified by a process which resembles—but is *different* from—iterative refinement for linear systems of equations. If we set $x^{(1)} = x_{L,\lambda}$, then the modified Tikhonov solutions are defined recursively as

$$x^{(k+1)} = x^{(k)} + (A^T A + \lambda^2 L^T L)^{-1} A^T (b - A x^{(k)}) \,, \quad k = 1, 2, \ldots \,. \tag{5.21}$$

This technique is usually called *iterated Tikhonov regularization*; see Eqs. (6.5)–(6.6) in the next chapter. We refer to [171, §5.2] for implementation details based on the bidiagonalization of $\bar{A}$. This technique gives slightly sharper filter factors than those for standard Tikhonov regularization.

Another modification of Tikhonov's method for achieving sharper filter factors was suggested by Rutishauser [298]. In the standard-form case this method amounts to solving the following system of equations:

$$\left(A^T A + \lambda^2 I_n + \lambda^2 (A^T A + \lambda^2 I_n)^{-1} \right) x = A^T b \,. \tag{5.22}$$

5.1. Tikhonov Regularization

The corresponding filter factors are

$$f_i = \frac{\sigma_i^2}{\sigma_i^2 + \lambda^2 + \lambda^2/(\sigma_i^2 + \lambda^2)} ,$$

and we have the approximations

$$f_i \approx \begin{cases} \sigma_i^2/(\lambda^2 + 1) , & \sigma_i < \lambda \\ \sigma_i^4/\lambda^2 , & \lambda < \sigma_i < \lambda^{1/2} \\ 1 , & \lambda^{1/2} < \sigma_i \end{cases} \quad \text{for} \quad \lambda \leq 1 ;$$

$$f_i \approx \begin{cases} \sigma_i^2/(\lambda^2 + 1) , & \sigma_i < \lambda \\ 1 , & \lambda < \sigma_i \end{cases} \quad \text{for} \quad \lambda \geq 1 .$$

We see that if $\lambda < 1$, then we achieve a sharper cut-off for singular values in the range $\lambda \leq \sigma_i \leq \lambda^{1/2}$. See [298] for implementation details.

For *symmetric positive definite* matrices A and L, Franklin[15] [133] suggested replacing Tikhonov's problem (5.1) with the problem

$$(A + \lambda L) x = b , \quad \lambda \geq 0 . \tag{5.23}$$

If $L = I_n$ then the solution to (5.23) can be expressed in terms of the SVD of A, i.e., in the form (4.3) with filter factors $f_i = \sigma_i/(\sigma_i + \lambda)$. If $L \neq I_n$ then the solution to (5.23) is most conveniently expressed in terms of the generalized eigenvalues and eigenvectors of (A, L). Specifically, if

$$W^T A W = \text{diag}(\alpha_1, \ldots, \alpha_n) \quad \text{and} \quad W^T L W = \text{diag}(\beta_1, \ldots, \beta_n)$$

and if $W = (w_1, \ldots, w_n)$, then the solution x_λ to (5.23) is given by

$$x_\lambda = \sum_{i=1}^n \frac{\alpha_i}{\alpha_i + \lambda \beta_i} \frac{w_i^T b}{\alpha_i} w_i .$$

Since the generalized eigenvalues and eigenvectors are generally not related to the GSVD of (A, L), this solution is generally not of the form (4.5).

One important application of Franklin's method is the "semidiscrete" approach to Tikhonov regularization of Fredholm integral equations of the form (1.2) with a discrete right-hand side, where the regularized solution is given by

$$f_{\text{reg}}(t) = \sum_{i=1}^n \xi_i k_i(t) . \tag{5.24}$$

Here, the solution vector ξ is computed by means of (5.23) with A's elements given by $a_{ij} = \int_0^1 k_i(t) k_j(t) \, dt$ and b consisting of the right-hand side data in (1.2).

---

[15] In the Russian literature, Lavrentiev's name is associated with this method.

O'Brien and Holt [264] presented a statistically based variant of Tikhonov regularization. Their approach is to determine the SVD filter factors f_i in (4.3) such that the expected value $\mathcal{E}(\|x^{\text{exact}} - x_{\text{reg}}\|_2)$ of the error norm is minimized. Assume that $b = b^{\text{exact}} + e$, $x_{\text{reg}} = A^\# b$, and that the covariance matrix for e is CC^T. Then we obtain

$$\begin{aligned}\mathcal{E}(\|x^{\text{exact}} - x_{\text{reg}}\|_2^2) &= \|x^{\text{exact}} - A^\# b^{\text{exact}}\|_2^2 + \text{trace}(A^\# C C^T (A^\#)^T) \\ &= \sum_{i=1}^n (1-f_i)^2 (v_i^T x^{\text{exact}})^2 + \sum_{i=1}^n f_i^2 \sigma_i^{-2} \|C^T u_i\|_2^2\,,\end{aligned}$$

and this quantity is minimized for

$$f_i = \frac{\sigma_i^2}{\sigma_i^2 + \|C^T u_i\|_2^2/(v_i^T x^{\text{exact}})^2}\,, \qquad i = 1,\ldots,n\,. \tag{5.25}$$

In particular, if $C = \sigma_0 I_m$, then the filter factors are

$$f_i = \frac{\sigma_i^2}{\sigma_i^2 + \sigma_0^2/(v_i^T x^{\text{exact}})^2}\,, \qquad i = 1,\ldots,n\,, \tag{5.26}$$

and if the solution's Fourier coefficients $v_i^T x^{\text{exact}}$ satisfy the discrete Picard condition, i.e., their absolute values decay, then the filter factors in (5.26) decrease faster than the Tikhonov filter factors $\sigma_i^2/(\sigma_i^2 + \sigma_0^2)$ as $\sigma_i \to 0$. But unless the quantities $|v_i^T x^{\text{exact}}|$ decay very rapidly, there is not much difference between the two sets of filter factors and Tikhonov regularization works almost as well as the method based on (5.26).

5.2. The Regularized General Gauss–Markov Linear Model

The general Gauss–Markov linear model is specially designed for least squares problems with an ill-conditioned or rank-deficient covariance matrix, which we again write as CC^T where $C \in \mathbb{R}^{m \times q}$ now has full column rank q. In this case, $\min \|C^\dagger (Ax - b)\|_2$ is *not* a valid formulation, and instead one should solve the problem

$$\min \|u\|_2 \quad \text{subject to} \quad Ax + Cu = b\,. \tag{5.27}$$

However, if A is ill conditioned, then we can expect the usual difficulties when solving (5.27) numerically; cf. [378, §1].

To overcome these difficulties, it was proposed in [378] to add stabilization to (5.27), similar to the way that Tikhonov's method adds stabilization to the ordinary least squares problem. The regularized version of the general Gauss–Markov linear model thus takes the form

$$\min \left\{ \|u\|_2^2 + \lambda^2 \|Lx\|_2^2 \right\} \quad \text{subject to} \quad Ax + Cu = b\,. \tag{5.28}$$

This equation involves three matrices A, C, and L, and the appropriate tool for its analysis is the restricted SVD (RSVD) from §2.1.4. In terms of the RSVD, and using the notation of Eq. (2.17), the solution $x_{L,C,\lambda}$ to (5.28) (cf. [378, Eq. (14)]) can be written as

$$x_{L,C,\lambda} = X F_\lambda \Sigma^\dagger X^T b , \qquad F_\lambda = \mathrm{diag}(f_1,\ldots,f_t,1,\ldots,1) , \qquad (5.29)$$

where $f_i = \sigma_i^2/(\sigma_i^2 + \lambda^2)$ for $i = 1,\ldots,t$. This relation is used in [378, §§3-4] to analyze the properties of the solution $x_{L,C,\lambda}$ and to derive the following perturbation bound.

Theorem 5.2.1. [378, Theorem 2]. *Let $\tilde{x}_{L,C,\lambda}$ denote the solution to (5.28) when b is replaced by $\tilde{b} = b + e$. If $\sigma_r \leq \lambda \leq \sigma_1$, then*

$$\frac{\|x_{L,C,\lambda} - \tilde{x}_{L,C,\lambda}\|_2}{\|x_{L,C,\lambda}\|_2} \leq \frac{\|\Sigma\|_2}{\min\{2\lambda,1\}} \mathrm{cond}(X)\,\mathrm{cond}(Z) \frac{\|e\|_2}{\|b_\lambda\|_2} , \qquad (5.30)$$

where $\|\Sigma\|_2 = \max\{\sigma_1, 1\}$ and $b_\lambda = A\, x_{L,C,\lambda}$.

As in the case of Tikhonov regularization, we see that λ controls the solution's sensitivity to errors in b. For more details we refer to [378], where an efficient algorithm based on a sequence of QR factorizations can also be found.

5.3. Truncated SVD and GSVD Again

A fundamental observation regarding Tikhonov regularization is that the ill-conditioning of A is circumvented by introducing a new problem with a new well-conditioned coefficient matrix $\binom{A}{\lambda L}$ with full rank. A different way to treat the ill-conditioning of A is to derive a new problem with a well-conditioned *rank-deficient* coefficient matrix. This is the philosophy behind the three methods TSVD, MTSVD (modified TSVD), and TGSVD, which were introduced in §3.2 in connection with numerically rank-deficient problems.

For problems with ill-determined numerical rank, it is not obvious that "brute force" truncation of the SVD/GSVD leads to a regularized solution. We have proved that, under suitable conditions, for any valid truncation parameter k there will always exist a regularization parameter λ such that the TGSVD solution (or the TSVD solution, if $L = I_n$) is close to the Tikhonov solution.

The TSVD and TGSVD methods are important in their own right, and they are perhaps even more useful for theoretical purposes because of their simplicity, which makes them much simpler to analyze than Tikhonov's method—although they produce practically the same solutions. For example, the filter factors for these methods simply consist of zeros and ones:

$$\text{TSVD:}\ f_i = \begin{cases} 1, & i \leq k , \\ 0, & i > k ; \end{cases} \qquad \text{TGSVD:}\ f_i = \begin{cases} 0, & i \leq n-k , \\ 1, & i > n-k , \end{cases} \qquad (5.31)$$

where k is the truncation parameter. The MTSVD solution $\hat{x}_{L,k}$ cannot be expressed merely in terms of n filter factors, and this can be seen from the following expression which follows immediately from (3.17):

$$\hat{x}_{L,k} = V \begin{pmatrix} I_k & 0 \\ \Theta_{21} & 0 \end{pmatrix} \Sigma^\dagger U^T b \, , \qquad \Theta_{21} = -(L\,V_k^o)^\dagger L\,V_k \, .$$

Some of the first papers that described the use of TSVD for regularization of discrete ill-posed problems are [18], [201], [246], and [352], and the relationships between TSVD/TGSVD and Tikhonov regularization in standard/general form are analyzed in [176], [179], [181], and [183]. To state our results, we recall that in terms of the GSVD of (A, L), the TGSVD and Tikhonov solutions can be written as $x_{L,k} = A_{L,k}^\# b$ and $x_{L,\lambda} = A_\lambda^\# b$, where

$$A_{L,k}^\# = X \begin{pmatrix} F_k \Sigma^\dagger & 0 \\ 0 & I_{n-p} \end{pmatrix} U^T \, , \qquad F_k = \mathrm{diag}(0, I_{p-k}) \tag{5.32}$$

and

$$A_\lambda^\# = X \begin{pmatrix} F_\lambda \Sigma^\dagger & 0 \\ 0 & I_{n-p} \end{pmatrix} U^T \, , \qquad F_\lambda = \mathrm{diag}(f_1, \ldots, f_p) \, , \tag{5.33}$$

in which $f_i = \gamma_i^2/(\gamma_i^2 + \lambda^2)$. The following theorems summarize our results (which agree with those derived in [176, Theorem 5.2] and [181, Theorem 3.2] when $L = I_n$).

Theorem 5.3.1. [183, Theorem 3]. *Let the matrices $A_{L,k}^\#$ and $A_\lambda^\#$ be given by (5.32) and (5.33), and let $\omega_k = \gamma_{p-k}/\gamma_{p-k+1}$. Then for any $\lambda > 0$ and any fixed $k < p$:*

$$\frac{\omega_k^{1/2}}{1 + \omega_k^{1/2}} \leq \min_\lambda \frac{\|L\,(A_\lambda^\# - A_{L,k}^\#)\|_2}{\|L\,A_{L,k}^\#\|_2} \leq \frac{\omega_k^{1/2}}{1 + \omega_k^{3/2}} \tag{5.34}$$

and

$$\min_\lambda \|A\,A_\lambda^\# - A\,A_{L,k}^\#\|_2 = \frac{\omega_k}{1 + \omega_k} . \tag{5.35}$$

The two minima are attained for

$$\lambda = \left(\left(\gamma_{p-k} \gamma_{p-k+1}^3 + \tfrac{1}{4}\gamma_{p-k}^2 (\gamma_{p-k+1} - \gamma_{p-k})^2 \right)^{1/2} \right.$$

$$\left. + \tfrac{1}{4}\gamma_{p-k}(\gamma_{p-k+1} - \gamma_{p-k}) \right)^{1/2}$$

$$\approx \left(\gamma_{p-k} \gamma_{p-k+1}^3 \right)^{1/4}$$

and $\lambda = (\gamma_{p-k} \gamma_{p-k+1})^{1/2}$, respectively.

Theorem 5.3.2. [183, Theorem 4]. *Let the right-hand side b satisfy the model (4.30); i.e.,*

$$u_i^T b = \begin{cases} \gamma_i^{\alpha+1}, & i = 1, \ldots, p, \\ \gamma_p^{\alpha+1}, & i = p+1, \ldots, n. \end{cases}$$

If $\gamma_{p-k} \leq \lambda \leq \gamma_{p-k+1}$ and $\omega_k = \gamma_{p-k}/\gamma_{p-k+1}$, then

$$\frac{\|L(x_{L,\lambda} - x_{L,k})\|_2}{\|L x_{L,k}\|_2} \leq \begin{cases} p^{1/2} \omega_k^{\alpha}, & \alpha < 0, \\ p^{1/2} (\gamma_{p-k+1}/\gamma_p)^{\alpha}, & 0 \leq \alpha < 2, \\ p^{1/2} (\gamma_{p-k+1}/\gamma_p)^2, & 2 \leq \alpha, \end{cases} \qquad (5.36)$$

and

$$\frac{\|A x_{L,\lambda} - A x_{L,k}\|_2}{\|b\|_2} \leq \begin{cases} p^{1/2} (\gamma_{p-k+1}/\gamma_p)^{\alpha+1}, & \alpha < 1, \\ p^{1/2} (\gamma_{p-k+1}/\gamma_p)^2, & 1 \leq \alpha. \end{cases} \qquad (5.37)$$

Whenever there is a distinct gap between γ_{p-k} and γ_{p-k+1} such that ω_k is small, then Theorem 5.3.1 guarantees that for every k there always exists a λ such that the two solutions $x_{L,k}$ and $x_{L,\lambda}$ are very similar. When there is no particular gap in the (generalized) singular value spectrum, then the discrete Picard condition (cf. §4.5) comes into play, and Theorem 5.3.2 shows that the faster the relative decay of the Fourier coefficients, the closer $x_{L,k}$ is to $x_{L,\lambda}$—and similarly with the residuals.

The similarities between standard-form Tikhonov regularization and TSVD can be illustrated graphically by means of ridge traces [329], i.e., plots of x_λ and x_k with λ and k as the parameters, respectively. As λ increases from 0 (corresponding to $k = n$, i.e., $x_\lambda = x_{\text{LS}}$) it is observed how x_λ moves towards x_{n-1}, then bends off and moves towards x_{n-2}, etc., until $\lambda \to \infty$ and $x_\lambda \to 0$ (corresponding to $k = 0$). See Fig. 5.1 for an illustration.

Perturbation bounds for the TSVD/TGSVD methods are given in §3.2.4. These methods are usually applied to discrete ill-posed problems with an "exact" coefficient matrix A where $\|E\|_2$ is of the same order as the machine precision or the approximation errors. The effective resolution limit (cf. §4.1), and hence the truncation parameter k, is therefore determined by the errors in the right-hand side b. For this reason, the requirement $\|E\|_2 < \sigma_k - \sigma_{k+1}$ in Theorem 3.2.1 is satisfied for these problems.

More details about the TSVD, MTSVD, and TGSVD methods as well as numerical examples can be found in [176], [179], [181], and [183].

5.4. Algorithms Based on Total Least Squares

As already mentioned in §3.2.2, most regularization methods assume that the errors in $A x \approx b$ are confined to the right-hand side b, and total least squares (TLS) is a good basis for designing methods that take both errors into account.

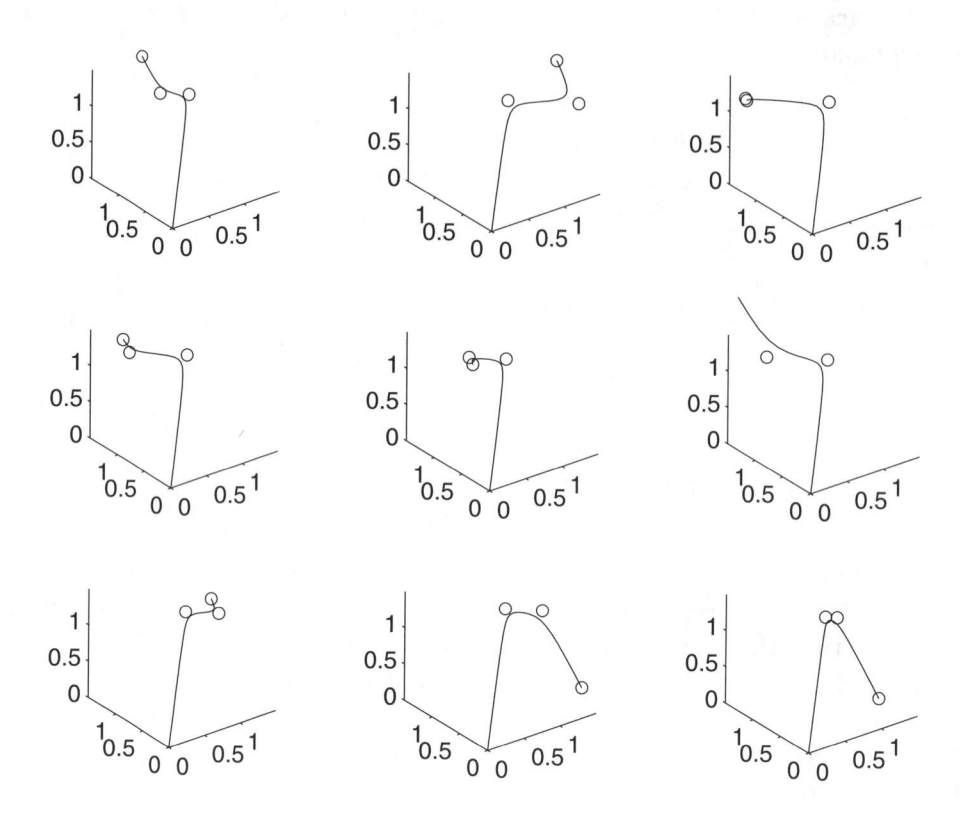

FIG. 5.1. *Ridge traces, i.e., plots of x_λ (solid lines) and x_k (circles) with λ and k as the parameters. The matrix A is 3×3 and fixed, $x^{\text{exact}} = (1,1,1)^T$, and the nine figures show the ridge traces for nine different perturbations of the system $b^{\text{exact}} = A\,x^{\text{exact}}$.*

In this section, we discuss two such methods for discrete ill-posed problems: truncated TLS (T-TLS) as defined in §3.2.2, and regularized TLS which is analogous to Tikhonov regularization. Our experience with these methods is that they can produce improved regularized solutions when the noise level in both A and b is large.

5.4.1. Truncated TLS Again

The T-TLS method, which was introduced in §3.2.2 in connection with numerically rank-deficient problems, is also suitable as a regularization method for discrete ill-posed problems; see [124]. For these problems, it may not be obvious why the T-TLS method yields a regularized solution. However, the filter factors for the T-TLS solution $\bar{x}_k$ show that the SVD components of $\bar{x}_k$ associated with the smallest singular values σ_i are indeed filtered out.

5.4. Algorithms Based on Total Least Squares

Theorem 5.4.1. [124, Theorems 3.6 and 3.7]. *If the nonzero singular values of A and (A, b) are simple, then the T-TLS solution $\bar{x}_k$ can be written in the form (4.3), and the filter factors corresponding to $u_i^T b \neq 0$ and $\sigma_i \neq 0$ are given by*

$$f_i = \sum_{j=1}^{k} \frac{\bar{v}_{n+1,j}^2}{\|\bar{V}_{22}\|_2^2} \left(\frac{\sigma_i^2}{\bar{\sigma}_j^2 - \sigma_i^2} \right) . \tag{5.38}$$

For $i \leq k$, these filter factors increase monotonically with i and satisfy

$$0 \leq f_i - 1 \leq \frac{\bar{\sigma}_{k+1}^2}{\sigma_i^2 - \bar{\sigma}_{k+1}^2} , \quad i \leq k , \tag{5.39}$$

while for $k < i \leq \mathrm{rank}(A)$ these filter factors satisfy

$$0 \leq f_i \leq \|\bar{V}_{22}\|_2^{-2} \frac{\sigma_i^2}{\bar{\sigma}_k^2 - \sigma_i^2} , \quad k < i \leq \mathrm{rank}(A) . \tag{5.40}$$

We emphasize that this theorem gives an expression for the T-TLS solution $\bar{x}_k$ in terms of the SVD of A, instead of the SVD of (A, b), as was previously the case.

An immediate consequence of the results in Theorem 5.4.1 is that $\|\bar{x}_k\|_2 \geq \|x_k\|_2$ for all k. From the theorem, we also obtain the following expressions for the first k filter factors:

$$1 \leq f_i \leq 1 + \frac{\bar{\sigma}_{k+1}^2}{\sigma_i^2} + \mathcal{O}\left(\frac{\bar{\sigma}_{k+1}^4}{\sigma_i^4}\right), \quad i \leq k ,$$

showing that the larger the ratio between σ_i and $\bar{\sigma}_{k+1}$, the closer the bound on f_i is to 1. Similarly, for the last $n - k$ filter factors, we obtain

$$0 \leq f_i \leq \|\bar{V}_{22}\|_2^{-2} \frac{\sigma_i^2}{\bar{\sigma}_k^2} \left(1 + \mathcal{O}\left(\frac{\sigma_i^2}{\bar{\sigma}_k^2}\right) \right) , \quad k < i \leq \mathrm{rank}(A) ,$$

showing that the smaller the ratio between σ_i and $\bar{\sigma}_k$, the closer f_i is to $\sigma_i^2/\bar{\sigma}_k^2$. Hence, Theorem 5.4.1 guarantees that the filter factors for $i \leq k$ are close to one and that those for $k < i \leq \mathrm{rank}(A)$ are approximately proportional to σ_i^2, even when there is no gap in the singular value spectrum. Hence, the T-TLS filter factors resemble the Tikhonov filter factors, showing that $\bar{x}_k$ is a filtered solution where k plays the role of the regularization parameter.

If $k = n$ then $f_i = \sigma_i^2/(\sigma_i^2 - \bar{\sigma}_{n+1}^2)$ for $i = 1, \ldots, n$ in agreement with [348, Theorem 2.7], and if the errors in A and b are small, then the difference $x_{\mathrm{LS}} - x_{\mathrm{TLS}}$ between the ordinary LS and TLS solutions is small [319]. Our experience is that the same is true for the difference $x_k - \bar{x}_k$ when $k < n$. When the noise is large, then $\bar{x}_k$ can be very different from x_k, and the filter factors f_i for $i \leq k$—and especially f_k—can differ somewhat from one (we have observed

$f_k \approx 1.2$ in our experiments). For more details and examples, see [124] where we compare T-TLS with TSVD and Tikhonov regularization, and illustrate that the T-TLS solution is indeed less sensitive to the perturbations of A and b when the noise level is large.

When the dimensions of A are not too large, then one can explicitly compute the SVD of (A, b) and experiment with different truncation parameters k. This is particularly useful when no a priori estimate of k is known. When the dimensions of A become large, then this approach becomes prohibitive due to the large complexity of the SVD algorithm. For such large-scale problems we propose in [124] an iterative algorithm based on Lanczos bidiagonalization. We postpone the discussion of this algorithm to §6.6.

5.4.2. Regularized TLS

An alternative approach to adding regularization to the TLS technique is based on the Tikhonov formulation in (5.14); see [145] and [197]. In the TLS setting, we add the bound $\|L x\|_2 \leq \alpha$ to the ordinary TLS problem (3.19), and the *regularized TLS* (R-TLS) problem thus becomes

$$\min \|(A, b) - (\tilde{A}, \tilde{b})\|_F \quad \text{subject to} \quad \tilde{b} = \tilde{A} x, \quad \|L x\|_2 \leq \alpha. \quad (5.41)$$

The R-TLS solution $\bar{x}_\alpha$ to (5.41) is characterized by the following theorem.

Theorem 5.4.2. [197, Theorem 2.1]. *The R-TLS solution $\bar{x}_\alpha$ to (5.41) is a solution to the problem*

$$(A^T A + \lambda_I I_n + \lambda_L L^T L) x = A^T b, \quad (5.42)$$

where the parameters λ_I and λ_L are given by

$$\lambda_I = -\frac{\|A x - b\|_2^2}{1 + \|x\|_2^2}, \quad (5.43)$$

$$\lambda_L = \frac{(b - A x)^T A x}{\alpha^2}. \quad (5.44)$$

Moreover, the TLS residual satisfies

$$\|(A, b) - (\tilde{A}, \tilde{b})\|_F^2 = -\lambda_I. \quad (5.45)$$

In the standard-form case, Eq. (5.42) simplifies to $(A^T A + \lambda_{IL} I_n) x = A^T b$ with $\lambda_{IL} = \lambda_I + \lambda_L$. If $0 \leq \alpha \leq \|x_{\text{LS}}\|_2$ then $\lambda_{IL} \geq 0$ and $\bar{x}_\alpha = x_{\text{LS}}$, while $\lambda_{IL} < 0$ and $\bar{x}_\alpha \neq x_{\text{LS}}$ when $\alpha > \|x_{\text{LS}}\|_2$. Hence, as long as $\alpha < \|x_{\text{LS}}\|_2$, which is normally the case in regularization problems where $\|x_{\text{LS}}\|_2$ is very large, the R-TLS solutions are identical to the Tikhonov solutions; i.e., replacing the LS residual with the TLS residual in (5.14) has no effect.

5.5. Mollifier Methods

In the general case ($L \neq I_n$), the R-TLS solution $\bar{x}_\alpha$ is different from the Tikhonov solution whenever the residual $A\bar{x}_\alpha - b$ is different from zero, since both λ_I and λ_L are nonzero. We show in [197] that λ_L is positive as long as $\alpha < \|L\,x_{\text{TLS}}\|_2$. On the other hand, λ_I is always negative, and thus adds some *deregularization* to the solution. Statistical aspects of a negative regularization parameter in Tikhonov's method are discussed in [214].

For a given α, there are usually several pairs of parameters λ_I and λ_L, and thus several solutions x, that satisfy relations (5.42)–(5.44), but only one of these satisfies the optimization problem (5.41). According to (5.45), this is the solution that corresponds to the smallest value of $|\lambda_I|$. If $\alpha < \|L\,x_{\text{TLS}}\|_2$ then the following relations hold:

$$\lambda_I < 0, \qquad \frac{\partial \lambda_I}{\partial \alpha} > 0, \qquad \lambda_L > 0.$$

If $\alpha \geq \|Lx_{\text{TLS}}\|_2$ then $\lambda_L = 0$ and $\bar{x}_\alpha = x_{\text{TLS}}$.

We note that if the matrix $\lambda_I I_n + \lambda_L L^T L$ is positive definite, then the R-TLS solution corresponds to a Tikhonov solution for which the seminorm $\|L\,x\|_2$ in (5.1) is replaced with the norm $(\lambda_I \|x\|_2^2 + \lambda_L \|L\,x\|_2^2)^{1/2}$. If $\lambda_I I_n + \lambda_L L^T L$ is indefinite or negative definite, then there is no equivalent interpretation.

Numerical experiments that compare the R-TLS method with Tikhonov regularization can be found in [145]. The conclusion from these experiments is that the two methods lead to almost identical solutions as long as the noise level is small. For large noise levels, R-TLS can lead to regularized solutions with smaller errors, provided that the deregularization with $\lambda_I I_n$ is suited for the given problem.

5.5. Mollifier Methods

Mollifier methods are different from the previous algorithms treated in this chapter in that the mollifier methods lead to explicit computation of a matrix $A^\#$ such that the regularized solution has the form $x_{\text{reg}} = A^\# b$. In the previous sections, the matrix $A^\#$ is not explicitly computed and acts primarily as an analysis tool, while here it is the goal of the computation.

To describe mollifier methods we return to the continuous setting and consider Fredholm integral equations of the form (1.2) with a discrete right-hand side, where we are given m measurements b_i which represent noisy values of m known functionals k_i on an unknown function f, i.e.,

$$\int_0^1 k_i(t)\,f(t)\,dt + \text{noise} = b_i, \qquad i = 1,\ldots,m. \tag{5.46}$$

The goal is now to estimate $f(t_0)$ at a given t_0 as a linear combination of the right-hand side data b_i, i.e.,

$$f_{\text{est}}(t_0) = \sum_{i=1}^{m} q_i(t_0) \, b_i = q(t_0)^T b \; . \qquad (5.47)$$

Here, we have introduced the vector $q(t_0) = (q_1(t_0), \ldots, q_m(t_0))^T$, and we see that $q(t_0)^T$ is simply a row in the matrix $A^{\#}$.

Let $f^{\text{exact}}(t)$ denote the exact solution. Then it is easy to see that in the absence of noise, $f_{\text{est}}(t_0)$ is given by

$$\begin{aligned}
f_{\text{est}}(t_0) &= \sum_{i=1}^{m} q_i(t_0) \int_0^1 k_i(t) \, f^{\text{exact}}(t) \, dt \\
&= \int_0^1 \sum_{i=1}^{m} q_i(t_0) \, k_i(t) \, f^{\text{exact}}(t) \, dt \\
&= \int_0^1 \mathcal{K}(t_0, t) \, f^{\text{exact}}(t) \, dt \; ,
\end{aligned}$$

where we have defined the *averaging kernel* at t_0 as the function

$$\mathcal{K}(t_0, t) = \sum_{i=1}^{m} q_i(t_0) \, k_i(t) \; . \qquad (5.48)$$

Notice that $\mathcal{K}(t_0, t)$ is simply a continuous version of the resolution matrix $\Xi = A^{\#} A$ defined in (4.20). Thus, $\mathcal{K}(t_0, t)$ determines the resolving power of the regularization method at t_0, and for $f_{\text{est}}(t_0)$ to be meaningful the function $\mathcal{K}(t_0, t)$ should be peaked around $t = t_0$. *Mollifier methods* are characterized as regularization methods in which $\mathcal{K}(t_0, t)$ is specified, from which the vector $q(t_0)$ is then computed. For this reason, mollifier methods are also referred to as optimally localized averaging methods.

The underlying idea, namely, to use the delta function for point evaluation of the solution and then to approximate the delta function by smooth regular functions, was mentioned by Courant in [63, p. 792].

5.5.1. The Target Function Method

In one variant of mollifier methods, $\mathcal{K}(t_0, t)$ is controlled by approximating it by a δ-like target function $T_\gamma(t_0, t)$ parametrized by γ. For example, Louis and Maass [240] suggest the target functions

$$T_\gamma(t_0, t) = \begin{cases} (2\,\gamma)^{-1} \, , & |t - t_0| \leq \gamma \, , \\ 0 \, , & \text{else} \, , \end{cases}$$

and

$$T_\gamma(t_0, t) = \frac{\gamma}{\pi} \frac{\sin \gamma \, (t - t_0)}{\gamma \, (t - t_0)} \, ,$$

5.5. MOLLIFIER METHODS

where γ controls the width of the δ-like function. Then $\mathcal{K}(t_0, t)$, and thus the vector $q(t_0)$, is determined from the problem

$$\min \|T_\gamma(t_0, t) - \mathcal{K}(t_0, t)\|_2 \ . \tag{5.49}$$

This problem can be solved numerically in different ways. For example, if we introduce the matrix $K \in \mathbb{R}^{m \times n}$ and the vector $g(t_0) \in \mathbb{R}^n$ consisting of "samples" of $k_i(t)$ and $T_\gamma(t_0, t)$ at the collocation points $t_1, \ldots, t_p$, i.e.,

$$K_{i\ell} = k_i(t_\ell) , \qquad i = 1, \ldots, m , \quad \ell = 1, \ldots, p , \tag{5.50}$$

$$g(t_0)_\ell = T_\gamma(t_0, t_\ell) , \qquad \ell = 1, \ldots, p , \tag{5.51}$$

then we obtain the least squares problem $\min \|K^T q(t_0) - g(t_0)\|_2$ to be solved for $q(t_0)$. Alternatively (see [240]), one can set up the normal equations $A\,q(t_0) = d(t_0)$ for computation of $q(t_0)$, where the elements of A and $d(t_0)$ are given by

$$a_{ij} = \int_0^1 k_i(t)\, k_j(t)\, dt , \qquad d(t_0)_i = \int_0^1 k_i(t)\, T_\gamma(t_0, t)\, dt \ . \tag{5.52}$$

Notice that only the right-hand side (i.e., either $g(t_0)$ or $d(t_0)$) depends on t_0, and thus only one factorization of either K or A is necessary.

In practice, it is necessary to add stabilization when both algebraic problems are solved numerically because the coefficient matrix (either K^T or A) becomes highly ill conditioned. Let $C\,C^T$ denote the right-hand side's covariance matrix. Then the system involving K naturally leads to Tikhonov regularization with $L = C^T$, while Franklin's method (5.23) with $L = C\,C^T$ is suited for the system involving A.

In both cases the estimate $f_{\text{est}}(t_0)$ is obtained by convolving a Tikhonov solution with the target function $T_\gamma(t_0, t)$. Using the discretization in (5.50) and (5.51), we have the relation $f_{\text{est}}(t_0) = g(t_0)^T x_\lambda$, where x_λ is the Tikhonov solution. Similarly, using the discretization in (5.52), we have the relation $f_{\text{est}}(t_0) = \int_0^1 T_\gamma(t_0, t)\, f_{\text{reg}}(t)\, dt$, where $f_{\text{reg}}(t)$ is the "semidiscrete" Tikhonov solution given by (5.24).

As suggested in [285] and [286], we can normalize the averaging kernel $\mathcal{K}(t_0, t)$ by incorporating the constraint

$$\int_0^1 \mathcal{K}(t_0, t)\, dt = 1 \ .$$

If the integral is computed by means of a quadrature rule with n abscissas t_j and weights w_j,

$$\int_0^1 \mathcal{K}(t_0, t)\, dt \approx \sum_{j=1}^n w_j\, \mathcal{K}(t_0, t_j) ,$$

then we can incorporate the constraint in the linear algebra setting as follows. We introduce the n-vector e of all ones, and the $n \times n$ diagonal matrix W whose diagonal elements are the quadrature weights,

$$e = (1, \ldots, 1)^T , \qquad W = \text{diag}(w_1, \ldots, w_n) \ . \tag{5.53}$$

Then the constraint takes the form $e^T W K^T q(t_0) = 1$, and we are now faced with a linearly constrained least squares problem

$$\min \|K^T q(t_0) - g(t_0)\|_2 \quad \text{subject to} \quad e^T W K^T q(t_0) = 1 \ . \quad (5.54)$$

Methods for solving (5.54) are described, e.g., in [154, §12.1.4] and [231, Chapters 20–22], while a detailed description of the complete algorithm is given in [230].

5.5.2. The Backus–Gilbert Method

In another variant of the mollifier methods, known as the *Backus–Gilbert method*, $\mathcal{K}(t_0, t)$ is controlled by specifying a positive δ^{-1}-like criteria function $J(t_0, t)$ and then solving

$$\min \int_0^1 J(t_0, t) \, (\mathcal{K}(t_0, t))^2 \, dt \quad \text{s.t.} \quad \int_0^1 \mathcal{K}(t_0, t) \, dt = 1 \ . \quad (5.55)$$

Examples of criteria functions are

$$J(t_0, t) = (t - t_0)^2$$

and

$$J(t_0, t) = 2\sigma^2 \left(1 - \exp\left(-\frac{(t - t_0)^2}{2\sigma^2}\right)\right) \ .$$

In the latter criteria function, σ controls the "width" of the δ-function. This method was originally presented in [15]; see [161, p. 5.6] or [287, §18.6] for more recent derivations of the method. Without the constraint in (5.55), we would obtain $\mathcal{K}(t_0, t) = 0$.

Again, in practice it is necessary to add regularization when the problem (5.55) is solved numerically, due to the ill-conditioning of the matrix problem that arises. The regularization term is not present in the original paper by Backus and Gilbert [15].

We now briefly summarize the computations in the Backus–Gilbert method when a quadrature rule with n abscissas t_j and weights w_j is used to compute the necessary integrals. For ease of exposition, we assume that the errors in b_i are uncorrelated and have unit standard deviation, such that the associated covariance matrix is I_m.

In addition to the vector e and the diagonal matrix W defined in (5.53), we need the $n \times n$ diagonal matrix $D(t_0)$ with diagonal elements

$$D(t_0)_{jj} = (J(t_0 - t_j))^{1/2} \ , \quad j = 1, \ldots, n \ .$$

Then $q(t_0)$ is given by

$$q(t_0) = \frac{p(t_0)}{e^T W K^T p(t_0)} \ , \quad (5.56)$$

5.5. Mollifier Methods

where K is defined in (5.50) and the vector $p(t_0)$ is given by

$$p(t_0) = \left(K\, W\, D(t_0)^2 K^T + \lambda^2 I_m \right)^{-1} K\, W\, e. \tag{5.57}$$

Here, the term $\lambda^2 I_m$ is necessary in order to stabilize the numerical computation of $p(t_0)$. See [161, §5.6] for a derivation of these equations.

The Backus–Gilbert method is computationally very expensive because the involved coefficient matrix depends on t_0, and therefore $\mathcal{O}(mn^2)$ flops are required for each value of t_0. Therefore, it is sometimes remarked that the Backus–Gilbert method should not necessarily be viewed as an inversion method in its own right, but rather as a method for assessing the significance of a solution; see [65, p. 97] and [277, p. 46]. However, there are also applications in which the Backus–Gilbert solution has advantages over other methods; see, e.g., [59] and [286].

An SVD analysis of the Backus–Gilbert method was presented in [188] for the case $J(t_0, t) = (t - t_0)^2$, and we summarize the main result here. If we assume that $t_0 \neq t_j$, $j = 1, \ldots, n$, then we can introduce the vector $\tilde{e} = D(t_0)^{-1} W^{1/2} e$. We also need the SVD of the matrix $K\, D(t_0)\, W^{1/2}$,

$$K\, D(t_0)\, W^{1/2} = \sum_{i=1}^{n} \hat{u}_i\, \hat{\sigma}_i\, \hat{v}_i^T,$$

and then $e^T W K^T p(t_0)$ and $p(t_0)^T b$ are given by

$$e^T W K^T p(t_0) = \sum_{i=1}^{n} f_i\, (\hat{v}_i^T \tilde{e})^2, \qquad p(t_0)^T b = \sum_{i=1}^{n} f_i\, \hat{v}_i^T \tilde{e}\, \frac{\hat{u}_i^T b}{\hat{\sigma}_i}, \tag{5.58}$$

where $f_i = \sigma_i^2/(\sigma_i^2 + \lambda^2)$ are the Tikhonov filter factors. The vector $\tilde{e}$ does not have decaying Fourier coefficients $\hat{v}^T \tilde{e}$, and therefore the discrete Picard condition is not satisfied for the system in (5.57).

To see where the filtering in the Backus–Gilbert method comes from, we note that the SVD of the matrix $K\, D(t_0)\, W^{1/2}$ inherits the characteristic features of the SVD of K. Hence, if K and b satisfy the discrete Picard condition, then the coefficients $\hat{u}_i^T b$ decay at least as fast as the singular values $\hat{\sigma}_i$. Due to the appearance of the filter factors f_i in (5.58), we therefore obtain a filtering of the noisy components of b in the Backus–Gilbert method which is similar to the filtering in Tikhonov's method.

For large-scale problems we can use Lanczos bidiagonalization to compute the largest singular values and corresponding vectors of $K\, D(t_0)\, W^{1/2}$, and then these quantities can be used to approximate $p(t_0)^T b$ and $e^T W K^T p(t_0)$ as follows:

$$p(t_0)^T b \approx \sum_{i=1}^{k} v_i^T \tilde{e}\, \frac{u_i^T b}{\sigma_i}, \qquad e^T W K^T p(t_0) \approx \sum_{i=1}^{k} (v_i^T \tilde{e})^2, \tag{5.59}$$

where k is the number of SVD components used in the approximation. See [188, §4] for an illustration of this approach.

5.6. Other Direct Methods

In this section we briefly mention a few other direct regularization methods of which we are aware. Some of these methods lead to nonlinear minimization problems.

Ekstrom and Rhoads [97] introduced a variant of Tikhonov's method for the case when A is Toeplitz and $L = I_n$. Their method is defined via the filter factors

$$f_i = \frac{\sigma_i}{\sigma_i + \rho_i}, \qquad i = 1, \ldots, n,$$

in the SVD expansion (4.3), and they suggest using either $\rho_i = \lambda$ (a constant) or $\rho_i = \|L\,v_i\|_2^2$, i.e., a measure of the "smoothness" of the ith singular vector v_i. The case $\rho_i = \lambda$ is an obvious way to generalize Franklin's method (5.23) to nonsymmetric A when $L = I_n$, and we shall refer to this method as *damped SVD*. A possible further extension to the case $L \neq I_n$ was proposed in [186], namely, to use the filter factors $f_i = \sigma_i/(\sigma_i + \lambda \mu_i)$ in the GSVD expansion (4.5). These filter factors decay more slowly than the Tikhonov filter factors and thus, in a sense, introduce less filtering.

Regularization in other norms than the 2-norm are also important, and Dax [76] considered problems of the general form

$$\min \left\{ \|A\,x - b\|_p + \lambda^2 \,\|x\|_s^s \right\}, \qquad (5.60)$$

where $1 < p < \infty$ and $1 < s < \infty$. Regarding the residual norm, the case $1 < p < 2$ is important in connection with robust estimation that takes into account possible "outliers" in the given data.

Regarding the solution norm, the 1-norm $\|x\|_1$ has achieved special attention in some applications, such as reflection seismology, where this norm is able to produce a "sparse spike train" solution, i.e., a solution that has the least number of nonzero components; cf. [232] and [303]. This feature of the 1-norm can be used to compute regularized solutions with steep gradients and even discontinuities, when $\|L\,x\|_2$ in Tikhonov's method (5.1) is replaced by the 1-norm (or some similar norm) of a derivative of x.

One example is image processing, where the *total variation* (TV) functional is useful as a measure of the "size" of the regularized solution. For a one-dimensional function u, the TV functional is defined as

$$\mathcal{J}_{\mathrm{TV}}(u) = \int_0^1 \left|\frac{du}{dt}\right| dt, \qquad (5.61)$$

and the following generalization can be used in the multidimensional case:

$$\mathcal{J}_{\mathrm{TV}}(u) = \int_\Omega |\nabla u|\, dt. \qquad (5.62)$$

For discrete one-dimensional problems, $\mathcal{J}_{\mathrm{TV}}$ is simply $\|L_1 x\|_1$, where L_1 is the discrete approximation to the first derivative operator (1.15). TV denoising and regularization are able to produce solutions with localized steep gradients without prior knowledge of the positions of these steep gradients; see [295], [359], and [361] for more details and algorithms.

By means of a variant of the MTSVD method (3.16), called PP-TSVD [195], we can compute a regularized solution that consists of *piecewise polynomials*, and in particular solutions that are piecewise constant or piecewise linear. We make the important assumption that the elements of the solution vector x are "samples" of the sought function that we wish to reconstruct. Then the idea is to replace the 2-norm in MTSVD by the 1-norm, to obtain the following problem:

$$\min \|L\, x\|_1 \quad \text{subject to} \quad \min \|A_k\, x - b\|_2\,, \tag{5.63}$$

where L approximates the $(n-p)$th derivative operator, A_k is the rank-k TSVD matrix given by (3.10), and k acts as the regularization parameter.

In analogy with the MTSVD method, we write the PP-TSVD solution as $\bar{x}_{L,k} = x_k - V_k^o z$ with z determined from $\min \|(L\, V_k^o) z - L\, x_k\|_1$, and we identify $L\, x$ as the residual vector of the linear ℓ_1-problem. The 1-norm forces this residual vector to have at least $p - (n-k)$ nonzero elements, since $L\, V_k^o$ is $p \times (n-k)$. Hence, we compute a solution vector $\bar{x}_{L,k}$ such that the vector $L\, \bar{x}_{L,k}$ is a "sparse spike train" with at most $p - (n-k)$ nonzero elements. Since this vector represents a sequence of delta functions, we conclude that the solution itself, represented by $\bar{x}_{L,k}$ and being the pth integral of the sequence of delta functions, is a piecewise polynomial of degree $n-p-1$ with at most $k - (n-p)$ break points. Moreover, the derivative of order $n-p-1$ is discontinuous across the break points.

For example, if $p = n-1$ such that $L = L_1$ approximates the first derivative operator, then $\bar{x}_{L,k}$ represents a piecewise constant function with at most k discontinuities, and if $p = n - 2$ such that $L = L_2$ approximates the second derivative operator, then $\bar{x}_{L,k}$ represents a continuous function consisting of at most $k - 1$ straight lines. The positions of the polynomial break points, represented by the nonzero elements in the vector $L\, \bar{x}_{L,k}$, are determined solely from the data—no a priori information is used. The parameter k controls the stabilization of the solution $\bar{x}_{L,k}$. The smaller the k, the "simpler" the model— at the expense of neglecting information from the right-hand side b.

As MTSVD is related to Tikhonov regularization in general form (5.1), the PP-TSVD method is related to TV regularization $\min\{\|A\, x - b\|_2^2 + \lambda \|L_1\, x\|_1\}$, but while the MTSVD and Tikhonov solutions are qualitatively very similar, the PP-TSVD and TV solutions are quite different.

The PP-TSVD method has interest in its own right, and it is also useful as a "preprocessor" for other algorithms, such as that from [310] in which the discontinuities must be incorporated explicitly via the modified regularization matrix L_Q (4.16). The PP-TSVD algorithm with $L = L_1$ can be used to detect and locate these discontinuities.

Maximum entropy regularization is used in image reconstruction and related applications where the discrete smoothing norm $\Omega(x)$ is not a (semi)norm. Here, $\Omega(x)^2$ is the negative of the entropy function:

$$\Omega(x)^2 = \sum_{i=1}^{n} x_i \log(w_i x_i) \,, \qquad (5.64)$$

where x_i are the elements of the positive vector x, and $w_1, \ldots, w_n$ are n weights. Equation (5.64) is the simplest form of the entropy function; an overview of different entropy functions can be found in [134]. See [3], [111, §5.3], and [311] for more details about maximum entropy regularization.

The mathematical justification for this particular choice of $\Omega(x)$ is that it yields a solution which is most "objective" or "maximally uncommitted" with respect to missing information in the right-hand side. Maximum entropy is an example of a linear regularization method for which filter factors cannot be used to express the regularized solution via (4.3) or (4.5).

Maximum entropy regularization is implemented in the package REGULARIZATION TOOLS by means of a nonlinear conjugate gradient algorithm applied to the function $\|A x - b\|_2^2 + \lambda^2 \, \Omega(x)^2$; see [186, §2.7.5] for details. The algorithm occasionally suffers from very slow convergence due to a flat minimum, and it seems necessary to supplement the algorithm with a suitable preconditioner. A different algorithm, based on convex programming, is given in [106].

Babolian and Delves [14], and later Belward [22], considered a regularization scheme based on the Galerkin discretization method (1.12) with Chebychev polynomials as basis functions (other orthonormal basis functions could also be used). In this method, the regularization is imposed by requiring that the expansion coefficients ξ_i satisfy

$$|\xi_i| \leq C \, i^{-r} \,, \qquad i = 1, \ldots, n \,,$$

where C and r are two real constants that act as regularization parameters. A cross-validation algorithm for determining C and r is given in [112]. Clearly, this method works well if the chosen basis functions resemble the appropriate SVD or GSVD basis vectors for the given problem.

Finally, we mention some methods related to the computation of linear functionals $w^T x$ defined on the solution x to discrete ill-posed problems, where $w \in \mathbb{R}^n$ is a given vector. In some applications, it is not the solution itself, but rather the linear functional $w^T x$ that is required, and the computation of the quantity $w^T x$ can be considerably less ill conditioned than the computation of x. To explain this, we use the SVD of A and obtain the expression

$$w^T x = \sum_{i=1}^{n} w^T v_i \, \frac{u_i^T b}{\sigma_i} \,,$$

showing that if the quantities $w^T v_i$ decay fast, then they have a filtering effect on the computation of $w^T x$. See [5] and [6] for more details and applications.

The linear functional $w^T x$ is of particular interest in connection with the computation of confidence intervals for the solution x, given certain bounds for the residual and the solution. The idea is to use the ordinary unit vectors as the vector w and to solve the following problems:

$$\min_{x \in \mathcal{S}} w^T x \quad \text{and} \quad \max_{x \in \mathcal{S}} w^T x \,, \tag{5.65}$$

where the subspace $\mathcal{S}$ defines the particular bounds. The three subspaces

$$\begin{aligned} \mathcal{S}_1 &= \{x \mid \|A x - b\|_2 \leq \delta \,,\, x_{\min} \leq x \leq x_{\max} \} \,, \\ \mathcal{S}_2 &= \{x \mid \|A x - b\|_2 \leq \delta \,,\, x \geq 0 \} \,, \\ \mathcal{S}_3 &= \{x \mid \|A x - b\|_2 \leq \delta \,,\, \|x - x^*\|_2 \leq \alpha \} \end{aligned} \tag{5.66}$$

are suggested in [284], [266], and [104], respectively, where algorithms can also be found. Further aspects of confidence intervals for discrete ill-posed problems are discussed in [204] and [297].

5.7. Characterization of Regularization Methods

We have already mentioned the lack of survey literature on numerical regularization methods. The situation is precisely the same regarding formal comparisons and characterizations of regularization methods (while purely numerical experiments with regularization methods for specific problems are not uncommon). In [235], Linz wrote:

> There are no general criteria by which different algorithms can be compared. Consequently, many methods are proposed which, on the evidence of a few special cases, are claimed to be effective.

One difficulty associated with formal comparisons of regularization methods is that two different regularized solutions can be very different even if their residual vectors are very similar. The following theorem quantifies this.

Theorem 5.7.1. [184, Theorem 3]. *Given two regularized solutions $x_{\text{reg}}^{(1)}$ and $x_{\text{reg}}^{(2)}$, both satisfying*

$$\|L x_{\text{reg}}^{(i)}\|_2 \leq \eta \,, \qquad \|A x_{\text{reg}}^{(i)} - b\|_2 \leq \delta \,, \qquad i = 1, 2 \,, \tag{5.67}$$

then, using the GSVD of (A, L), the difference $x_{\text{reg}}^{(1)} - x_{\text{reg}}^{(2)}$ satisfies

$$\|x_{\text{reg}}^{(1)} - x_{\text{reg}}^{(2)}\|_2 \leq 2 \|X\|_2 \sqrt{\delta^2 + \eta^2} \,, \tag{5.68}$$

$$\|L (x_{\text{reg}}^{(1)} - x_{\text{reg}}^{(2)})\|_2 \leq 2 \min\{ \delta/\gamma_1 \,,\, \eta \} \,, \tag{5.69}$$

$$\|A (x_{\text{reg}}^{(1)} - x_{\text{reg}}^{(2)})\|_2 \leq 2 \min\{ \delta \,,\, \eta \gamma_p \} \,. \tag{5.70}$$

We see that a small common upper bound δ for the two residual norms ensures that the residual vectors are close, but it does not ensure that the solutions $x_{\text{reg}}^{(1)}$ and $x_{\text{reg}}^{(2)}$ are close, neither does it ensure that $L\,x_{\text{reg}}^{(1)}$ and $L\,x_{\text{reg}}^{(2)}$ are close.

Common for many direct and iterative regularization methods is that they minimize some norm of the residual $r_{\text{reg}} = A\,x_{\text{reg}} - b$ subject to one or more constraints on x_{reg} and its "size" $\Omega(x_{\text{reg}})$. It is therefore possible to give a characterization of these regularization methods [196], showing exactly how the measurement of "size" is dependent on the particular regularization method chosen. Below, we summarize the results from [196].

Since it is always possible to transform a problem in general form into one in standard form (see §2.3), we limit our discussion here to standard-form problems. Tikhonov regularization in the form (5.14) is obviously in this above-mentioned class of methods, and so is maximum entropy regularization when formulated as

$$\min \|A\,x - b\|_2 \quad \text{subject to} \quad \sum_{i=1}^n x_i \log(w_i\,x_i) \leq \alpha \;. \tag{5.71}$$

The same is true for MTSVD (3.16) and the 1-norm variant PP-TSVD (5.63), as well as regularized TLS (5.41).

The conjugate gradient (CG) method, treated in §6.3, is also in this general class of regularization methods since the CG iterate $x^{(k)}$, after k iterations, satisfies

$$\min \|A\,x - b\|_2 \quad \text{subject to} \quad x \in \mathcal{K}_k(A^T A, A^T b) \;, \quad \|x\|_2 \leq \alpha \;. \tag{5.72}$$

Here, $\mathcal{K}_k(A^T A, A^T b) = \text{span}\{A^T b, A^T A\,A^T b, \ldots, (A^T A)^{k-1} A^T b\}$ is the Krylov subspace associated with applying the CG method to the normal equations $A^T A\,x = A^T b$, and k is determined by the choice of α.

Other methods can also be characterized in this framework if we use the p-norms in the SVD basis, denoted by $\|\cdot\|_{\underline{p}}$, and work with

$$\|x\|_{\underline{p}} = \|V^T x\|_p \quad \text{and} \quad \|r\|_{\underline{p}} = \|U^T r\|_p + \|(I_m - U\,U^T)\,b\|_p \;. \tag{5.73}$$

Then the TSVD solution x_k is characterized by

$$\min \|A\,x - b\|_{\underline{1}} \quad \text{subject to} \quad \|x\|_{\underline{1}} \leq \alpha \tag{5.74}$$

with a real parameter α, and we have the relations

$$\|x_k\|_{\underline{1}} = \sum_{i=1}^n f_i\,|u_i^T b|/\sigma_i \;, \tag{5.75}$$

$$\|A\,x_k - b\|_{\underline{1}} = \sum_{i=1}^n (1 - f_i)\,|u_i^T b| + \|(I_m - U\,U^T)\,b\|_1 \;, \tag{5.76}$$

5.7. CHARACTERIZATION OF REGULARIZATION METHODS

where it can be shown that the filter factors f_i are given by

$$f_i = \begin{cases} 1, & i = 1, \ldots, k, \\ \sigma_{k+1}/|u_{k+1}^T b| \left(\alpha - \sum_{j=1}^{k} |u_j^T b|/\sigma_j \right), & i = k+1, \\ 0, & i = k+2, \ldots, n. \end{cases} \quad (5.77)$$

Similarly, we can define a new method based on the norm $\|\cdot\|_\infty$,

$$\min \|A x - b\|_\infty \quad \text{subject to} \quad \|x\|_\infty \leq \alpha, \quad (5.78)$$

with filter factors given by

$$f_i = \min\{\, 1 \,,\, \alpha\, \sigma_i/|u_i^T b| \,\}, \quad i = 1, \ldots, n. \quad (5.79)$$

While the TSVD method (obtained for $p = 1$) has no components in directions corresponding to the small singular values, and the Tikhonov solution (obtained for $p = 2$) has small components in these directions, the solution defined in (5.78) has components in these directions which are comparable to those corresponding to the large singular values.

From this discussion we see that the choice of regularization method is a choice of an appropriate pair of functions for measuring the "size" of the solution and the corresponding residual (possibly supplemented with a requirement about the subspace that the solution must lie in). These functions determine precisely how the errors are damped in the regularized solution, and different functions lead to regularized solutions with different properties. For example, the number p in the norm $\|\cdot\|_p$ (5.73) controls the damping of the SVD components corresponding to small singular values. The 1-norm in $\|L x\|_1$, combined with an L that approximated the dth derivate operator, yields solutions with discontinuities in the $(d-1)$th derivative. The TLS residual $\|(A, b) - (\tilde{A}, \tilde{b})\|_F$ yields solutions that are less sensitive to perturbations of A than methods using the least squares residual $\|A x - b\|_2$. See [265] for an experimental study of a variety of functions for measuring the solution's "size" in image reconstruction problems.

Another way to characterize the properties of regularization methods for convolution operators was proposed by Anderssen and Prenter [7]. They use Fourier analysis and show that for convolution operators, the regularization process can be characterized by means of Wiener filtering in the Fourier domain. Write the right-hand side g in (1.1) as a sum of the exact right-hand side g^{exact} and the noise η, and let $\mathcal{F}$ denote the Fourier transform. Then the ideal Wiener filter for g is given by

$$\mathcal{F}(g^{\text{exact}})/(\mathcal{F}(g^{\text{exact}}) + \mathcal{F}(\eta)),$$

and this filter gives the optimal filtering of the noise, in the least squares sense—but it requires specific knowledge of the noise η.

In this framework, different regularization methods lead to different approximations to the ideal Wiener filter. One result of the analysis in [7] is that Tikhonov regularization (§5.1) inherits the potential of Wiener filtering for achieving a realistic separation of signal from noise, because the Tikhonov filter is of the same form as the ideal Wiener filter. Landweber iteration (see §6.2), on the other hand, corresponds to an "ideal low-pass filter" (i.e., a Heaviside step function) which, from a filtering point of view, is a poor approximation to the optimal Wiener filter. Via the Wiener filtering framework in [7], it is also possible to give conditions in which the Backus–Gilbert method (§5.5.2) is equivalent to Tikhonov regularization.

5.8. Direct Regularization Methods in Action

We conclude this chapter with five numerical examples that illustrate several of the direct regularization methods. The first example uses a simple 2×2 test matrix. The test problem used in the next three examples is **shaw** from §1.4.3, discretized with $m = n = 64$ and with a perturbed right-hand side $b = b^{\text{exact}} + e$ with noise level $\|e\|_2 / \|b^{\text{exact}}\|_2 = 0.01$. This problem has a smooth solution. The test problem used in the last example is **wing** from §1.4.4, which has a discontinuous solution.

5.8.1. A Geometric Perspective

In our first example we illustrate the geometry of standard-form Tikhonov regularization and TSVD by means of a small 2×2 example with

$$A = \begin{pmatrix} 0.41 & 1.00 \\ -0.15 & 0.06 \end{pmatrix}, \qquad x^{\text{exact}} = \begin{pmatrix} 1 \\ 1 \end{pmatrix},$$

and $b^{\text{exact}} = A\,x^{\text{exact}}$. Tikhonov regularization in standard-form minimizes the function $\|A\,x - b\|_2^2 + \lambda^2\,\|x\|_2^2$, and to illustrate the behavior of this function we show four contour plots of the Tikhonov function for $\lambda = 0,\ 0.2,\ 0.6$, and 1.5 in the top part of Fig. 5.2. The contours are ellipses centered at the Tikhonov solution x_λ (shown as the large cross) and with semiaxes whose lengths and orientations are determined by the two eigenvalues and eigenvectors of the matrix $A^T A + \lambda I_n$, namely, $\sigma_i^2 + \lambda^2$ and v_i for $i = 1, 2$.

For $\lambda = 0$, the Tikhonov function is $\|A\,x - b\|_2^2$, and the contour ellipses are fairly eccentric ellipses, due to the ill-conditioning of A, centered at $x^{\text{exact}} = (1, 1)^T$. For $\lambda \to \infty$, the Tikhonov function approaches $\lambda^2\,\|x\|_2^2$, whose contours are circles centered at $(0, 0)^T$; if an a priori estimate x^* is included, then the contours are circles centered as x^*. For intermediate values of λ, the eccentricity of the ellipses decrease as λ increases.

The solid line in the bottom right figure is the ridge trace of x_λ as λ varies from 0 to ∞, and we see that the corresponding Tikhonov solution assumes a smooth transition from $(1, 1)^T$ to $(0, 0)^T$.

5.8. Direct Regularization Methods in Action

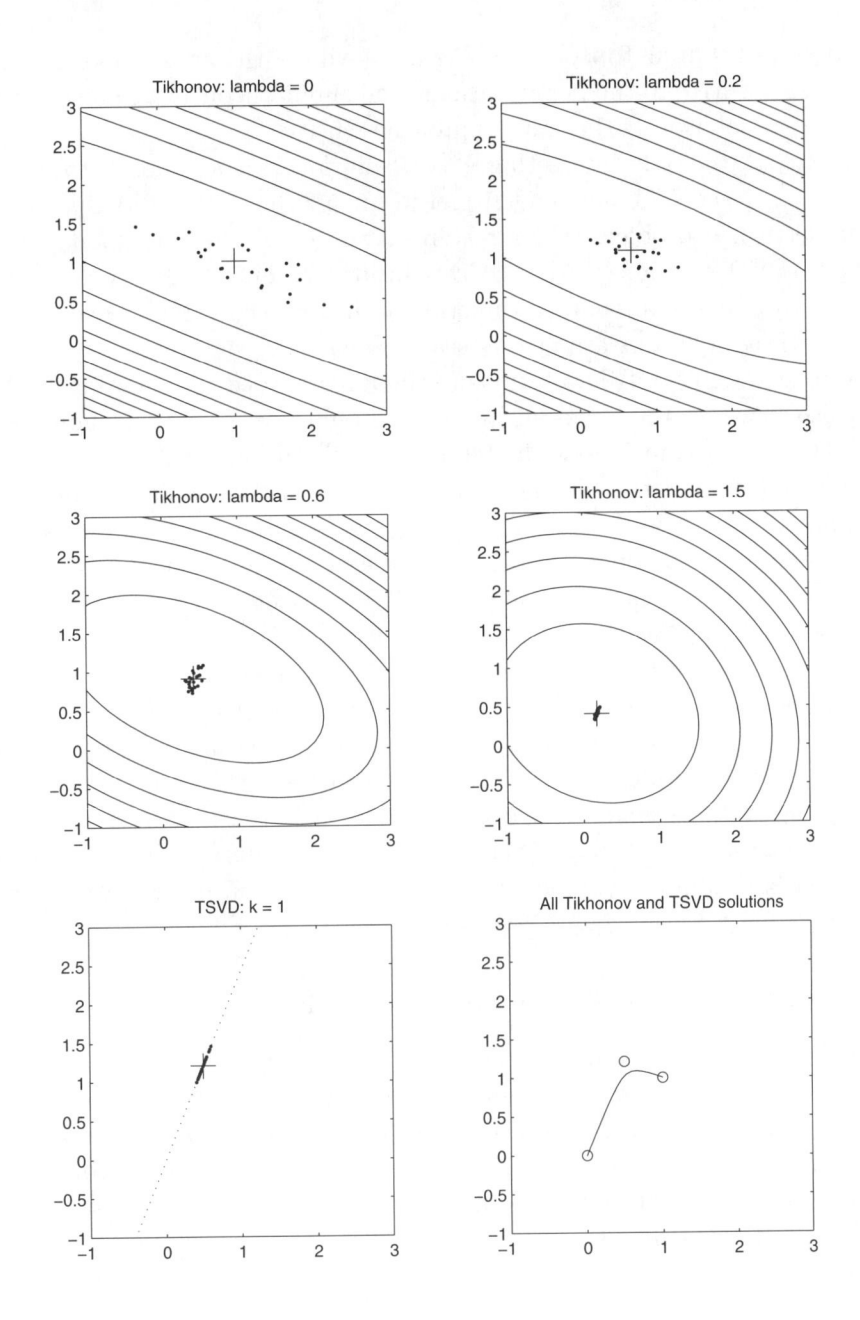

FIG. 5.2. *Geometric interpretation of Tikhonov regularization and TSVD for the small* 2×2 *test problem. See the text for details.*

To illustrate the sensitivity of the Tikhonov solution x_λ as a function of λ, we generated 25 random perturbations $b = b^{\text{exact}} + e$ with noise level $\|e\|_2/\|b\|_2 = 0.15$. For each value of λ, we computed the corresponding 25 perturbed Tikhonov solutions, and these solutions are shown as the 25 large dots in each of the four top plots in Fig 5.2. We see that as λ increases, x_λ becomes less sensitive to the perturbations, and simultaneously x_λ moves away from the exact solution $(1,1)^T$ and approaches $(0,0)^T$.

When $n = 2$, the only interesting TSVD solution x_k corresponds to $k = 1$, i.e., $x_k = \sigma_1^{-1} u_1^T b\, v_1$. The bottom left plot in Fig 5.2 shows this TSVD solution to the unperturbed problem (the large cross) as well as the corresponding 25 perturbed TSVD solutions. All TSVD solutions lie on the dotted line that intersects $(0,0)^T$ and points in the direction of the first right singular vector v_1, because all these solutions are merely scaled versions of this vector.

The three possible TSVD solutions to the unperturbed problem, for $k = 0$, 1, 2 are also shown as the circles in the bottom right part of Fig 5.2. For $k = 2$ the TSVD solution coincides with the Tikhonov solution for $\lambda = 0$, namely, x^{exact}, and for $k = 0$ the TSVD solution equals $(0,0)^T$, which is identical to the limit of x_λ as $\lambda \to \infty$. In this plot, we see that as λ increases, the Tikhonov solution first approaches the TSVD solution for $k = 1$, until it starts to approach the TSVD solution for $k = 2$. The same behavior was illustrated in the ridge traces in Fig. 5.1.

The importance of the right singular vectors v_1 and v_2 can also be illustrated by this example. Recall that these two vectors determine the orientation of the semiaxes of all the contour ellipses. We clearly see that the largest component of the perturbation of x_λ lies in the direction of the long semiaxis, corresponding to the second SVD component $\sigma_2^{-1} u_2^T e\, v_2$. As λ increases, the second filter factor $f_2 = \sigma_2^2/(\sigma_2^2+\lambda^2)$ decreases, and therefore the v_2-component of the perturbation error also decreases. For $\lambda \gtrsim 0.5$ the perturbation error lies mainly along the first singular vector v_1.

5.8.2. From Oversmoothing to Undersmoothing

Now we illustrate the influence of the regularization parameter on the regularized solution. We show how the standard-form Tikhonov solution x_λ to the test problem shaw changes as λ varies from large values (leading to oversmoothing) to small values (leading to undersmoothing). The same transition from oversmoothing to undersmoothing is also illustrated for the TSVD method applied to the same test problem, where increasing values of the truncation parameter k introduces less smoothing in the TSVD solution x_k.

Figure 5.3 illustrates how x_λ and x_k vary with λ and k. In the two top figures, we show the L-curves corresponding to the regularization parameters used in this example, namely,

$$10^{-3} \leq \lambda \leq 3 \quad \text{and} \quad k = 1, 2, \ldots, 9\, .$$

In both figures we also show the full Tikhonov L-curve as the dotted line. For large λ and small k the solutions are overregularized: the solution is "smooth" and its norm is small, at the cost of a large residual norm. The reverse is true

5.8. DIRECT REGULARIZATION METHODS IN ACTION

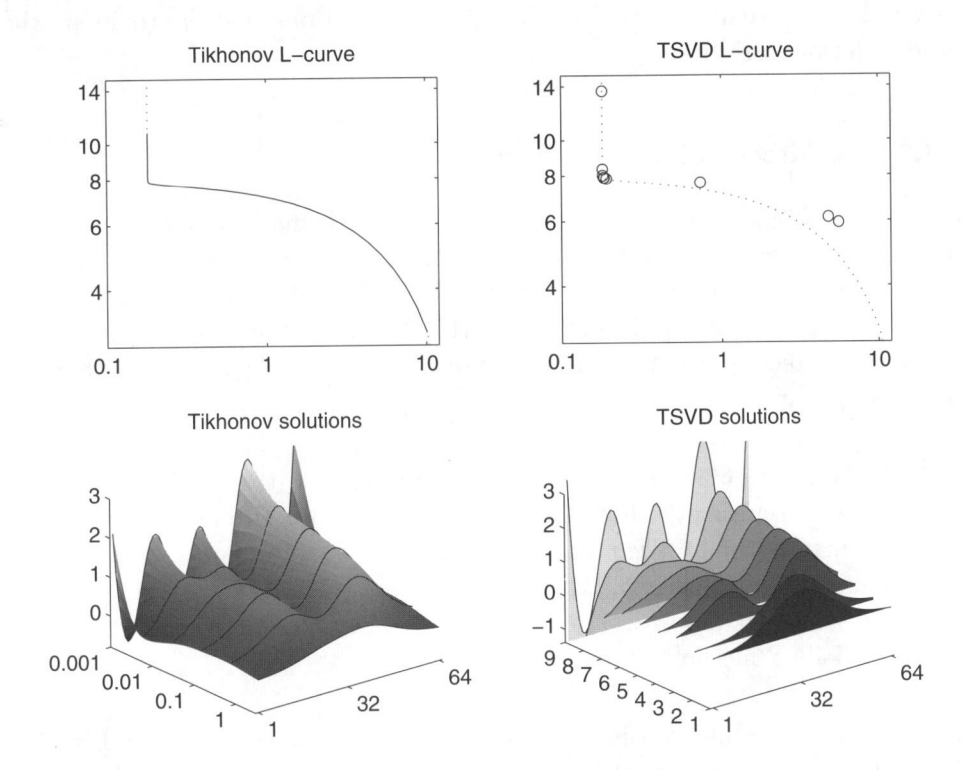

FIG. 5.3. *Illustration of the influence of the regularization parameter on the regularized solution. The dotted line in the two top figures is the full Tikhonov L-curve. The solid line in the upper left figure is the part of the L-curve that corresponds to $10^{-3} \leq \lambda \leq 3$, and the circles in the upper right figure represent the TSVD L-curve for $k = 1, 2, \ldots, 9$. The bottom left figure shows the Tikhonov solution x_λ as a continuous function of λ, with certain solutions emphasized for clarity. The bottom right figure shows the first nine TSVD solutions x_k.*

for small λ and large k, where the solutions are underregularized: the residual norm is small, but the solution has oscillations with large amplitudes, and the solution norms are therefore large.

In the bottom part of Fig. 5.3 we have plotted the solutions x_λ and x_k as functions of λ and k, respectively. Since λ is a continuous parameter, we see a continuous variation of x_λ from oversmoothing to undersmoothing (selected solutions as shown as solid lines for clarity). In the TSVD method, the truncation parameter k is a discrete parameter, and in the nine TSVD solutions x_k, we again see the transition from oversmoothing to undersmoothing.

To summarize, for $\lambda = 3$ and $k = 1$, the solutions are much too smooth. For $\lambda = 10^{-1}$ and $k = 9$, the solutions are dominated by large oscillating contributions from the errors in the right-hand side. In between these extreme values of λ and k, we are able to compute regularized solutions that approximate the exact solution fairly well.

5.8.3. Six Direct Solutions to shaw

In our second example, we compute six different regularized solutions to the shaw test problem computed by direct methods. For each method, we choose the regularization parameter for the regularized solution x_{reg} such that the error norm $\|x^{\text{exact}} - x_{\text{reg}}\|_2$ is minimized; this illustrates the particular method's capability to reconstruct the exact solution. The six regularization methods in this example are

1. Tikhonov regularization (5.1) in standard form with $L = I_n$,
2. Rutishauser's method (5.22),
3. the method by O'Brien and Holt (5.26),
4. maximum entropy regularization (5.64),
5. the TSVD method (3.11),
6. the T-TLS method (3.22),

and the corresponding "optimal" regularized solutions are plotted in Fig. 5.4. All of the regularized solutions approximate x^{exact} quite well, and none of the six solutions stand out as being distinctly better than the others.

Figure 5.5 shows the values of $(\|A\,x_{\text{reg}} - b\|_2\|_2, \|x_{\text{reg}}\|_2)$ relative to the Tikhonov L-curve for the six "optimal" regularized solutions in Fig. 5.4. The Tikhonov solution obviously lies on the L-curve, while the other five solutions lie above the L-curve and fairly close to the Tikhonov solution.

5.8.4. The Backus–Gilbert Solution to shaw

In our third example, we compute the Backus–Gilbert solution to the test problem shaw at t_0 equal to the n collocation points $t_i = (i - 0.5)\,\pi/n$, $i = 1,\ldots,n$. We selected a range of regularization parameters λ in (5.57), and the "optimal" solution was computed as above using the same[16] λ at each t_0. This solution is shown as the solid line in the left part of Fig. 5.6. The agreement with the exact solution, shown as the dotted line, is not very good for this test problem. In the right part of Fig. 5.6 we show the resolution matrix $\Xi = A^\# A$ corresponding to the "optimal" Backus–Gilbert solution. For clarity, each row of Ξ is plotted as a solid line. The resolution matrix resembles, to some extent, the identity matrix, and the figure clearly illustrates the localized nature of the Backus–Gilbert solution.

---

[16] In principle, we can use a different λ at each t_0, but we do not pursue this aspect here.

5.8. Direct Regularization Methods in Action

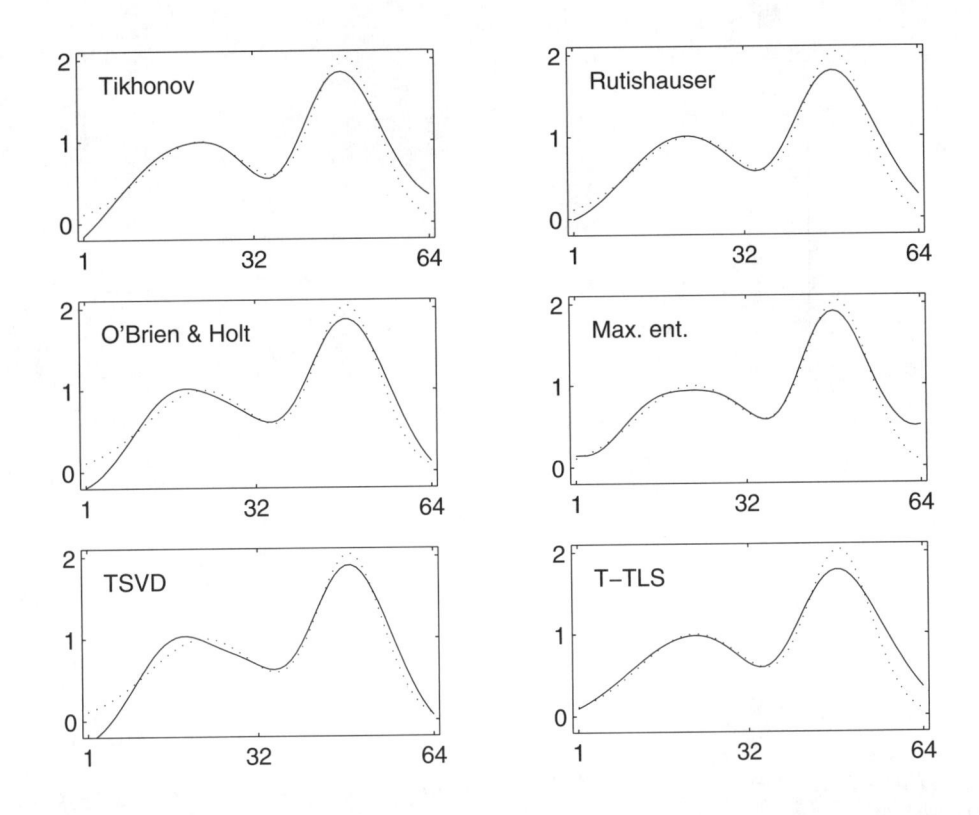

FIG. 5.4. *The solid lines are the "optimal" regularized solutions x_{reg} to the test problem* shaw *for six different regularization methods. Each solution minimizes the error norm $\|x^{\text{exact}} - x_{\text{reg}}\|_2$. The exact solution is shown as the dotted lines.*

5.8.5. Discontinuous Solutions to wing

In our final example we use the TSVD, MTSVD, and PP-TSVD methods to compute solutions to the test problem wing discretized with $m = n = 64$ and noise level $\|e\|_2/\|b^{\text{exact}}\|_2 = 10^{-5}$. The truncation parameters are $k = 2, 3, 4, 5$ and all the solutions are shown in Fig. 5.7. Neither the TSVD method nor the MTSVD method is able to reconstruct the two discontinuities, and the same is true for other methods such as Tikhonov and maximum entropy regularization.

For $k = 4$ the PP-TSVD is able to reproduce the exact solution well with two discontinuities at almost the right positions, plus a small discontinuity in the far right part of the solution. For $k = 2$ and $k = 3$, the solution is oversmoothed and it does not reproduce x^{exact}. For $k = 5$, all three methods produce undersmoothed solutions with too much influence from the right-hand side errors, and we see that the PP-TSVD is precisely as sensitive to these errors as the other two methods.

132   5. Direct Regularization Methods

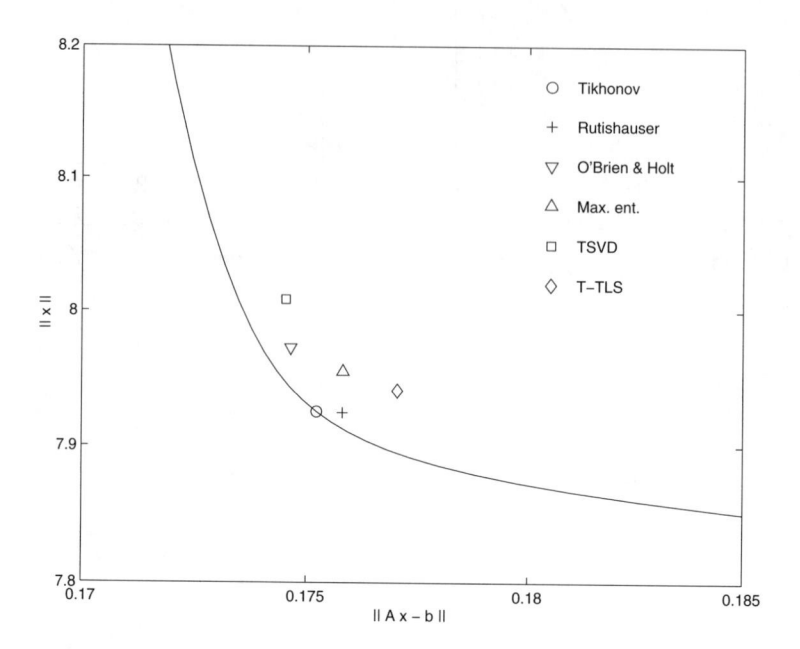

FIG. 5.5. *A close-up view of the Tikhonov L-curve and the points* ($\|A x_{\mathrm{reg}} - b\|_2\|_2$, $\|x_{\mathrm{reg}}\|_2$) *for the six "optimal" regularized solutions x_{reg} from the previous figure.*

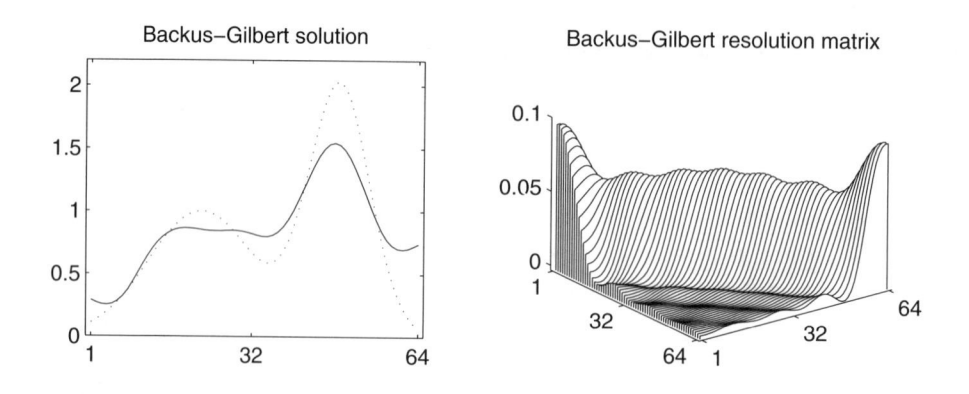

FIG. 5.6. *Left: the "optimal" Backus–Gilbert solution (solid line) and the exact solution x^{exact} (dotted line) to the* shaw *test problem. Right: the corresponding resolution matrix $\Xi = A^{\#} A$; each row of Ξ is plotted as a solid line.*

5.8. DIRECT REGULARIZATION METHODS IN ACTION 133

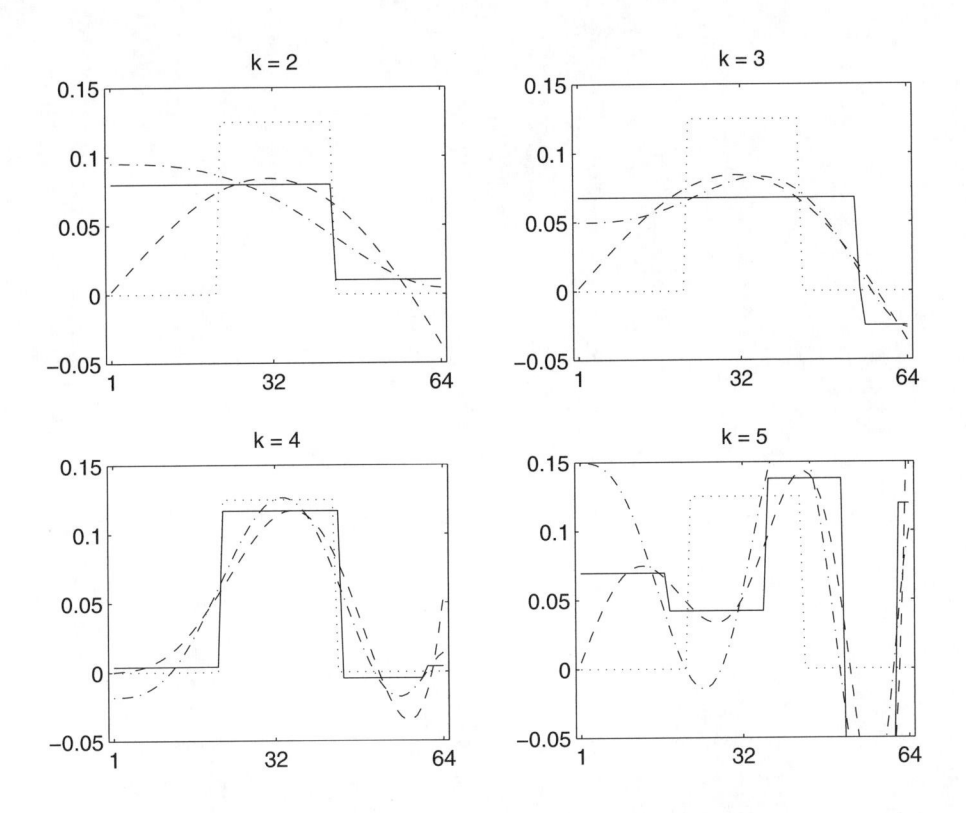

FIG. 5.7. *Regularized solutions to test problem* wing *computed by TSVD (dashed lines), MTSVD (dash-dotted lines), and PP-TSVD (solid lines), for truncation parameters $k = 2, 3, 4, 5$. The exact solution is shown by the dotted lines.*

6

Iterative Regularization Methods

Iterative methods for linear systems of equations and linear least squares problems are based on iteration schemes that access the coefficient matrix A only via matrix-vector multiplications with A and A^T, and they produce a sequence of iteration vectors $x^{(k)}$, $k = 1, 2, \ldots$, that converge to the desired solution. Iterative methods are preferable to direct methods when the coefficient matrix is so large that it is too time-consuming or too memory-demanding to work with an explicit decomposition of A.

Obviously, we can apply iterative methods to the symmetric positive definite Tikhonov system $(A^T A + \lambda^2 L^T L)\, x = A^T b$ or the equivalent least squares problem, and in this way compute a regularized solution [47], [273]. However, this approach requires a new iteration process for each new value of λ, and for general matrices it is difficult to construct efficient preconditioners.

Instead, we are interested in iterative regularization methods in which each iteration vector $x^{(k)}$ can be considered as a regularized solution, with the iteration number k playing the role of the regularization parameter. Hence, we need iteration schemes with the intrinsic property that they, when applied to discrete ill-posed problems, initially pick up those singular value decomposition (SVD) components $(u_i^T b / \sigma_i)\, v_i$ corresponding to the *largest* singular values (which are the ones that are desired in a regularized solution), in such a way that the iteration number k can be considered as a regularization parameter. This phenomenon is sometimes referred to as *semiconvergence* [259, p. 89], because the iteration vector $x^{(k)}$ initially approaches a regularized solution and then, in later stages of the iterations, converges to some other undesired vector—often the least squares solution $x_{\mathrm{LS}} = A^\dagger b$.

Below, we first describe some classical iterative methods that have been studied extensively in the literature, and then we focus on the use of the conjugate gradient (CG) algorithm and related algorithms based on Lanczos bidiagonalization. We give a detailed description of the regularizing properties of these algorithms in terms of the associated filter factors, and we discuss the influence of finite-precision arithmetic. Finally, we describe how a hybrid method can be obtained by performing "inner" regularization in each step of an iterative regularization algorithm. All these methods have been studied often in operator form (see, e.g., [111], [169]) but, as in the previous chapter, we restrict our presentation to the linear algebra setting. The results in §§6.4.1 and 6.4.3 have not been published before, and therefore we include the proofs—in contrast to the rest of the book.

6.1. Some Practicalities

For large-scale problems, the iterative regularization methods can be favorable alternatives to the direct methods for the following reasons.

- The matrix A is never altered, but only "touched" via the matrix-vector products with A and A^T, while matrix factorizations—and in particular orthogonal factorizations such as the QR factorization and the SVD—destroy any sparsity or structure of A. Hence, iterative methods are suited whenever one can take advantage of the sparsity or structure of A in the matrix-vector multiplications. Recall that a matrix-vector multiplication with A or A^T requires $m\,\eta$ flops if A is sparse and η is the average number of nonzeros per row, and $15\,p\,\log_2 p$ flops if A has Hankel or Toeplitz structure and p is the smallest power of 2 satisfying $p = 2^t \geq 2\max(m,n) - 1$; cf. [351, §4.2.4]. See also [171, §6.2]. Sparse matrix-vector products are discussed in [36, 7.1.2].

- The "atomic operations" of the iterative methods are the matrix-vector products with A and A^T plus saxpy operations (i.e., $y \leftarrow \alpha\,x + y$), vector 2-norms, and possibly backsolves. These operations are fairly simple to parallelize, and it is known how to overlap the computations with communication on message-passing parallel computers [83], [328].

- Iterative methods are the only methods of choice for problems where the matrix-vector products $A\,x$ and $A^T z$ are defined through operators acting on the vectors x and z, and where matrix representations of these operators are not available.

- Since the iteration number k plays the role of the regularization parameter, iterative regularization methods produce a sequence of regularized solutions $x^{(k)}$ and corresponding residuals $r^{(k)} = b - A\,x^{(k)}$ whose properties can be monitored as k increases. This is helpful, e.g., when deciding when to stop the iterations.

In connection with the use of iterative methods for solving symmetric positive definite systems of linear equations $A\,x = b$ it is common to use some form of preconditioning $M\,A\,x = M\,b$ that improves the convergence of the method by manipulating the spectrum of the coefficient matrix A. In particular, it is important to choose the preconditioner M such that the condition number of $M\,A$ is smaller than that of A.

6.1. SOME PRACTICALITIES

The situation is different for discrete ill-posed problems. Here, there is no point in improving the condition of A (and thus the condition of $A^T A$) because we are interested in a regularized solution that essentially consists of a fraction of all the SVD components. We would rather prefer to use a preconditioner that improves a part of the singular value spectrum of A—namely, those singular values σ_1 through σ_k that contribute most to the regularized solution—and leaves the remaining singular values unchanged. One possibility is to use a classical stationary iterative method (§6.2) as a preconditioner in a CG method (§6.3); see, e.g., [38]. Preconditioners for discrete ill-posed problems is a subject of current research.

One instance where a preconditioner with the above property can be constructed is when the coefficient matrix A is Toeplitz or block Toeplitz with Toeplitz blocks, in which case a suited circulant preconditioner can be computed efficiently; cf. [173] and [258]. These matrices arise in image deblurring problems with spatially invariant blur. Efficient preconditioners for spatially variant blur are considered by Nagy and O'Leary in [258].

It may be possible to construct multilevel preconditioners that are useful for discrete ill-posed problems. King [224] has proposed multilevel preconditioners for both $A A^T + \lambda^2 I_m$ and $A A^T$, but the latter requires an estimate of the condition number of $M A A^T$ which is often impractical.

In §2.3.2 we showed how a general-form regularization problem can always be transformed into a standard-form problem, and we showed that if L is a banded matrix then the necessary multiplications with $L_A^\dagger$ and $(L_A^\dagger)^T$ can always be implemented efficiently. Thus, there is no practical hindrance to using a matrix $L \neq I_n$ in connection with iterative methods. However, for ease of presentation, throughout this chapter we will assume without loss of generality that any necessary standard-form transformation is "built into" the given coefficient matrix, keeping in mind the simple relation (2.37) between the generalized SVD (GSVD) of (A, L) and the SVD of $A L_A^\dagger$. For practical details, see the model implementations in REGULARIZATION TOOLS; cf. Chapter 8.

When the multiplications with $L_A^\dagger$ and $(L_A^\dagger)^T$ are "built into" the iterative schemes, they act as a kind of "preconditioner," and we can work directly with $x^{(k)}$ and avoid the back-transformation from the standard-form vector $\bar{x}^{(k)}$ to $x^{(k)}$. To see this, we note that if $x^{(0)} = 0$ then $\bar{x}^{(k)}$ can always be written as a polynomial $\mathcal{P}_k$ in $\bar{A}^T \bar{A}$ of degree $k - 1$ times the vector $\bar{A}^T \bar{b}$:

$$\bar{x}^{(k)} = \mathcal{P}_k(\bar{A}^T \bar{A}) \bar{A}^T \bar{b} .$$

If we insert $\bar{A} = A L_A^\dagger$ and $\bar{b} = b - A x_0$ into this expression, we obtain

$$\bar{x}^{(k)} = \mathcal{P}_k\left((L_A^\dagger)^T A^T A (L_A^\dagger)\right) (L_A^\dagger)^T A^T (b - A x_0) .$$

Using Eqs. (2.34) and (2.35) for $L_A^\dagger$ and x_0 together with the GSVD it is straightforward to show that $(L_A^\dagger)^T A^T A x_0 = 0$. Thus, by inserting the above expressions for $\bar{x}^{(k)}$ into the back-transformation $x^{(k)} = L_A^\dagger \bar{x}^{(k)} + x_0$, we obtain

$$x^{(k)} = \mathcal{P}_k(L_A^\dagger (L_A^\dagger)^T A^T A) \, L_A^\dagger (L_A^\dagger)^T A^T b + x_0 \; . \tag{6.1}$$

From this relation we see that we can consider the matrix $L_A^\dagger (L_A^\dagger)^T$ a "preconditioner" for the iterative methods, and we emphasize that the purpose of this "preconditioner" is not to improve the condition of the iteration matrix but rather to ensure that the "preconditioned" iteration vector $x^{(k)}$ lies in the correct subspace and thus minimizes $\|L\,x^{(k)}\|_2$. Minimization of the correct residual norm is ensured by Eq. (2.39).

6.2. Classical Stationary Iterative Methods

One of the classical iterative methods is *Landweber iteration*[17], which takes the form

$$x^{(k)} = x^{(k-1)} + \omega\, A^T r^{(k-1)} \;, \qquad k = 1, 2, \ldots, \tag{6.2}$$

where $x^{(0)}$ is the starting vector (often $x^{(0)} = 0$), ω is a real parameter satisfying $0 < \omega < 2\,\|A^T A\|_2^{-1}$, and $r^{(k)} = b - A\,x^{(k)}$ is the residual vector corresponding to $x^{(k)}$. This method was generalized by Strand [327] to a scheme of the form

$$x^{(k)} = x^{(k-1)} + \mathcal{F}(A^T A)\, A^T r^{(k-1)} \;, \qquad k = 1, 2, \ldots, \tag{6.3}$$

where $\mathcal{F}$ is a rational function of $A^T A$. Classical Landweber iteration thus corresponds to $\mathcal{F}(A^T A) = \omega$. If $A = \sum_{i=1}^n u_i \sigma_i v_i^T$ is the SVD of A, then the eigenvalue decomposition of $\mathcal{F}(A^T A)$ is given by

$$\mathcal{F}(A^T A) = \sum_{i=1}^n v_i\, \mathcal{F}(\sigma_i^2)\, v_i^T \;,$$

and it is easy to show that the filter factors $f_i^{(k)}$ for the kth iteration vector $x^{(k)}$ in (6.3) are given by

$$f_i^{(k)} = 1 - \left(1 - \sigma_i^2\, \mathcal{F}(\sigma_i^2)\right)^k \;, \qquad i = 1, \ldots, n \;. \tag{6.4}$$

In this way, the iteration number k determines the filter factors and thus it plays the role of a regularization parameter. In particular, the filter factors for (6.2) become

$$f_i^{(k)} = 1 - (1 - \omega\,\sigma_i^2)^k \;, \qquad i = 1, \ldots, n \;,$$

in which case $f_i^{(k)} \approx k\,\omega\,\sigma_i^2$ for $\sigma_i \ll \omega^{-1/2}$ while $f_i^{(k)} \approx 1$ for the large σ_i; cf. [171, §6.1]. See also Fig. 6.1.

---

[17]According to [365], at least five names are associated with this algorithm: Richardson, Landweber, Fridman, Picard, and Cimino.

6.2. CLASSICAL STATIONARY ITERATIVE METHODS

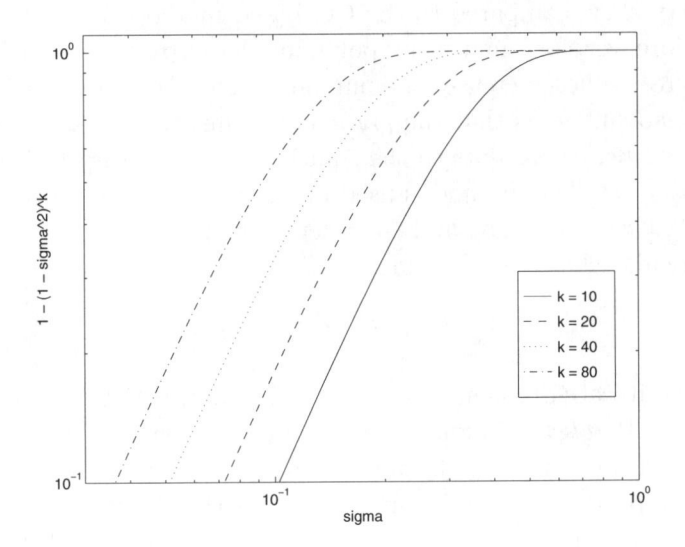

FIG. 6.1. *Plots of the function $1 - (1 - \omega \sigma^2)^k$, which defines the Landweber filter factors, for $\omega = 1$ and $k = 10, 20, 40, 80$.*

Strand [327] and Graves and Prenter [157] studied various cases where $\mathcal{F}$ is chosen such that the function $1 - (1 - t \mathcal{F}(t))^k$ associated with the filter factors $f_i^{(k)}$ in (6.4) approximates the Heaviside step function

$$H_\tau(t) = \begin{cases} 0, & 0 \leq t < \tau, \\ 1, & \tau \leq t. \end{cases}$$

After suitably many iterations the filter factors $f_i^{(k)}$ corresponding to singular values $\sigma_i \geq \sqrt{\tau}$ will be close to 1, while the remaining filter factors will be small. See [161, §5.3], [157], and [327] for more details.

Fleming [128] showed that if $m \geq n$ and M is an arbitrary nonsingular matrix, then there is a one-to-one correspondence between the iteration $x^{(k)} = x^{(k-1)} + M A^T r^{(k-1)}$ and Tikhonov regularization (5.1) with a square nonsingular L and a weighted residual norm $\|W(Ax - b)\|_2$ where W is nonsingular. More general results can be found in [128] and [300].

In [143], the basic Landweber iteration with $\mathcal{F}(A^T A) = \omega$ is augmented with a projection,

$$x^{(k)} = \mathbf{P}\left(x^{(k-1)} + \omega A^T r^{(k-1)}\right), \qquad k = 1, 2, \ldots,$$

where the projection $\mathbf{P}$ represents so-called "hard constraints" by which additional knowledge about the desired regularized solution (such as positivity or band-limitation) is incorporated.

The main disadvantage of Landweber iteration is that the convergence rate is often very slow, compared to the CG-based methods presented in the next section. Moreover, for small τ any polynomial approximation to H_τ must have a high degree, which increases the number of computations per iteration. Yet another disadvantage is that the parameter τ must be chosen in advance. On the positive side, constraints to the solution are fairly easy to incorporate via the projection **P**. The method is used in signal and image processing; cf., e.g., [127], [220], [281], and [299] and the references therein.

A particular choice of $\mathcal{F}$, namely,

$$\mathcal{F}(A^T A) = (A^T A + \lambda^2 I_n)^{-1}, \tag{6.5}$$

leads to the iterative scheme $x^{(k)} = (A^T A + \lambda^2 I_n)^{-1}(A^T b + \lambda^2 x^{(k-1)})$ often referred to as *iterated Tikhonov regularization* when $x^{(0)} = 0$. With this choice of $\mathcal{F}$, the iteration scheme (6.3) is identical to the iterative refinement process in (5.21). In particular, the first iterate $x^{(1)}$ is the Tikhonov solution x_λ, and after one more step we obtain $x^{(2)} = x_\lambda + (A^T A + \lambda^2 I_n)^{-1} A^T (b - A x_\lambda)$. An early reference to this method is [294]; see also [144]. With this choice of $\mathcal{F}$, each iteration step involves the computation of a Tikhonov regularized solution to a system with fixed coefficient matrix A and right-hand side $r^{(k)}$, and the filter factors take the form

$$f_i^{(k)} = 1 - \left(1 - \frac{\sigma_i^2}{\sigma_i^2 + \lambda^2}\right)^k, \qquad i = 1, \ldots, n. \tag{6.6}$$

This method has been studied extensively; see, e.g., [225]. Some applications of the method are described in [304] and [309]. Implementation aspects, based on the bidiagonalization (5.5) of A, are described in [171, §5.2].

There have been several attempts to accelerate Landweber iteration by replacing $x^{(k-1)}$ in (6.3) by a weighted average of some of the last iterates and allowing ω to depend on k. One of the more successful methods in this class of nonstationary methods is the so-called ν-*method* by Brakhage [43], which is a two-step procedure of the form

$$x^{(k)} = \mu_k x^{(k-1)} + (1 - \mu_k) x^{(k-2)} + \omega_k A^T r^{(k-1)}, \qquad k = 1, 2, \ldots \tag{6.7}$$

with coefficients μ_k and ω_k given by

$$\mu_k = 1 + \frac{(k-1)(2k-3)(2k+2\nu-1)}{(k+2\nu-1)(2k+4\nu-1)(2k+2\nu-3)}, \tag{6.8}$$

$$\omega_k = \frac{4(2k+2\nu-1)(k+\nu-1)}{(k+2\nu-1)(2k+4\nu-1)}, \tag{6.9}$$

where ν is a prescribed real constant satisfying $0 < \nu < 1$. The ν-method does not converge if $\|A\|_2 > 1$, and the convergence is fastest if $\|A\|_2$ is slightly smaller than 1. For more details, see [43] and [166]. However, even with this acceleration the ν-method is generally much slower than the CG method discussed below; see [166, §11] for numerical examples.

Finally, we wish to mention the row-action methods [46] that access the coefficient matrix A one row at a time, which may be advantageous for very large problems. A classical—and simple—example of a row-action method is *Kaczmarz's method*, originally described in [219]. This method is often used in connection with computerized tomography, where it is also known as the algebraic reconstruction technique (ART); cf. [259, Chapter 5]. In Kaczmarz's method, each iteration involves a sweep through the rows a_i^T of A of the form

$$x \leftarrow x + \frac{b_i - a_i^T x}{\|a_i\|_2^2} a_i , \qquad i = 1, \ldots, m , \qquad (6.10)$$

where b_i is the ith component b. Each new iteration vector is the projection of the old iteration vector on the hyperplane $\{z : a_i^T z = b_i\}$. In [38, §3] it is shown that Kaczmarz's method is equivalent to the Gauss–Seidel method for the problem $x = A^T y$, $A A^T y = b$, and a symmetric successive overrelaxation (SSOR) method for the same problem is discussed. An iterative algorithm, based on Kaczmarz's method, for incorporating equality and inequality constraints into Tikhonov regularization is described in [77]. Although successful in certain applications such as computerized tomography, the row-action methods can exhibit very slow convergence in other applications.

6.3. Regularizing CG Iterations

Currently, there is a lot of interest in iterative regularization methods based on the CG method. This method was originally designed for solving large sparse systems of equations with a symmetric positive definite coefficient matrix, and there is a wealth of literature about this method, its implementation, and its convergence properties; see, e.g., [12], [154, §10.2–10.3], [345], and the references therein. An understanding of the CG method's behavior in finite-precision arithmetic is also emerging; cf. [158].

In connection with least squares problems and regularization problems, the CG method is applied to the normal equations $A^T A x = A^T b$ whose coefficient matrix $A^T A$ is symmetric and positive semidefinite.

An essential property of the CG iterates $x^{(k)}$ with residual vectors $r^{(k)} = b - A x^{(k)}$ is that the corresponding residual vectors $A^T r^{(k)} = A^T b - A^T A x^{(k)}$ for the normal equations are orthogonal. An important consequence of this is that if the starting vector $x^{(0)}$ is zero, then the solution norm $\|x^{(k)}\|_2$ increases monotonically with k. This follows from the following theorem due to Hestenes and Stiefel, formulated here in terms of the normal equations.

Theorem 6.3.1. [206, Theorem 6:1]. *Let $\eta(x^{(k)})$ denote the CG error function*

$$\eta(x) = (x_{\mathrm{LS}} - x^{(k)})^T A^T A (x_{\mathrm{LS}} - x^{(k)}) . \quad (6.11)$$

Then the CG iterate $x^{(k)}$ can be written as

$$x^{(k)} = x^{(0)} + \sum_{i=0}^{k-1} \frac{\eta(x^{(i)}) - \eta(x^{(k)})}{\|A^T r^{(i)}\|_2^2} A^T r^{(i)} , \quad (6.12)$$

and if $x^{(0)} = 0$ then

$$\|x^{(k)}\|_2^2 = \sum_{i=0}^{k-1} \left(\frac{\eta(x^{(i)}) - \eta(x^{(k)})}{\|A^T r^{(i)}\|_2} \right)^2 . \quad (6.13)$$

Since the error function $\eta(x^{(k)})$ decreases monotonically with k, we conclude that if the CG starting vector is zero then each of the terms in (6.13) increases with k, and therefore the solution norm $\|x^{(k)}\|_2$ increases monotonically with k. The residual norm $\|r^{(k)}\|_2$, on the other hand, decreases monotonically with k if $x^{(0)} = 0$; this is an immediate consequence of the Krylov subspace formulation in (6.20) below. The monotonic behavior of both $\|x^{(k)}\|_2$ and $\|r^{(k)}\|_2$ is important in connection with the use of the L-curve criterion (§7.5) as a stopping rule for regularizing CG iterations.

The CG method often produces iteration vectors in which the spectral components associated with the large eigenvalues tend to converge faster than the remaining components. In connection with discrete ill-posed problems, the same behavior is observed when the CG algorithm is applied to the normal equations $A^T A x = A^T b$. Since the eigenvalues of $A^T A$ are simply σ_i^2, this means that the SVD components associated with the large singular values tend to converge faster than the remaining SVD components, in which case the CG algorithm has an intrinsic regularizing effect when stopped long before the convergence to the least squares solution $x_{\mathrm{LS}} = A^\dagger b$ sets in.

Hanke's book [169] contains a wealth of background material about the CG method applied to $A x = b$ (for symmetric semidefinite A) and to the normal equations $A^T A x = A^T b$ (for general problems), with focus on the regularizing aspects of these methods in the Hilbert space setting. Hanke also discusses the regularizing effects of the CG method applied to the problem $x = A^T y$, $A A^T y = b$, as well as the minimal residual (MINRES) algorithm [271] which solves problems $A x = b$ with a symmetric indefinite A without the transition to the normal equations, thus avoiding multiplications with A^T. An application of MINRES in image restoration is described in [172].

6.3.1. Implementation Issues

In a model implementation of the standard CG algorithm for symmetric positive definite matrices, the loop consists of merely five statements; see, e.g., [154,

6.3. REGULARIZING CG ITERATIONS

Algorithm 10.2.1]. There are several mathematically equivalent implementations of this algorithm when it is applied to the system $A^T A x = A^T b$. Elfving [105] found experimentally that the most stable implementation—on average—is the one which is often[18] referred to as CGLS; see, e.g., [36, §7.4.1]. The heart of this algorithm, which is due to Hestenes and Stiefel [206], also consists of five statements, to be executed for $k = 1, 2, \ldots$:

$$
\begin{aligned}
\alpha_k &= \|A^T r^{(k-1)}\|_2^2 / \|A\, d^{(k-1)}\|_2^2 , \\
x^{(k)} &= x^{(k-1)} + \alpha_k\, d^{(k-1)} , \\
r^{(k)} &= r^{(k-1)} - \alpha_k\, A\, d^{(k-1)} , \\
\beta_k &= \|A^T r^{(k)}\|_2^2 / \|A^T r^{(k-1)}\|_2^2 , \\
d^{(k)} &= A^T r^{(k)} + \beta_k\, d^{(k-1)} ,
\end{aligned}
\qquad (6.14)
$$

where $r^{(k)}$ is the residual vector $r^{(k)} = b - A x^{(k)}$ and $d^{(k)}$ is an auxiliary n-vector. The CGLS algorithm is initialized with the starting vector $x^{(0)}$, residual $r^{(0)} = b - A x^{(0)}$, and $d^{(0)} = A^T r^{(0)}$.

A variant of the CGLS implementation, in which the normal-equation residual vector $A^T r^{(k)}$ is recurred instead of $r^{(k)}$, is proved in [39] to be less stable than CGLS.

The CG algorithm applied to the normal equations with $x^{(0)} = 0$ can also be implemented by means of the Lanczos bidiagonalization algorithm (cf. [154, §9.3.3–4]), which leads to another class of methods of which LSQR [273], [274] is generally believed to be the most stable (cf. [39]). The Lanczos bidiagonalization algorithm with starting vector b produces a bidiagonal matrix $B_k \in \mathbb{R}^{(k+1) \times k}$ and two matrices $\hat{U}_{k+1} \in \mathbb{R}^{m \times (k+1)}$ and $\hat{V}_k \in \mathbb{R}^{n \times k}$ with orthonormal columns such that

$$
A \hat{V}_k = \hat{U}_{k+1} B_k , \qquad \hat{u}_1 = \beta_1^{-1} b . \qquad (6.15)
$$

The columns $\hat{u}_1, \ldots, \hat{u}_{k+1}$ of $\hat{U}_{k+1}$ and the columns $\hat{v}_1, \ldots, \hat{v}_k$ of $\hat{V}_k$ are called the *left and right Lanczos vectors*. Then it follows that $\hat{U}_{k+1}^T b = \beta_1 e_1^{(k+1)}$, where $e_1^{(k+1)} = (1, 0, \ldots, 0)^T$ has length $k+1$.

The Lanczos bidiagonalization algorithm with starting vector b is initialized with

$$
\beta_1 = \|b\|_2 , \qquad \hat{u}_1 = b/\beta_1 , \qquad \hat{v}_0 = 0 ,
$$

and the loop takes the form, for $k = 1, 2, \ldots$,

---

[18]Unfortunately, the naming conventions in the literature are not entirely consistent. Some authors refer to this implementation as CGNE.

$$\begin{aligned}
p^{(k)} &= A^T \hat{u}_k - \beta_k \hat{v}_k , \\
\alpha_k &= \|p^{(k)}\|_2 , \\
\hat{v}_k &= p^{(k)}/\alpha_k , \\
q^{(k)} &= A \hat{v}_k - \alpha_k \hat{u}_k , \\
\beta_{k+1} &= \|q^{(k)}\|_2 , \\
\hat{u}_{k+1} &= q^{(k)}/\beta_{k+1} ,
\end{aligned} \qquad (6.16)$$

where $p^{(k)}$ and $q^{(k)}$ are two auxiliary vectors of lengths n and m, respectively. The $(k+1) \times k$ bidiagonal matrix B_k is given by

$$B_k = \begin{pmatrix} \alpha_1 & & & \\ \beta_2 & \alpha_2 & & \\ & \beta_3 & \ddots & \\ & & \ddots & \alpha_k \\ & & & \beta_{k+1} \end{pmatrix} . \qquad (6.17)$$

In step k of the LSQR algorithm, the iteration vector $x^{(k)}$ is then defined as $\hat{V}_k \xi^{(k)}$, where $\xi^{(k)}$ solves the least squares problem

$$\min \|B_k \, \xi^{(k)} - \beta_1 e_1^{(k+1)}\|_2 ; \qquad (6.18)$$

i.e., $x^{(k)}$ is formally given by $x^{(k)} = \beta_1 \hat{V}_k B_k^\dagger e_1^{(k+1)}$. In practice, $x^{(k)}$ is computed via a QR factorization of B_k which can be updated efficiently in each stage. The result is that $x^{(k-1)}$ can be updated efficiently to $x^{(k)}$, at the cost of one additional iteration vector of length n, and neither $\hat{U}_{k+1}$ nor $\hat{V}_k$ needs to be stored. The key to the efficiency is the particular choice of starting vector $\hat{u}_1 = \beta_1^{-1} b$ such that the right-hand side in (6.18) takes the special form $\beta_1 e_1^{(k+1)}$. See [36, §7.6.3] and [273] for details.

For ill-conditioned full-rank problems that do not require regularization, it was experimentally found in [273] that LSQR is superior to CGLS and related methods. For discrete ill-posed problems, where the iterations are stopped long before the final convergence, LSQR does not give significantly more accurate solutions than CGLS. However, only the LSQR algorithm allows for applying "inner regularization" in each iteration step, which leads to a more general class of regularization methods; see §6.6.

We mention in passing that LSQR, as well as the recently developed extended Craig algorithm [305], includes the possibility of solving the Tikhonov problem (5.1) with $L = I_n$ and a *fixed* parameter λ using very little overhead. However, we are here interested in the intrinsic regularizing properties of the CG algorithm.

In the REGULARIZATION TOOLS package, CGLS and LSQR are implemented as functions cgls and lsqr, respectively. These implementations correspond to standard-form regularization; for general-form problems with $L \neq I_n$, the "preconditioned" versions pcgls and plsqr incorporate the standard-form transformation from §2.3.2. In all four functions, full reorthogonalization can be applied in order to "simulate"—for many practical purposes—exact arithmetic: in LSQR, one can reorthogonalize the two sets of Lanczos vectors, and in CGLS one can reorthogonalize the normal-equation residual vectors $A^T r^{(k)}$. Two reorthogonalization techniques are available: modified Gram–Schmidt and Householder transformations (of which the Householder approach is the most expensive and also the most accurate).

6.3.2. The Regularizing Effects of CG Iterations

In 1950, Lanczos [228, p. 270] made the following note regarding the Krylov subspace associated with $A A^T b$ on a 12 × 12 problem:

> The strong grading of the successive eigenvectors has the consequence that in k minimized iterations essentially only the highest k vibrational modes will come into evidence. This is of eminent practical value, since it allows us to dispense with the calculation of the very low eigenvalues, which are often of little physical interest [...]

The regularizing effect of CG iterations has been rediscovered several times over the years. Tal [330] proved in 1966 that the steepest descent algorithm initially filters out solution components corresponding to small eigenvalues (singular values) and noted that experimental results showed similar behavior for CG, although fewer iterations were required. Similar experimental results were later obtained by Squire [315] and Nolet [262]. Scales and Gersztenkorn [306] used the intrinsic regularizing effects of CG in a "regularized" iteratively weighted least squares algorithm. In 1979 Björck and Eldén [37] wrote about regularizing CG iterations:

> More research is needed to tell for which problems this approach will work, and what stopping criterion to choose.

Since then, some progress has been made (see [169], [346], and [357]), but a complete mathematical understanding of the intrinsic regularizing effect of the CG iterations, describing the convergence of the individual SVD components of $x^{(k)}$, is still lacking.

The following discussion holds only in infinite precision; for finite-precision effects, see §6.5. Since the vectors $x^{(k)} - x^{(0)}$ are identical to the iteration vectors for CG applied to the system $A^T A x = A^T (b - A x^{(0)})$, it is no restriction to assume $x^{(0)} = 0$ in our analysis. The CG algorithm is a Krylov subspace method [36, §7.4.1]. In the kth step the iteration vector $x^{(k)}$ minimizes the error function $\eta(x^{(k)})$ in (6.11) over all vectors in the *Krylov subspace*

$$\mathcal{K}_k(A^T A, A^T b) \equiv \mathrm{span}\{A^T b, A^T A A^T b, \ldots, (A^T A)^{k-1} A^T b\} . \tag{6.19}$$

Alternatively, we can use the orthogonality of $\hat{U}_{k+1}$ and $\hat{V}_k$ to show that the kth CG iterate can be written as

$$x^{(k)} = \mathrm{argmin} \|A\,x - b\|_2 \qquad \text{subject to} \qquad x \in \mathcal{K}_k(A^T A, A^T b)\,. \qquad (6.20)$$

The regularizing properties of the CG algorithm are due to the fact that in certain applications the Krylov subspace $\mathcal{K}_k(A^T A, A^T b)$ can be considered as an approximation to the subspace $\mathrm{span}\{v_1, \ldots, v_k\}$ spanned by the first k right singular vectors. Hence, $x^{(k)}$ can be considered as an approximation to the truncated SVD (TSVD) solution x_k (3.11)—or the TGSVD solution $x_{L,k}$ (3.15) if the CG process is "preconditioned" with $L_A^\dagger$.

To obtain more insight into the regularizing properties of the CG method, it is necessary to link the CG algorithm with $x^{(0)} = 0$ to the Lanczos bidiagonalization algorithm applied to A with starting vector b (or, equivalently, to the Lanczos tridiagonalization algorithm applied to $A^T A$). After k iterations, the Lanczos bidiagonalization algorithm has produced the $(k+1) \times k$ bidiagonal matrix B_k in (6.17) in such a way that $B_k^T B_k$ is identical to the symmetric positive definite tridiagonal matrix produced by the Lanczos tridiagonalization process applied to $A^T A$. Hence, the singular values of B_k are identical to the square roots of the *Ritz values* $\theta_j^{(k)}$, $j = 1, \ldots, k$, defined as the eigenvalues of $B_k^T B_k$.

As shown, e.g., in [345, Property 2.8], the corresponding CG iterate $x^{(k)}$ is given by $x^{(k)} = (1 - \mathcal{R}_k(A^T A))\,x_{\mathrm{LS}}$, where $\mathcal{R}_k$ is the so-called *Ritz polynomial*

$$\mathcal{R}_k(\theta) = \prod_{j=1}^{k} \frac{\theta_j^{(k)} - \theta}{\theta_j^{(k)}}\,, \qquad (6.21)$$

which satisfies $\mathcal{R}_k(0) = 1$ and $\mathcal{R}_k(\theta_j^{(k)}) = 0$, $j = 1, \ldots, k$. Hence, the filter factors associated with the kth iterate $x^{(k)}$ are given by

$$f_i^{(k)} = 1 - \prod_{j=1}^{k} \frac{\theta_j^{(k)} - \sigma_i^2}{\theta_j^{(k)}}\,, \qquad i = 1, \ldots, n\,. \qquad (6.22)$$

Notice that $f_i^{(k)}$ is a polynomial in σ_i^2 of degree k with constant term equal to zero. We emphasize that $f_i^{(k)}$ depends on both A and b via the Ritz values—in contrast to, e.g., Tikhonov regularization, TSVD, and TGSVD where the filter factors are independent of b.

We see from (6.22) that if the square root of a Ritz value $\theta_j^{(k)}$ has converged to a singular value σ_i of A, i.e., if $\theta_j^{(k)} = \sigma_i^2$ for some pair (i,j), then the corresponding filter factor $f_i^{(k)} = 1$. Filter factors corresponding to multiple singular values of A are identical and, in particular, if $\theta_j^{(k)}$ has converged to a multiple singular value then all the corresponding filter factors are one. For a

6.3. REGULARIZING CG ITERATIONS

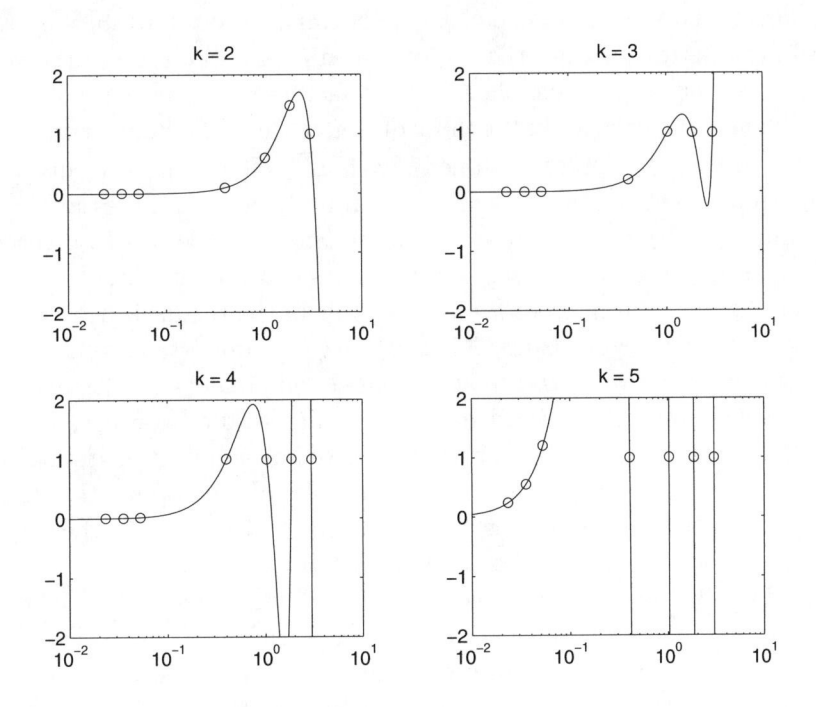

FIG. 6.2. *The polynomial* $1 - \prod_{j=1}^{k}(\theta_j^{(k)} - \sigma^2)/\theta_j^{(k)}$ *as a function of the real parameter* σ *(solid line), and the corresponding filter factors* f_i *versus the singular values* σ_i *(circles), for test problem* shaw *with* $m = n = 8$ *and computed with full reorthogonalization in LSQR.*

real parameter σ lying *between* the converged Ritz values $\theta_j^{(k)}$, the polynomial $1 - \prod_{j=1}^{k}(\theta_j^{(k)} - \sigma^2)/\theta_j^{(k)}$ oscillates and takes on very large positive and negative values; see Fig. 6.2 for an illustration.

Since $\mathcal{R}_k$ is monotonic in the interval from 0 to $\theta_k^{(k)}$ (the smallest Ritz value in the kth iteration), all filter factors corresponding to $\sigma_i^2 \leq \theta_k^{(k)}$ are positive and less than 1; i.e.,

$$0 \leq f_i^{(k)} \leq 1 \quad \text{for} \quad \sigma_i^2 \leq \theta_k^{(k)}. \tag{6.23}$$

Moreover, if $\sigma_i^2 \ll \theta_k^{(k)}$, then the corresponding filter factor $f_i^{(k)}$ is proportional to σ_i^2; more precisely:

$$f_i^{(k)} = \sigma_i^2 \sum_{j=1}^{k} \frac{1}{\theta_j^{(k)}} + \mathcal{O}\left(\frac{\sigma_i^4}{\theta_k^{(k)} \theta_{k-1}^{(k)}}\right) \quad \text{for} \quad \sigma_i^2 \ll \theta_k^{(k)}. \tag{6.24}$$

We see that the spectral filtering of the CG process is entirely controlled by the Ritz values whose convergence, in turn, is related to the number k of iterations. If the Ritz values tend to approximate the eigenvalues σ_i^2 in their natural order, starting from the largest, then k essentially plays the role of a regularization parameter and the CG filter factors resemble those of standard-form Tikhonov regularization (or those of general-form Tikhonov regularization if the CG process is "preconditioned" with $L_A^\dagger$). This, in turn, explains the semiconvergent nature of CG, because $x^{(k)}$ initially approximates x^{exact} as long as all converged Ritz values are fairly large, while $x^{(k)}$ starts to diverge from x^{exact} again when small converged Ritz values start to appear.

Since the iteration number k plays the role of the regularization parameter in connection with regularizing CG iterations, it is important to use a reliable stopping rule—otherwise, too many spectral components will appear in $x^{(k)}$. In this connection, we emphasize an important result by Eicke, Louis, and Plato [96], namely, that an a priori choice of the number of iterations as a function of the error in the data is not possible. We postpone the discussion of stopping rules for regularizing CG iterations to Chapter 7, where we treat methods for choosing the regularization parameter.

The filter factors cannot be computed without explicit knowledge of the singular values of A, but they can be computed without knowledge of the Ritz values by means of a three-term recurrence. This approach to computing the $f_i^{(k)}$ is used in functions cgls, lsqr, pcgls, and plsqr in REGULARIZATION TOOLS. The following theorem from [357] gives the details (we include the proof since [357] was never published).

Theorem 6.3.2. *The CG iteration vector $x^{(k)}$ computed by means of the CGLS algorithm in (6.14) with starting vector $x^{(0)} = 0$ satisfies*

$$x^{(k)} = \mathcal{P}_k(A^T A)\, A^T b \,, \qquad (6.25)$$

where the CG polynomial $\mathcal{P}_k$ is a polynomial of degree $k-1$ satisfying the recurrence relation

$$\mathcal{P}_k(\theta) = \left(1 - \theta\,\alpha_k + \frac{\alpha_k \beta_{k-1}}{\alpha_{k-1}}\right)\mathcal{P}_{k-1}(\theta) - \frac{\alpha_k \beta_{k-1}}{\alpha_{k-1}}\mathcal{P}_{k-2}(\theta) + \alpha_k \qquad (6.26)$$

and with initial conditions $\mathcal{P}_0(\theta) = \mathcal{P}_{-1}(\theta) = 0$. The filter factors for $x^{(k)}$ are given by

$$f_i^{(k)} = \sigma_i^2\, \mathcal{P}_k(\sigma_i^2)\,, \qquad i = 1,\dots,n\,. \qquad (6.27)$$

Proof. In addition to $\mathcal{P}_k$ we introduce the polynomial $\hat{\mathcal{P}}_k$ such that the vector $d^{(k)}$ in CGLS can be written as

$$d^{(k)} = \hat{\mathcal{P}}_k(A^T A)\, A^T b\,.$$

It is easy to show that $\mathcal{P}_k$ and $\hat{\mathcal{P}}_k$ satisfy the recurrence relations

$$\begin{aligned}
\mathcal{P}_k(\theta) &= \mathcal{P}_{k-1} + \alpha_k\, \hat{\mathcal{P}}_{k-1}(\theta)\,, \\
\hat{\mathcal{P}}_k(\theta) &= 1 - \theta\, \mathcal{P}_k(\theta) + \beta_k\, \hat{\mathcal{P}}_{k-1}(\theta)
\end{aligned}$$

with initial conditions $\mathcal{P}_0(\theta) = 0$ and $\hat{\mathcal{P}}_0(\theta) = 1$. Eliminating $\hat{\mathcal{P}}_{k-1}$ and $\hat{\mathcal{P}}_k$ from these relations we obtain (6.26). To derive the expression for the filter factors, we use that $A^T b = A^T A\, x_{\mathrm{LS}}$ and therefore $V^T x^{(k)} = \mathcal{P}_k(\Sigma^2)\, \Sigma^2 V^T x_{\mathrm{LS}}$, which leads to (6.27). □

The polynomial $\mathcal{P}_k$ in (6.25) is related to the Ritz polynomial $\mathcal{R}_k$ from Eq. (6.21) as $\theta \mathcal{P}_k(\theta) = 1 - \mathcal{R}_k(\theta)$.

6.4. Convergence Properties of Regularizing CG Iterations

From the above discussion we see that a fundamental task in connection with a theoretical investigation of the regularizing properties of CG iterations is to determine the convergence properties of the Ritz values, given the particular features of discrete ill-posed problems mentioned in §2.1. This is a difficult problem, and it has not been solved satisfactorily yet.

6.4.1. Convergence of the Ritz Values

To heuristically explain why the Lanczos bidiagonalization process tends to capture the largest singular values of A first, we switch to the normal equations $A^T A x = A^T b$ and to the basis of the right singular vectors v_i of A, and again we assume without loss of generality that $x^{(0)} = 0$. Then a basis for the Krylov subspace $\mathcal{K}_k(\Sigma^2, \Sigma U^T b)$ is given by the columns of the $n \times k$ matrix $D W_k$ with

$$D = \mathrm{diag}(\sigma_i\, u_i^T b) \quad \text{and} \quad W_k = \begin{pmatrix} 1 & \sigma_1^2 & \cdots & \sigma_1^{2k-2} \\ 1 & \sigma_2^2 & \cdots & \sigma_2^{2k-2} \\ \vdots & \vdots & & \vdots \\ 1 & \sigma_n^2 & \cdots & \sigma_n^{2k-2} \end{pmatrix}, \quad (6.28)$$

and the jth column of $D W_k$ has elements $\sigma_i^{2j-1} u_i^T b$, $i = 1, \ldots, n$. As long as the discrete Picard condition is satisfied, the elements of each column decay, on the average, with index i. Hence, the first unit vectors $e_1^{(n)}, e_2^{(n)}, \ldots$, which are eigenvectors of Σ^2, can be approximated well by the columns of $D W_k$.

The largest Ritz values $\theta_1^{(k)}, \theta_2^{(k)}, \ldots$, obtained by restricting Σ^2 to the range of $D W_k$, are therefore approximations to the largest eigenvalues $\sigma_1^2, \sigma_2^2, \ldots$ of Σ^2. The Krylov subspace $\mathcal{K}_k(A^T A, A^T b)$, which is spanned by the columns of $V D W_k$, is therefore primarily spanned by the first "smooth" right singular vectors $v_1, v_2, \ldots$ of A. This is not true for general problems; it is a genuine property of discrete ill-posed problems that satisfy the discrete Picard condition.

As already mentioned, the particular starting vector $\hat{u}_1 = b/\|b\|_2$ for the Lanczos bidiagonalization process makes the LSQR algorithm so efficient because the iteration vector $x^{(k)}$ can be updated in each stage. As we shall see in §6.5, this starting vector is also essential for the numerical stability of the

LSQR algorithm. Yet another advantage of this particular starting vector—in connection with discrete ill-posed problems—is that (as noted in [35]) it is guaranteed to be rich in the first SVD components.

More insight into the convergence properties of the Ritz values has been obtained by Hansen, O'Leary, and Stewart [198], and their hitherto unpublished analysis is summarized below. This analysis is purely algebraic and therefore it does not fully capture the behavior of the CG process, which is strongly connected to the Ritz polynomial. Partition the matrices Σ, D, and W_k according to

$$\Sigma = \begin{pmatrix} \Sigma_1 & 0 \\ 0 & \Sigma_2 \end{pmatrix}, \qquad D = \begin{pmatrix} D_1 & 0 \\ 0 & D_2 \end{pmatrix}, \qquad W_k = \begin{pmatrix} W_{k1} \\ W_{k2} \end{pmatrix}$$

such that Σ_1, D_1, and W_{k1} have k rows. Another basis for $\mathcal{K}_k(\Sigma^2, \Sigma U^T b)$ is then given by the columns of

$$S_k = \begin{pmatrix} I_k \\ \Delta_k \end{pmatrix} \qquad \text{with} \qquad \Delta_k = D_2 W_{k2} W_{k1}^{-1} D_1^{-1}. \qquad (6.29)$$

The Ritz values $\theta_1^{(k)}, \ldots, \theta_k^{(k)}$ are therefore the eigenvalues of the generalized eigenvalue problem $S_k^T \Sigma^2 S_k z = \theta S_k^T S_k z$ or, equivalently, of

$$\left(\Sigma_1^2 + \Delta_k^T \Sigma_2^2 \Delta_k\right) z = \theta \left(I_k + \Delta_k^T \Delta_k\right) z. \qquad (6.30)$$

The two matrices in (6.30) are symmetric positive definite, so they form a regular pair. We can therefore use the first-order perturbation theory for the generalized eigenvalue problem from [326, Theorem 2.2] and consider $\Delta_k^T \Sigma_2^2 \Delta_k$ and $\Delta_k^T \Delta_k$ as perturbations of Σ_1^2 and I_k, respectively. Then we obtain

$$\theta_i^{(k)} = \frac{\sigma_i^2 + \|\Sigma_2 \delta_i^{(k)}\|_2^2}{1 + \|\delta_i^{(k)}\|_2^2} + \mathcal{O}\left(\|e_i^{(k)} - z_i^{(k)}\|_2^2\right), \qquad (6.31)$$

where $\delta_i^{(k)}$ is the ith column of Δ_k and $z_i^{(k)}$ is the eigenvector corresponding to $\theta_i^{(k)}$. Unfortunately, we have not been able to derive a useful upper bound for the norm $\|e_i^{(k)} - z_i^{(k)}\|_2$. The residual vector corresponding to the approximate eigenpair $(\theta, z) = (\sigma_i^2, e_i^{(k)})$ in (6.30) satisfies

$$\|\Delta_k^T \Sigma_2^2 \delta_i^{(k)} - \sigma_i^2 \Delta_k^T \delta_i^{(k)}\|_2 \leq (\sigma_i^2 + \sigma_{k+1}^2) \|\Delta_k\|_2 \|\delta_i^{(k)}\|_2. \qquad (6.32)$$

The smaller the right-hand sides in (6.31) and (6.32), the better $\theta_i^{(k)}$ approximates σ_i^2, and the norm of $\delta_i^{(k)}$ plays an important role.

6.4. Convergence Properties of Regularizing CG Iterations

Theorem 6.4.1. [198]. *The norm of the ith column $\delta_i^{(k)}$ of the matrix Δ_k in (6.29) satisfies*

$$\|\delta_i^{(k)}\|_\infty \leq \frac{\sigma_{k+1}}{\sigma_i} \frac{\max_{j=k+1}^n |u_j^T b|}{|u_i^T b|} |\mathcal{L}_i^{(k)}(0)|, \qquad (6.33)$$

where $\mathcal{L}_i^{(k)}$ is the ith Lagrange polynomial and

$$|\mathcal{L}_i^{(k)}(0)| = \prod_{\substack{j=1 \\ j \neq i}}^k \frac{\sigma_j^2}{|\sigma_j^2 - \sigma_i^2|}.$$

Proof. Since $D_1^{-1} e_i^{(k)} = (\sigma_i u_i^T b)^{-1} e_i^{(k)}$, we get

$$\delta_i^{(k)} = \Delta_k e_i^{(k)} = (\sigma_i u_i^T b)^{-1} D_2 W_{k2} W_{k1}^{-1} e_i^{(k)},$$

and thus

$$\|\delta_i^{(k)}\|_\infty \leq \frac{\sigma_{k+1}}{\sigma_i} \frac{\max_{j=k+1}^n |u_j^T b|}{|u_i^T b|} \|W_{k2} W_{k1}^{-1} e_i^{(k)}\|_\infty.$$

The ith column of W_{k1}^{-1} consists of the coefficients of the ith Lagrange polynomial

$$\mathcal{L}_i^{(k)}(\lambda) = \prod_{\substack{j=1 \\ j \neq i}}^k \frac{\lambda - \sigma_j^2}{\sigma_j^2 - \sigma_i^2}$$

that interpolates the elements of $e_i^{(k)}$ at the abscissas $\sigma_1^2, \ldots, \sigma_k^2$. Consequently, the ith column of $W_{k2} W_{k1}^{-1}$ is

$$W_{k2} W_{k1}^{-1} e_i^{(k)} = \left(\mathcal{L}_i^{(k)}(\sigma_{k+1}^2), \ldots, \mathcal{L}_i^{(k)}(\sigma_n^2) \right)^T, \qquad (6.34)$$

and since $\mathcal{L}_i^{(k)}(\lambda)$ is monotonic for $\lambda \leq \sigma_k^2$, the elements of $W_{k2} W_{k1}^{-1} e_i^{(k)}$ are bounded in size by $|\mathcal{L}_i^{(k)}(0)|$. Thus, we obtain (6.33). $\square$

To obtain more insight into the quantity $|\mathcal{L}_i^{(k)}(0)|$ for $i = 1, \ldots, k$, we computed $|\mathcal{L}_1^{(k)}(0)|, \ldots, |\mathcal{L}_k^{(k)}(0)|$ for a model singular value spectrum given by

$$\sigma_i = \rho^{1-i}, \qquad i = 1, \ldots, k$$

for various values of ρ. The results are shown in Fig. 6.3. We see that for a fixed i, $|\mathcal{L}_i^{(k)}(0)|$ decays monotonically with k. Moreover, as long as the decay of the singular values is not too gentle, $\max_i |\mathcal{L}_i^{(k)}(0)| = \mathcal{O}(1)$ such that $\|\Delta_k\|_2 = \mathcal{O}(1)$. We also see that for small i, we have $|\mathcal{L}_i^{(k)}(0)| \ll 1$ after relatively few CG iterations k.

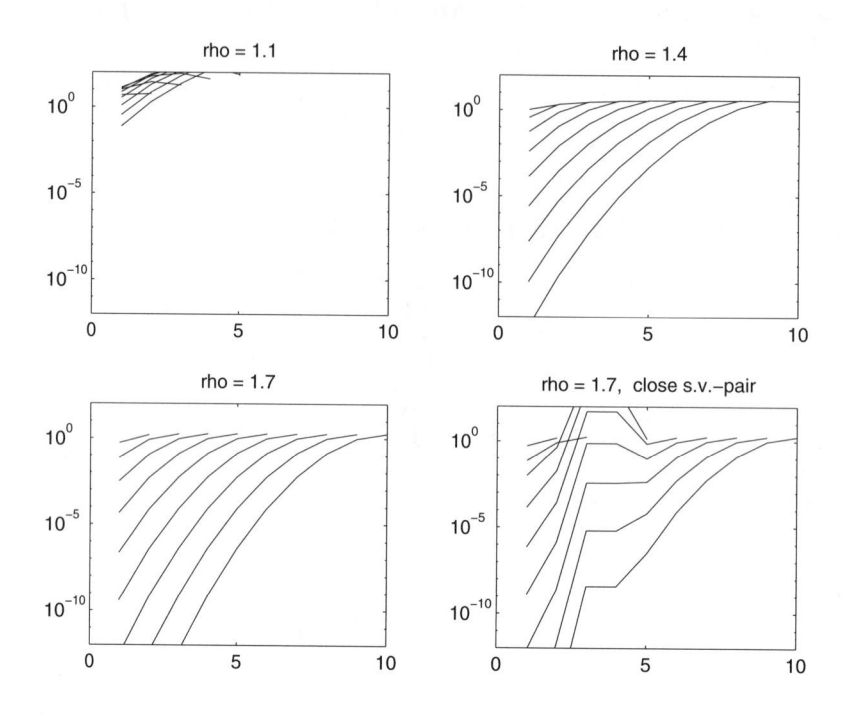

FIG. 6.3. *Plots of $|\mathcal{L}_i^{(k)}(0)|$ versus i for the model singular value spectrum $\sigma_i = \rho^{1-i}$. Each curve corresponds to a fixed k for $k = 1, \ldots, 10$. In the bottom right figure, we have set $\sigma_4 = \sigma_3/1.001$ such that the matrix has one pair of close singular values.*

Figure 6.3 also shows the effect of a pair of close singular values. When k and i are larger than the index of the pair, then $|\mathcal{L}_i^{(k)}(0)|$ is greater than its counterpart in the case of no close singular values. For certain k and i the value $|\mathcal{L}_i^{(k)}(0)|$ is very large, of the order 10^4, and for these k the Lanczos process captures an average of the two close singular values. However, as k increases, the huge values of $|\mathcal{L}_i^{(k)}(0)|$ disappear, and this happens when the Lanczos process captures both singular values of the close pair.

We conclude from (6.31)–(6.33) that if the first $k+1$ singular values of A are well separated and do not decay too slowly, then the accuracy of the Ritz value $\theta_i^{(k)}$ is largely determined by $\sigma_i^{-1}\|\delta_i^{(k)}\|_2$, and therefore the first Ritz values approximate the largest σ_i^2 in their natural order. Moreover, the larger the gap between σ_k and σ_{k+1}, the better the approximations—provided that b satisfies the discrete Picard condition such that b has no large components $|u_j^T b|$ for $j > k$. When A has close singular values and k is small, then the bounds in (6.31)–(6.33) become large, and in this sense our theory agrees with common experience that the symmetric single-vector Lanczos process is not

6.4.2. Convergence Rates for the CG Solution

We now turn to the convergence of the CG iterates $x^{(k)}$. Let CG^{exact} denote the CG method applied to a problem with exact right-hand side $b = b^{\text{exact}}$ in infinite precision, in which case CG^{exact} always converges to the least squares solution $x^{\text{exact}} = A^\dagger b^{\text{exact}}$. Convergence results for this "easy case"—compared to the case where b is contaminated by errors—were obtained by Brakhage and Louis, who related the convergence of CG^{exact} to the decay of the exact Fourier coefficients $u_i^T b^{\text{exact}}$.

To express their results, which are formulated in terms of operators in Hilbert spaces, in our linear algebra notation it is instructive to again consider the model problem (4.30) from §4.5 where the coefficients $u_i^T b^{\text{exact}}$ are related to the singular values σ_i as follows:

$$u_i^T b^{\text{exact}} = \sigma_i^{1+\alpha}, \qquad i = 1, \ldots, n, \tag{6.35}$$

where α is a real number. Notice that α controls the decay of the Fourier coefficients relative to the decay of the singular values, and that $v_i^T x^{\text{exact}} = \sigma_i^\alpha$. The Fourier coefficients decay faster than the singular values when $\alpha > 0$.

Assuming the model (6.35), Brakhage [43, Theorem 2] showed that

$$\|x^{(k)} - x^{\text{exact}}\|_2 = \mathcal{O}(k^{-\alpha}). \tag{6.36}$$

In comparison, the convergence rate for Landweber iteration (6.2) is only $\mathcal{O}(k^{-\alpha/2})$; cf. [43, p. 166]. Thus, (6.36) confirms that the CG method is typically much faster than Landweber iteration.

Assuming the same model (6.35), Louis [238], [239, §4.3] proved the following results:

$$\|A\,x^{(k)} - b^{\text{exact}}\|_2 \leq \sigma_{k+1}^{1+\alpha} \|P_k^\perp x^{\text{exact}}\|_2 \tag{6.37}$$

and

$$\|x^{(k)} - x^{\text{exact}}\|_2 \leq c \min\left\{\sigma_{k+1}^\alpha \|P_k^\perp x^{\text{exact}}\|_2, k^{-\alpha}\|x^{\text{exact}}\|_2\right\}, \tag{6.38}$$

where $P_k^\perp$ is the orthogonal projection onto the subspace span$\{v_{k+1}, \ldots, v_n\}$ and c is an unspecified constant. These bounds imply that in the absence of noise and with $\alpha \approx 0$ the CG^{exact}-iterate $x^{(k)}$ behaves similarly to the TSVD solution x_k (3.11) for which we have $\|A\,x_k - b^{\text{exact}}\|_2 \leq \sigma_{k+1}\|P_k^\perp x^{\text{exact}}\|_2$ and $\|x_k - x^{\text{exact}}\|_2 = \|P_k^\perp x^{\text{exact}}\|_2$.

Van der Sluis and van der Vorst [346, §7] presented similar results for CG^{exact}. In our linear algebra notation,[19] they assume essentially the same

---

[19] In [346], $\|x^{(k)} - x^{\text{exact}}\|_2^2$ is approximated by $\int_0^1 \mathcal{R}_k(t)^2 \rho(t)\,dt$, where $\mathcal{R}_k$ is the Ritz polynomial satisfying $A^T b - A^T A x^{(k)} = \mathcal{R}_k(A^T A)\,A^T b$, and $\mathcal{R}_k(\sigma_i^2) = 1 - f_i^{(k)}$. Thus, $\rho(\sigma_i^2)\,\sigma_i^2 \approx (u_i^T b)^2\,\sigma_i^{-2}$, and (6.35) leads to $\rho(t)\,t = t^{2\alpha}$. Then the model problem (7.3) in [346], $\rho(t) = a'\,t^a\,(1-t)^\beta \approx a'\,t^a$ for $t \ll 1$, together with [346, Theorem 7.24] leads to (6.39).

model problem as above, but they restrict the assumption (6.35) to holding for the small singular values only, and prove that

$$\|x^{(k)} - x^{\text{exact}}\|_2 \approx \bar{c}\, k^{-\alpha}\, , \qquad \bar{c} = \text{const.} \tag{6.39}$$

The key result is that for this model problem, the convergence of the CG^{exact} process is almost independent of the behavior of the exact Fourier coefficients corresponding to the large singular values.

The results (6.36)–(6.39), which were derived in the noise-free case, do not immediately carry over to problems with perturbed data, because the CG process depends in a nonlinear way on both the matrix A and the right-hand side b.

The general case where errors are present in *both* A and b was analyzed by Nemirovskii [261] (the presentation in [169, §3.2] is recommendable). These results are also presented in terms of Hilbert space operators, and again we interpret them by means of the above model (6.35). In this framework Nemirovskii proved that initially the residual norm and error norm decay like

$$\|A\,x^{(k)} - b\|_2 = \mathcal{O}(k^{-(\alpha+1)})\, , \qquad \|x^{(k)} - x^{\text{exact}}\|_2 = \mathcal{O}(k^{-\alpha})\, , \tag{6.40}$$

and eventually these norms stagnate. If A is error-free then stagnation occurs when the residual norm levels off at $\|b - b^{\text{exact}}\|_2$ and, simultaneously, the solution $x^{(k)}$ reaches an accuracy that corresponds to the resolution limit $\mathcal{O}(\|b - b^{\text{exact}}\|_2^{\alpha/(1+\alpha)})$; cf. (4.1) and (4.35).

These results are important because they show that the initial convergence of the CG process, when applied to a perturbed problem, is similar to the convergence of CG^{exact}. As Nemirovskii concludes:

> Roughly speaking, the presence of errors in the initial data limits the range of accuracies with which we can reduce the solution [...], but does not increase the time in which we can reduce the solution with permissible accuracy.

Moreover, since the CG solution $x^{(k)}$ at some stage k reaches an accuracy corresponding to the resolution limit, the CG method is able to produce regularized solutions of the same quality as the direct methods.

6.4.3. Filter Factors for the CG Solution

None of the results mentioned above say anything about the convergence of the individual SVD components. Such an analysis requires knowledge about the individual filter factors $f_i^{(k)}$.

So far, the only available results about the CG filter factors are based on Theorem 6.4.1. Assuming again that $x^{(0)} = 0$ and that the first k singular values of A are simple, our analysis gives the following characterization of the filter factors $f_i^{(k)}$ for the CG iterate $x^{(k)}$.

6.4. Convergence Properties of Regularizing CG Iterations

Theorem 6.4.2. [198]. *The first k filter factors $f_i^{(k)}$ for CG satisfy*

$$|f_i^{(k)} - 1| \leq \frac{\sigma_{k+1}}{\sigma_i} \frac{\|(U_k^o)^T b\|_2 + \sigma_{k+1}\|\Delta_k\|_2 \|x^{(k)}\|_2}{|u_i^T b|} \|\delta_i^{(k)}\|_2 , \qquad (6.41)$$

where $U_k^o = (u_{k+1}, \ldots, u_n)$ and $\delta_i^{(k)}$ is the ith column of Δ_k in (6.29). The last $n - k$ filter factors satisfy

$$0 \leq f_i^{(k)} \leq \frac{\sigma_i^2}{\sigma_k^2} C_k , \qquad i = k+1, \ldots, n , \qquad (6.42)$$

where the constant C_k is given by

$$C_k = |\mathcal{L}_k^{(k)}(0)| \sum_{j=1}^{k} f_j^{(k)} . \qquad (6.43)$$

Proof. Write the CG iterate as $x^{(k)} = V S_k w^{(k)}$, where V is the right singular matrix of A, and the vector $w^{(k)}$ is determined from the equation

$$S_k^T \Sigma^2 S_k w^{(k)} = S_k^T \Sigma U^T b .$$

If we partition $U = (U_k, U_k^o)$ such that U_k has k columns, and insert $S_k^T = (I_k, \Delta_k^T)$, then we obtain

$$\Sigma_1^2 w^{(k)} + \Delta_k^T \Sigma_2^2 \Delta_k w^{(k)} = \Sigma_1 U_k^T b + \Delta_k^T \Sigma_2 (U_k^o)^T b$$

or, by multiplying with Σ_1^{-2} and reordering the terms,

$$w^{(k)} - \Sigma_1^{-1} U_k^T b = \Sigma_1^{-2} \Delta_k^T \Sigma_2 \left((U_k^o)^T b - \Sigma_2 \Delta_k w^{(k)}\right) .$$

Writing this equation elementwise, we obtain for $i = 1, \ldots, k$

$$w_i^{(k)} - \frac{u_i^T b}{\sigma_i} = \sigma_i^{-2} (\delta_i^{(k)})^T \Sigma_2 \left((U_k^o)^T b - \Sigma_2 \Delta_k w^{(k)}\right) .$$

Here, $\sigma_i^{-1} u_i^T b$ are the coefficients to the unregularized least squares solution, and from the relation

$$x^{(k)} = V S_k w^{(k)} = V \begin{pmatrix} w^{(k)} \\ \Delta_k w^{(k)} \end{pmatrix}$$

we see that the elements of $w^{(k)}$ are the first k SVD coefficients of the CG iterate $x^{(k)}$. Hence, for $i = 1, \ldots, k$, the quantity $w_i^{(k)}/(\sigma_i^{-1} u_i^T b)$ is the ith filter factor $f_i^{(k)}$ for $x^{(k)}$, and we obtain

$$f_i^{(k)} - 1 = \frac{1}{\sigma_i u_i^T b} (\delta_i^{(k)})^T \Sigma_2 \left((U_k^o)^T b - \Sigma_2 \Delta_k w^{(k)}\right) .$$

Equation (6.41) is then obtained by taking norms and using the fact that $\|w^{(k)}\|_2 \leq \|x^{(k)}\|_2$. The left part of (6.42) is Eq. (6.23). The elements of $\Delta_k w^{(k)}$ are the last $n-k$ SVD coefficients of $x^{(k)}$, and therefore the elements of $\Sigma_2^2 D_2^{-1} \Delta_k w^{(k)}$ are the corresponding filter factors. If we insert the expression (6.29) for Δ_k, then we obtain

$$\begin{aligned}\Sigma_2^2 D_2^{-1} \Delta_k w^{(k)} &= \Sigma_2^2 W_{k2} W_{k1}^{-1} D_1^{-1} w^{(k)} \\ &= \Sigma_2^2 W_{k2} W_{k1}^{-1} \Sigma_1^{-2} \left(\Sigma_1^2 D_1^{-1} w^{(k)} \right),\end{aligned}$$

in which we recognize $\Sigma_1^2 D_1^{-1} w^{(k)}$ as the first k filter factors. It follows from (6.34) that the ith row of $W_{k2} W_{k1}^{-1}$ is given by

$$(e_i^{(n-k)})^T W_{k2} W_{k1}^{-1} = \left(\mathcal{L}_1^{(k)}(\sigma_{k+i}^2), \ldots, \mathcal{L}_k^{(k)}(\sigma_{k+i}^2) \right),$$

and therefore we obtain, for $i = k+1, \ldots, n$,

$$f_i^{(k)} = \sigma_i^2 \sum_{j=1}^k \mathcal{L}_j^{(k)}(\sigma_i^2) \sigma_j^{-2} f_j^{(k)} \leq \frac{\sigma_i^2}{\sigma_k^2} |\mathcal{L}_k^{(k)}(0)| \sum_{j=1}^k f_j^{(k)},$$

which establishes the right part of (6.42). □

From Theorems 6.4.1 and 6.4.2 we make the following conclusions about the CG filter factors, assuming again that the discrete Picard condition is satisfied. The smaller the index i, the smaller the norms $\|\delta_i^{(k)}\|_2$, and therefore the more accurate the Ritz approximations. Hence, the first filter factors $f_i^{(k)}$, $i = 1, 2, \ldots$, are close to 1. On the other hand, the filter factors for $i > k$ are approximately proportional to σ_i^2, as with the Tikhonov filter factors. For $i = k$ and i slightly smaller than k, the bound in (6.41) becomes less sharp, reflecting the fact that the corresponding filter factors can indeed differ somewhat from 1 (which is also observed in practice).

Our theory does not fully describe the behavior of regularizing CG iterations. The CG method also works well in the case of multiple or very close eigenvalues of $A^T A$, as long as at least one of the Ritz values $\theta_j^{(k)}$ approximates the clustered eigenvalues, for then (6.22) ensures that all filter factors $f_i^{(k)}$ corresponding to the cluster are close to 1. A refined analysis should reflect this.

In spite of our lack of a full theoretical understanding of the convergence of the Ritz values, numerical experiments give a fairly clear picture of what goes on in regularizing CG iterations. If the initial decay of the coefficients $|u_i^T b|/\sigma_i$ is not too slow, then the CG process captures the singular values in their natural order, starting with the largest.

As an example, we applied LSQR with full reorthogonalization to the one-dimensional image reconstruction test problem shaw from §1.4.3 with $m = n = 100$. The filter factors for $k = 4, 8, 11$, and 14 are shown in Fig. 6.4. We see

6.5. The LSQR Algorithm in Finite Precision

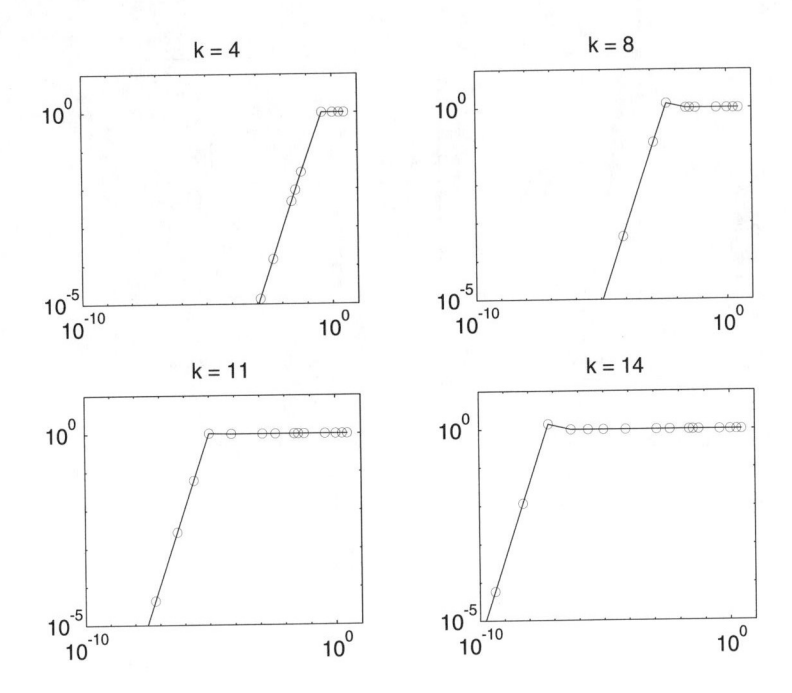

FIG. 6.4. *Filter factors $f_i^{(k)}$ as functions of the singular values σ_i for test problem* shaw *with $m = n = 100$, $k = 4, 8, 11, 14$ iterations, and full reorthogonalization in LSQR.*

that the first k filter factors $f_i^{(k)}$ corresponding to the largest k singular values are close to 1, while the filter factors for singular values $\sigma_i < \sigma_k$ decay as σ_i^2.

For $k = 8$ and $k = 14$ we notice that $f_k^{(k)}$ is slightly larger than one. This is not due to rounding errors; it is allowed by Eq. (6.22). The same behavior is observed in Fig. 6.2 for $k = 2$ and $k = 5$.

The influence of a double singular value on the CG iterate $x^{(k)}$ is illustrated in Fig. 6.5, where we plot $f_i^{(k)}$ versus i. The test problem is the same as before, except that $\sigma_3 = \sigma_4$ and therefore, as predicted by (6.22), $f_3^{(k)} = f_4^{(k)}$. Thus, after k iterations with $k \geq 3$, the CG process has roughly captured $k + 1$ SVD components.

6.5. The LSQR Algorithm in Finite Precision

In finite-precision arithmetic the Lanczos bidiagonalization algorithm loses orthogonality among the Lanczos vectors, i.e., the columns of the matrices $\hat{U}_{k+1}$ and $\hat{V}_k$ in (6.15). Similarly, the finite-precision CGLS algorithm loses orthogonality among the vectors $A^T r^{(k)}$. Detailed studies of the CG and Lanczos

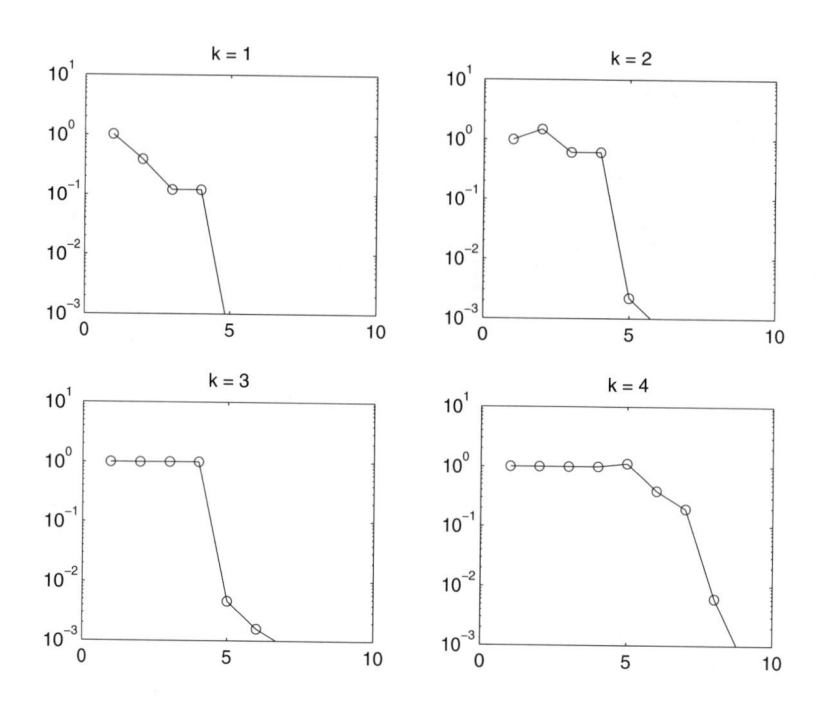

FIG. 6.5. *Filter factors $f_i^{(k)}$ versus i, computed with full reorthogonalization in LSQR. Same test problem as in the previous figure, except that σ_4 is replaced with σ_3 such that A has a double singular value.*

algorithms in finite precision have recently appeared [68], [69], [70], [158], but these studies do not focus on the regularizing properties of CG.

The consequences of the loss of orthogonality in the Lanczos process are extremely important in connection with procedures for computing eigenvalues and singular values, and the subject has been studied extensively; see, e.g., [71] and the references therein. We emphasize the similarity between the Lanczos bidiagonalization process in the LSQR algorithm and the reorthogonalization-free Lanczos procedures for computing the SVD in [71, Chapter 5]. However, as we describe in the following, the loss of orthogonality among the Lanczos vectors in LSQR is not nearly as harmful when solving least squares problems as when computing singular triplets.

Recall that the singular values of the bidiagonal matrix B_k in (6.15) are the square roots of the Ritz values $\theta_j^{(k)}$. One consequence of finite-precision arithmetic is that, as the number k of iterations increases, those singular values of B_k which have converged to isolated (possibly multiple) singular values of A become multiple,[20] and the multiplicity increases with k. The appearance of

---

[20]Theoretically, the singular values of an unreduced bidiagonal matrix are distinct, but in

6.5. THE LSQR ALGORITHM IN FINITE PRECISION

such multiple singular values of B_k is usually taken as a sign that a singular value of A has been accurately captured by a Ritz value [71, p. 124].

As a singular value of B_k converges to a singular value σ_j of A, the corresponding SVD component $v_j^T x^{(k)}$ of the LSQR iterate $x^{(k)}$ converges to $\sigma_j^{-1} u_j^T b$, corresponding to a unit filter factor $f_j^{(k)}$. We will now show that in most situations the multiple singular values of B_k can only have a minor deteriorating effect on the converged SVD components of $x^{(k)}$, compared to the iterate computed in infinite precision (provided that both processes capture the same SVD components). Investigations of finite-precision effects in the algorithm SQMR (which is an alternative version of LSQR) can be found in [68] and [70] (see also [69]), but our analysis of LSQR seems to be new.

To simplify our notation in this section, the quantities $\hat{U}_{k+1}$, B_k, $\hat{V}_k$, and $x^{(k)}$ associated with LSQR denote the *computed quantities* in finite precision. Write the exact SVD of the computed B_k as

$$B_k = H_k \Gamma_k Q_k^T = \sum_{i=1}^{k} h_i \gamma_i q_i^T \qquad (6.44)$$

(where we have skipped an index k in the singular values and vectors), and let $\mathcal{I}_j$ denote the set of indices for which γ_i has converged to σ_j, i.e.,

$$\gamma_i \doteq \sigma_j , \qquad i \in \mathcal{I}_j . \qquad (6.45)$$

Here $\doteq$ denotes a close approximation, and note that $\mathcal{I}_j$ may be empty for some j.

The key to our analysis is the fact that in finite precision, Eq. (6.15) still holds within machine precision [273, p. 47], i.e., $A \hat{V}_k \doteq \hat{U}_{k+1} B_k$. If we insert (6.44) and (6.45) into this relation, then we obtain

$$A \hat{V}_k q_i \doteq \sigma_j \hat{U}_{k+1} h_i , \qquad i \in \mathcal{I}_j ,$$

from which we see that the left and right Ritz vectors $\hat{U}_{k+1} h_i$ and $\hat{V} q_i$ are proportional to u_j and v_j, respectively:

$$\left. \begin{array}{rcl} \hat{U}_{k+1} h_i & \doteq & c_i u_j \\ \hat{V}_k q_i & \doteq & c_i v_j \end{array} \right\} \qquad i \in \mathcal{I}_j , \qquad (6.46)$$

where c_i is a constant. Next, we write the right-hand side b as

$$b = \beta_1 \hat{U}_{k+1} H_k w_k , \qquad (6.47)$$

where $w_k^T = (h_{11}, \ldots, h_{1,k+1})$ is the first *row* of H_k. Using (6.46) we thus obtain

$$u_j^T b \doteq \beta_1 \sum_{i \in \mathcal{I}_j} c_i h_{1i} + \delta_j , \qquad (6.48)$$

---

practice, numerically-multiple singular values occur.

where δ_j is a (small) contribution coming from left Ritz vectors $\hat{U}_{k+1} h_i$ that have not converged yet—and their component along u_j is rarely large. Similarly, for $x^{(k)}$, we obtain

$$x^{(k)} = \beta_1 \hat{V}_k B_k^\dagger e_1^{(k+1)} \doteq \beta_1 \hat{V}_k Q_k \Gamma_k^{-1} w_k = \beta_1 \hat{V}_k \sum_{i=1}^k \frac{h_{1i}}{\gamma_i} q_i \qquad (6.49)$$

and, using (6.48) and (6.49),

$$\begin{aligned} v_j^T x^{(k)} &\doteq \beta_1 \sum_{i \in \mathcal{I}_j} c_i \gamma_i^{-1} h_{1i} + \epsilon_j \\ &\doteq \sigma_j^{-1} u_j^T b - \sigma_j^{-1} \delta_j + \epsilon_j \,, \end{aligned} \qquad (6.50)$$

where ϵ_j is due to right Ritz vectors $\hat{V}_k q_i$ that have not converged yet.

We see from (6.50) that once γ_i has converged to a singular value of A, the appearance of multiple copies of γ_i in B_k does not seriously deteriorate the component of $x^{(k)}$ along v_j, as long as $|\delta_j|$ and $|\epsilon_j|$ are small and σ_j is not too small, and thus preventing $|\sigma_j^{-1} \delta_j|$ from blowing up. We conclude that in these circumstances multiple singular values in B_k do not deteriorate $x^{(k)}$ dramatically. Their main consequence is a modest loss of accuracy in the converged SVD components of $x^{(k)}$ and a slow-down of the convergence, since $x^{(k)}$ stays almost unchanged in those steps where a new multiple singular value of B_k appears. And as a result, the residual norm $\|r^{(k)}\|_2$ exhibits plateaus when multiple singular values appear; see [68] and [69] for more on this subject.

The choice of starting vector $\hat{u}_1 = \beta_1^{-1} b$ in LSQR is crucial, not only for the efficiency of the algorithm, but also for its numerical stability in finite precision. When $\hat{u}_1 = \beta_1^{-1} b$, we use the formula $\hat{U}_{k+1}^T b = \beta_1^{-1} e_1^{(k+1)}$ to avoid the influence of the loss of orthogonality among the columns of $\hat{U}_{k+1}$. With all other choices of $\hat{u}_1$, we must compute $\hat{U}_{k+1}^T b$ explicitly, and this vector suffers from the loss of orthogonality among the Lanczos vectors. Consequently $v_j^T x^{(k)}$ can differ considerably from $\sigma_j^{-1} u_j^T b$ when we compute $x^{(k)}$ via the expression $x^{(k)} = \hat{V}_k B_k^\dagger \hat{U}_{k+1}^T b$. Hence, reorthogonalization is needed if we wish to use any other starting vector $\hat{u}_1$ than the normalized right-hand side.

Another consequence of the finite-precision arithmetic is the appearance of spurious singular values,[21] i.e., singular values of B_k that do not converge to any of the singular values of A, but instead either disappear or reappear in another part of the spectrum. These spurious singular values require special attention when Lanczos bidiagonalization without reorthogonalization is used to approximate the SVD of A [71, §§4.5 and 5.4] but, as we shall see, they do not impose any difficulties in connection with the LSQR algorithm.

---

[21]Spurious singular values of B_k are completely analogous to the spurious eigenvalues that arise in connection with Lanczos tridiagonalization without reorthogonalization; cf. [71, §4.5].

6.5. The LSQR Algorithm in Finite Precision

Consider the triplet (γ_i, h_i, q_i) of the bidiagonal matrix B_k. If the first component of h_i (and q_i) is pathologically small[22], then the corresponding Ritz vector $\hat{U}_{k+1} h_j$ has no component along the starting vector $\hat{u}_1$. Hence, either the triplet (γ_i, h_i, q_i) has converged to a singular triplet of A for which $\hat{U}_{k+1} h_i$ is orthogonal to $\hat{u}_1$ and therefore also orthogonal to b (which is not possible in exact arithmetic), or γ_i is a spurious singular value. In both cases, γ_i does not *correctly* contribute to the LSQR iterate $x^{(k)}$ since $h_{1i} \doteq 0$ in (6.49).

We note in passing that the identification test for spurious eigenvalues used by Cullum and Willoughby [71, §4.5] is closely related to the above. Let T denote a $k \times k$ symmetric tridiagonal matrix, let (λ, w) be an eigenpair of T, and let $\hat{T} = T(2{:}k, 2{:}k)$ (in MATLAB notation). Then Cullum and Willoughby identify λ as spurious if λ is isolated and also an eigenvalue of $\hat{T}$. The first element of w is given by [279, Corollary (7-9-4)]:

$$w_1^2 = \det(\hat{T} - \lambda I_{k-1}) / \xi'(\lambda), \qquad \xi = \det(T - \lambda I_k),$$

and since $\xi'(\lambda) \neq 0$ for an isolated λ, we see that $w_1 \doteq 0$ if λ is a spurious eigenvalue. Similarly, we can show that $h_{1i} \doteq 0$ if γ_i is a spurious singular value of B_k.

A third consequence of finite-precision arithmetic is that there is no guarantee that finite-precision Lanczos bidiagonalization approximates the same sequence of singular values of A as exact Lanczos bidiagonalization does. In other words, during the iterations the exact Lanczos process may miss certain large singular values of A, while the finite-precision Lanczos process may miss other singular values of A. Hence, it is generally difficult to relate the CGLS and LSQR iterates in infinite and finite precision.

The situation is somewhat simpler in those circumstances where the Lanczos processes in finite and infinite precision pick up the singular values of A in their natural order (i.e., there are no "missing" Ritz values). In this case, the appearance of multiple singular values of B_k slows down the convergence of $x^{(k)}$, and since spurious singular values do not contribute to $x^{(k)}$, it follows that the finite-precision LSQR iterate $x^{(k)}$ is practically identical to the iterate $\hat{x}^{(\ell_k)}$ after ℓ_k steps of the infinite-precision LSQR algorithm,

$$x^{(k)} \approx \hat{x}^{(\ell_k)}, \qquad k \geq \ell_k. \tag{6.51}$$

Hence, the loss of orthogonality in finite-precision arithmetic does not deteriorate the iteration vector dramatically, it merely slows down the convergence process, compared to exact arithmetic. In some iteration steps the finite-precision iterate $x^{(k)}$ stagnates, the corresponding finite-precision residual $\|r^{(k)}\|_2$ exhibits a plateau, and more iterations are needed to compute a regularized solution with a desired number of converged SVD components.

---

[22]Theoretically, if B_k is unreduced then the first component of h_i and k_i cannot be exactly zero [279, Theorem (7-9-5)], but in practice numerically-zero components occur.

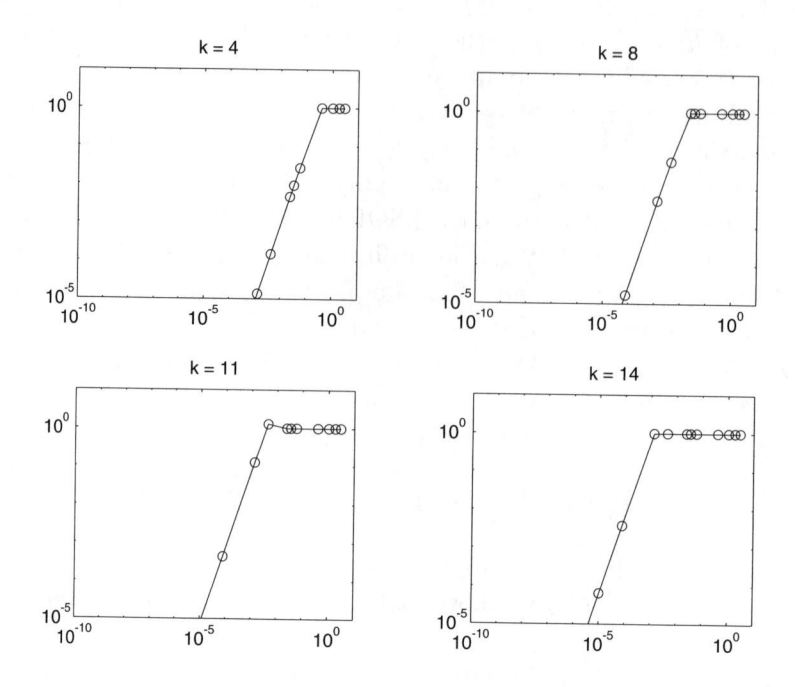

FIG. 6.6. *Filter factors $f_i^{(k)}$ as functions of the singular values σ_i. Same test problem* shaw *as in the figures in §6.4, but computed without reorthogonalization in LSQR.*

Figure 6.6 illustrates this point. With our test problem shaw, in which there are no multiple singular values in A, LSQR with and without reorthogonalization indeed picks up the singular values of A in their natural order. With reorthogonalization, exactly k SVD components are captured in the kth iteration (see Fig. 6.4), while without reorthogonalization the number of captured SVD components for $k = 4, 8, 11, 14$ is $\ell_k = 4, 7, 8, 9$, respectively. That is, in this example we have

$$x^{(4)} \approx \hat{x}^{(4)}, \quad x^{(8)} \approx \hat{x}^{(7)}, \quad x^{(11)} \approx \hat{x}^{(8)}, \quad x^{(14)} \approx \hat{x}^{(9)}.$$

This behavior of regularizing CG iterations is likely to be observed in large problems as well, as long as the singular values of A are well separated and the discrete Picard condition is satisfied; cf. our theory in §6.4. Numerical examples that illustrate these issues are presented in §§6.7.3 and 6.7.4.

6.6. Hybrid Methods

Although the CG process itself has a regularizing effect due to the order in which the Ritz values converge, there is no guarantee that this particular con-

6.6. Hybrid Methods

vergence always takes place—in particular in finite precision. Hence, it is sometimes convenient to combine the Lanczos bidiagonalization process in LSQR with an "inner regularization" algorithm applied to the bidiagonal $(k+1) \times k$ least squares problem $\min \|B_k\, \xi_k - \beta_1\, e_1^{(k+1)}\|_2$ (6.18), and with an "inner regularization parameter" that depends on the iteration number k. In this way, we can effectively filter out small Ritz values that may converge before all the large singular values of A have been captured. This is sometimes called a "hybrid method" [171, §7]. The cost of this feature is that we must save the left Lanczos vectors, i.e., the matrix V_k; this is not necessary in the LSQR algorithm.

These ideas were first explored by O'Leary and Simmons in their bidiagonalization-regularization procedure [267], and later described by Björck [35]. In these papers, TSVD and Tikhonov regularization are considered as "inner regularization" methods. The main difference between the approaches in the two papers is that [267] uses a general start vector for the Lanczos bidiagonalization process, while [35] uses the right-hand side b as start vector and explicitly uses the LSQR algorithm based on the relation $\hat{U}_{k+1}^T b = \beta_1\, e_1^{(k+1)}$.

The use of the "inner regularization" has an additional effect. When the iteration number k is large enough, and all large singular values—plus, possibly, some of the small—have been captured, then ideally the iteration vector $x^{(k)}$ does not deteriorate again, because of the "inner regularization." Thus, a hybrid method does not depend as critically on the stopping rule as CGLS and LSQR do, because the parameter choice is effectively moved from the iteration number k to the "inner regularization parameter."

Another motivation for adding "inner regularization" in LSQR is to compute (or approximate) regularized solutions with properties other than the CG-iterates $x^{(k)}$. For example, this idea is used in the Lanczos truncated total least squares (T-TLS) algorithm proposed in [124], where the bidiagonal least squares problem (6.18) in LSQR is replaced by a TLS problem:

$$\min \|(B_k,\, \beta_1\, e_1^{(k+1)}) - (\tilde{B}_k,\, \tilde{e}_k)\|_F \quad \text{subject to} \quad \tilde{B}_k\, \bar{\xi}^{(k)} = \tilde{e}_k\,, \quad (6.52)$$

in which $\tilde{B}_k$ and $\tilde{e}_k$ are generally full. In this way, we use the Lanczos bidiagonalization algorithm to compute approximations $\bar{x}^{(k)} = \hat{V}_k\, \bar{\xi}^{(k)}$ to the T-TLS solution $\bar{x}_k$ (3.22). In [124, Theorem 4.1] it is proved that the norm $\|\bar{x}_k\|_2$ is a nondecreasing function of k, while the TLS residual norm $\|(A,\, b) - U_{k+1}\, (\tilde{B}_k,\, \tilde{e}_k)\, V_k^T\|_F$ is a nonincreasing function of k.

The central idea in a hybrid method is that we reduce the large $m \times n$ problem to a much smaller $(k+1) \times k$ problem by projecting A onto suitably chosen left and right subspaces $\mathcal{U}_{k+1}$ and $\mathcal{V}_k$. This idea is also the basis for the TSVD and TGSVD methods, where $\mathcal{U}_{k+1}$ and $\mathcal{V}_k$ are spanned by the left and right singular vectors of A, in which case the subspaces are perfectly "tuned" to the problem. The hybrid method avoids the demanding SVD/GSVD computations by constructing the subspaces $\mathcal{U}_{k+1}$ and $\mathcal{V}_k$, as well as the projection B_k, iteratively by means of the Lanczos process such that $\mathcal{U}_{k+1}$ and $\mathcal{V}_k$ approximate the SVD subspaces—or the GSVD subspaces if A is "preconditioned" with $L_A^\dagger$. There are other algorithms for constructing subspaces $\mathcal{U}_{k+1}$ and $\mathcal{V}_k$ with the desired properties, such as subspace iterations. However, the Lanczos process gives the superior convergence; see the comparison in [362].

Once the subspaces $\mathcal{U}_{k+1}$ and $\mathcal{V}_k$ are large enough to capture all the desired information in the right-hand side b—plus, possibly, some undesired components associated with smaller Ritz values—then we can use sophisticated regularization algorithms to treat the bidiagonal problem with small dimensions. For example, computing the SVD of the bidiagonal $(k+1) \times k$ matrix B_k is not restrictive. Neither is it restrictive to apply a mollifier method of the form (5.49) to the $(k+1) \times k$ problem. On the other hand, hybrid methods do not seem to be suited to regularization methods that impose special constraints on the solution such as, e.g., the positivity constraint in maximum entropy regularization (see §5.6).

We emphasize again that without reorthogonalization, for stability reasons (cf. §6.5) the starting vector must be the normalized right-hand side, $\hat{u}_1 = \beta_1 b$, such that the right-hand side in the bidiagonal problem (6.18) can take the particularly simple form $\beta_1 e_1^{(k+1)}$.

In practice, it is desirable to keep the number of Lanczos steps small, because of the storage requirements for the right Lanczos matrix V_k. One way to achieve this is to use complete reorthogonalization among the Lanczos vectors, at the cost of an increasing number of computations as k increases. In [40], this computational overhead is reduced by repeated implicit restarts of the Lanczos process. An implicit restart after k Lanczos steps is mathematically equivalent to restarting the Lanczos process with the starting vector $(A A^T)^k b$, but the implicit restart is essential for the numerical stability. As the number of iterations and restarts increases, the Lanczos subspaces $\mathcal{U}_{k+1}$ and $\mathcal{V}_k$ converge to the SVD subspaces span$\{u_1, \ldots, u_{k+1}\}$ and span$\{v_1, \ldots, v_k\}$. Hence, if the "inner regularization method" is TSVD—as is the case in [40]—then the iteration vector from the hybrid method approaches the TSVD solution.

6.7. Iterative Regularization Methods in Action

As in the previous chapter, we conclude this chapter with numerical examples that illustrate the behavior of some of the iterative regularization methods, with special focus on regularizing CG iterations. The test problem used throughout the examples is again shaw from §1.4.3, discretized with $m = n = 64$.

6.7.1. Error Histories

The *error history* for an iterative method with iterates $x^{(k)}$ is a plot of the relative error norm $\|x^{\text{exact}} - x^{(k)}\|_2 / \|x^{\text{exact}}\|_2$ versus the iteration number k.

6.7. Iterative Regularization Methods in Action

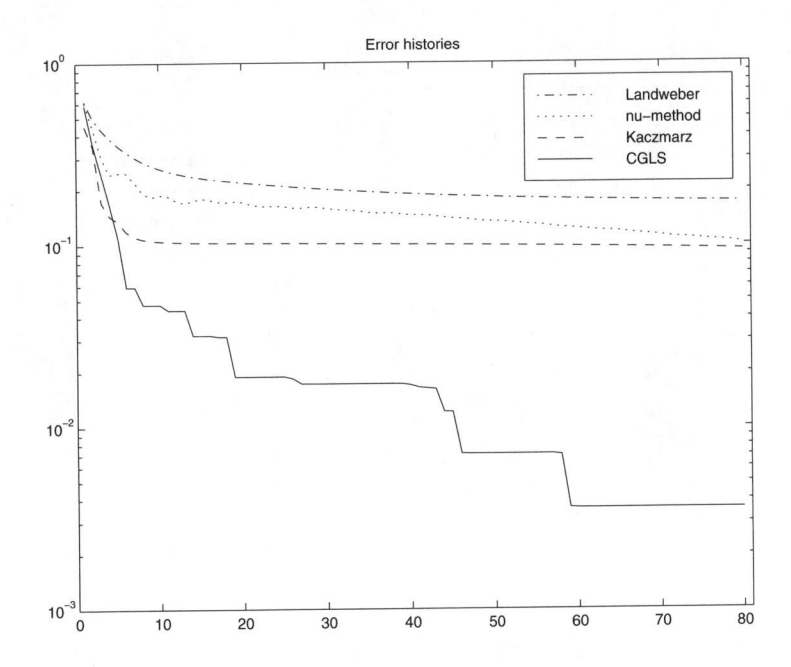

FIG. 6.7. *Error histories, i.e.,* $\|x^{\text{exact}} - x^{(k)}\|_2 / \|x^{\text{exact}}\|_2$ *versus k, for four iterative methods applied to the* shaw *test problem with exact right-hand side.*

Figure 6.7 shows the error histories for classical Landweber iteration (6.2) with $\omega = 0.15$, the ν-method (6.7), Kaczmarz's method (6.10), and CGLS (6.14) applied to the shaw test problem with exact right-hand side. Notice the nonmonotonic behavior of the error for the ν-method.

All four methods have an initial stage in which the convergence is faster than in the remaining iterations. In this initial stage, the solution's SVD components corresponding to the largest singular values are picked up. Overall, the first three methods converge much more slowly than CGLS, and the example clearly illustrates the potential power of regularizing CG iterations.

6.7.2. Four Iterative Solutions to shaw

In analogy to the six direct solutions to shaw from §5.8.3, we compute four iterative solutions to the same test problem with the same perturbed right-hand side $b = b^{\text{exact}} + e$ with noise level $\|e\|_2 / \|b^{\text{exact}}\|_2 = 0.01$. The four iterative methods are the same as above.

Figure 6.8 shows the error histories for the four methods. CGLS computes the "optimal" regularized solution (i.e., the one that minimizes the error norm $\|x^{\text{exact}} - x^{(k)}\|_2$) in only five iterations. Kaczmarz's method reaches its minimal error after seven iterations, but notice that this minimum is much larger

Error histories

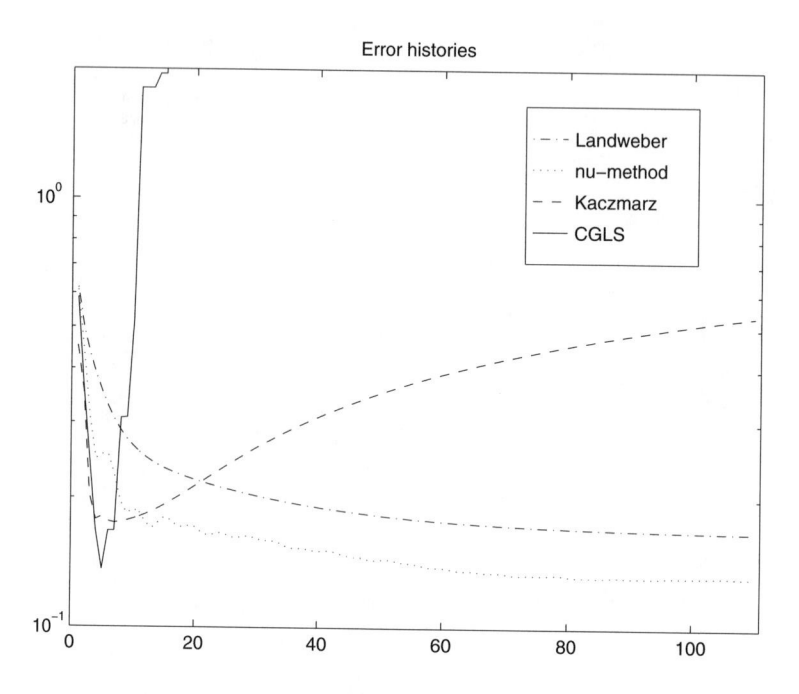

FIG. 6.8. *Error histories similar to the previous figure, except that the right-hand side is now perturbed. CGLS, Kaczmarz's method, and the ν-method reach their minimum error in 5, 7, and 94 iterations, respectively. Landweber's method did not reach a minimum within 3,000 iterations.*

than those for the other methods. The ν-method achieves its minimum after 94 iterations, while Landweber's method converges very slowly (no minimum was found in 3,000 iterations). Notice the rapid increase of the CG error as soon as k exceeds its optimal value.

The corresponding "optimal" solutions are shown as the solid lines in Fig. 6.9 (for Landweber's method, we show $x^{(120)}$) where, for reference, the exact solutions are also shown by the dotted lines. The solutions computed by both CGLS and the ν-method are of the same quality as the direct solutions in Fig. 5.4, while the other two solutions are poorer approximations to the exact solution.

6.7.3. Ritz Plots for Lanczos Bidiagonalization

In this example, we apply Lanczos bidiagonalization (6.16) to the matrix A from the shaw test problem. If the starting vector for the Lanczos bidiagonalization process is the normalized right-hand side, $\hat{u}_1 = b/\|b\|_2$, then we obtain the LSQR algorithm. We use the following three starting vectors:

6.7. Iterative Regularization Methods in Action

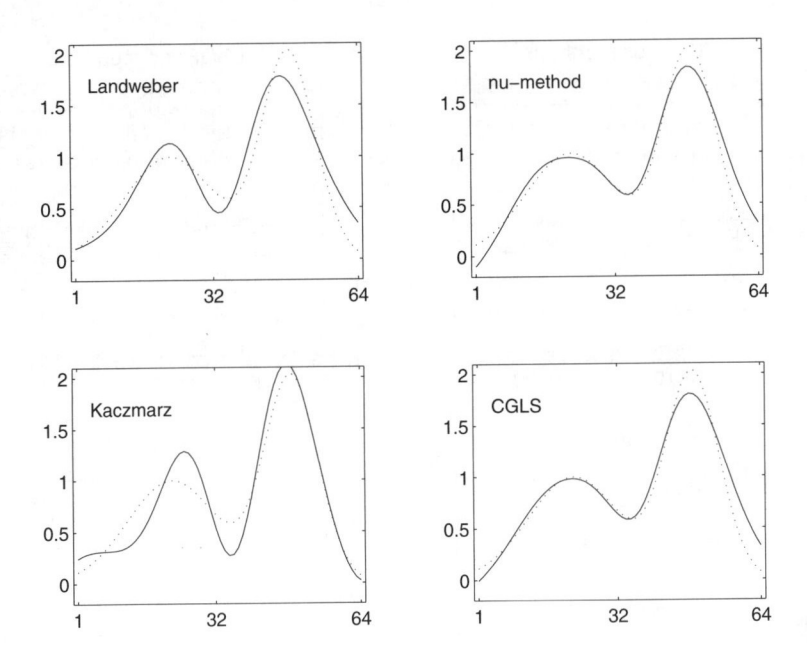

FIG. 6.9. *"Optimal" regularized solutions to the test problem* shaw *with noise for the four iterative regularization methods. Each solution minimizes the error norm* $\|x^{\text{exact}} - x^{(k)}\|_2$.

1. the normalized right-hand side $b/\|b\|_2$,
2. a normalized vector with unbiased normally distributed elements,
3. the "constant" vector $n^{-1/2}(1, \ldots, 1)^T$.

In addition, we apply the Lanczos bidiagonalization process with full reorthogonalization to the first starting vector, in order to illustrate the influence of the loss of orthogonality among the Lanczos vectors.

Figure 6.10 shows plots of all the singular values of the bidiagonal matrix B_k (shown as circles) versus the iteration number k. We recall that the Ritz values $\theta_j^{(k)}$ are the squares of these singular values—hence the name Ritz plots. In addition, the singular values of the matrix A are represented by the solid horizontal lines. The figure illustrates how an increasing number of singular values of A are being approximated by the singular values of B_k as the iteration number k increases.

The top left part of Fig. 6.10 illustrates that with full reorthogonalization (which simulates computations in exact arithmetic) we capture, within the accuracy of the plot, at least $k - 1$ singular values in the kth stage.

The top right part of Fig. 6.10 illustrates the influence of the loss of orthogonality among the Lanczos vectors, which becomes significant for $k > 6$;

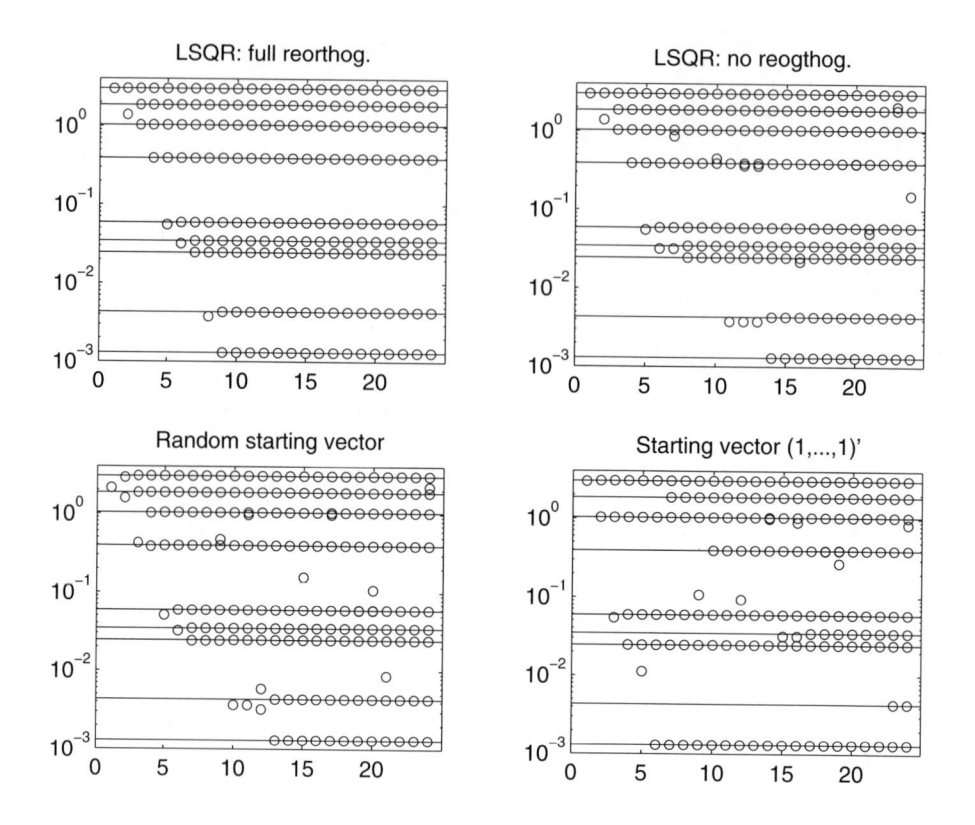

FIG. 6.10. *Ritz plots, i.e., plots of the singular values of the bidiagonal matrix B_k (circles) versus the iteration number k, for the* shaw *test problem. The solid lines represent the singular values of the matrix A. The two top figures, in which the starting vector is $\hat{u}_1 = b/\|b\|_2$, correspond to the LSQR algorithm. The two bottom figures correspond to a random and a "constant" starting vector.*

eight iterations are necessary to capture the first seven singular values, eleven iterations are needed to capture the eighth singular value, and an additional three iterations are needed to capture the ninth singular value. We also see occasional spurious singular values for k equal to 7, 10, 16, 21, 23, and 24.

The figure does not show the multiplicity of the singular values of B_k which—as mentioned in §6.5—is closely connected with the convergence of a Ritz value. After 25 iterations, the multiplicities of the first 7 singular values of B_{25} are 5, 4, 3, 3, 1, 2, and 3, respectively, and we recall that the fifth singular value (with multiplicity 1) is a spurious singular value. As discussed in §6.5, these multiple and spurious singular values do not lead to any significant deterioration of the iteration vector $x^{(k)}$, compared to computations in exact arithmetic.

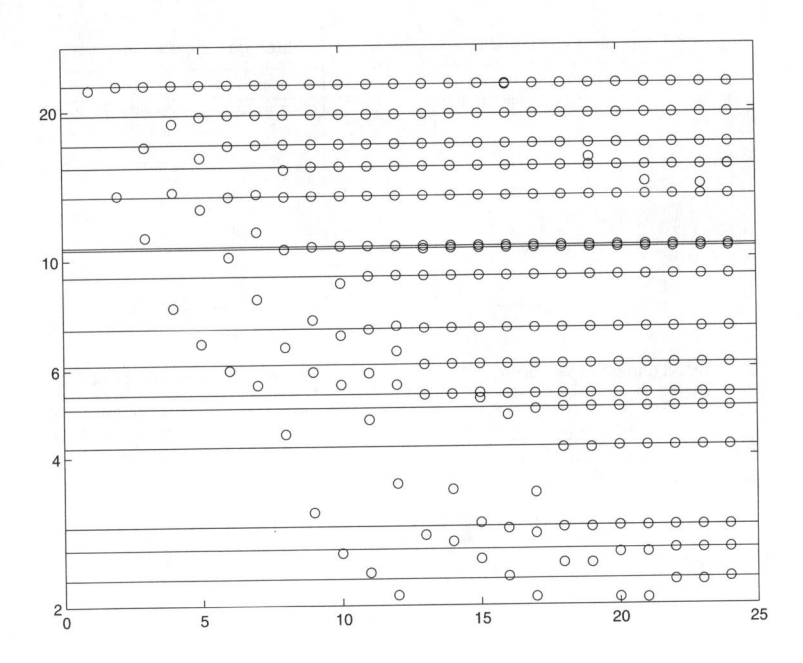

6.7. Iterative Regularization Methods in Action

FIG. 6.11. *Ritz plot of the singular values of B_k (circles) for the 256×256 image deblurring problem in the next chapter. The singular values of A are shown as solid lines.*

The influence of starting vectors other than the normalized right-hand side is shown in the bottom part of Fig. 6.10 (no reorthogonalization is used). The random vector gives an iteration sequence which resembles that for LSQR, while the "constant" starting vector gives a completely different sequence: every second SVD component of the vector $(1, \ldots, 1)^T$ is small, and the corresponding Ritz values converge much more slowly than the other Ritz values.

The above Ritz plot for LSQR without reorthogonalization illustrates how an increasing number of singular values of A are approximated as the iteration number k increases, but it does not give a realistic picture of the convergence behavior for large-scale problems. Therefore, we also applied LSQR without reorthogonalization to the 256×256 image deblurring problems from §7.7.2 in the next chapter. Figure 6.11 shows the Ritz plot for the first 24 iterations. For this problem, many iterations are needed before a new singular value of A is approximated within the precision of the plot, and the singular values are not captured in their natural order (e.g., σ_5 is captured before σ_4). The sixth singular value of B_k initially converges to the average of the close pair σ_6 and σ_7 of A, until the 13th iteration where the Ritz value splits into two. At $k = 24$, the first 4 singular values of B_{24} have multiplicity 2.

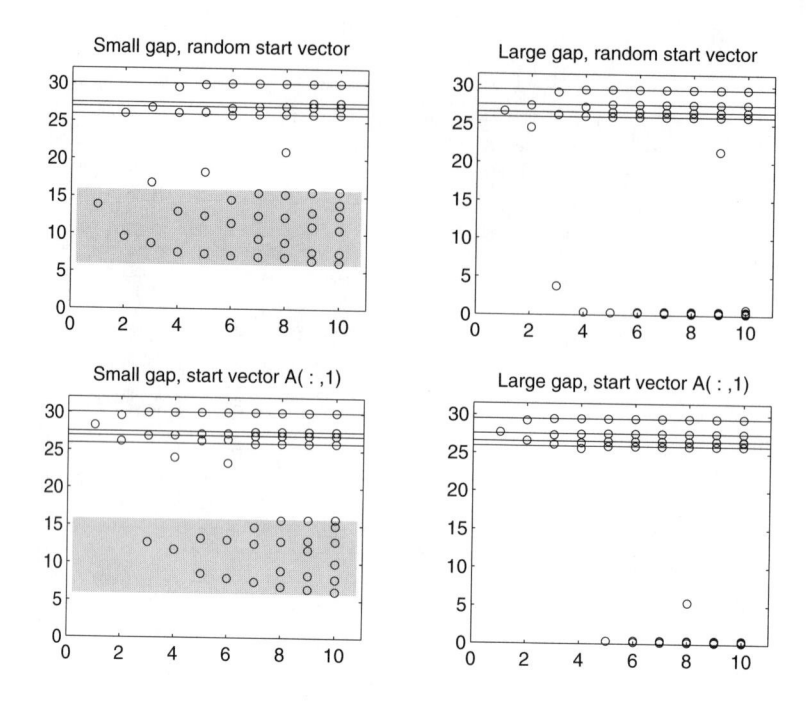

FIG. 6.12. *Ritz plots of the singular values of B_k (circles) for the 97×32 Hankel matrix representing two sinusoids in noise. The four large singular values of A are shown as solid lines, and the remaining singular values of A lie in the shaded regions in the two left plots and close to zero in the two right plots. Some spurious singular values are present.*

To illustrate the behavior of the Lanczos bidiagonalization algorithm in the presence of a gap in the singular value spectrum, we applied this algorithm to the Hankel matrix A (1.18) from the signal processing example in §1.4.1. The matrix has four large singular values, and the size of the remaining singular values depends on the noise level. We use two noise levels, leading to two problems with a small and a large gap in the singular value spectrum. Moreover, we use two different starting vectors for the Lanczos process, namely, a random vector and the first column of a noise-free version of A. The first vector has quite large components along all the left singular vectors u_i of A, and the second has four large components along $u_1, \ldots, u_4$, while the remaining components are small.

According to the theory in Theorem 6.4.1, the Lanczos process in infinite precision is guaranteed to capture the largest p singular values of A in their natural order if the gap between σ_p and σ_{p+1} is large and all the Fourier coefficients $u_i^T b$ are small for $i > p$. The Ritz plots shown in Fig. 6.12 agree with this theory. When the gap is small, or when the starting vector has large

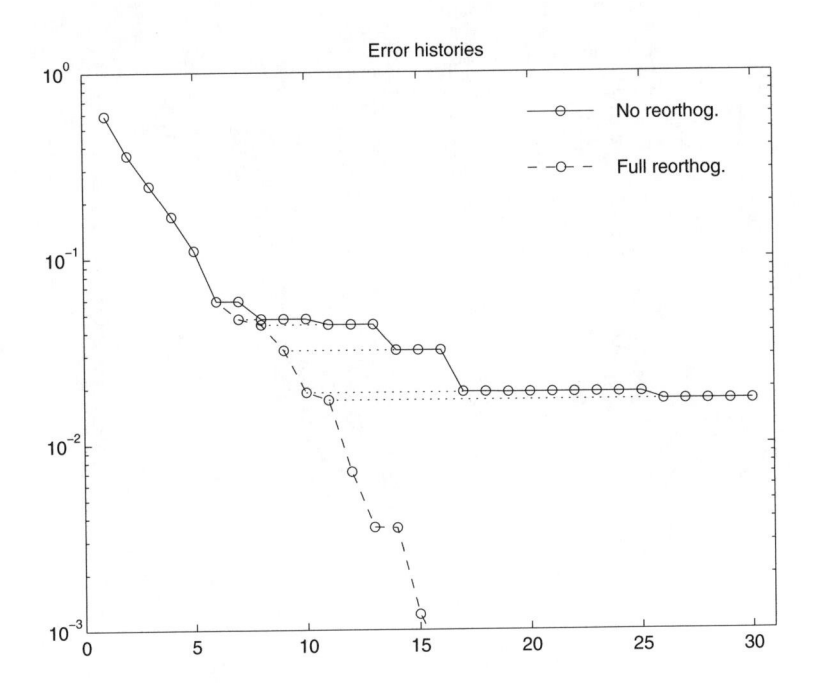

6.7. Iterative Regularization Methods in Action

FIG. 6.13. *Error histories for LSQR with and without reorthogonalization applied to the* shaw *test problem with exact right-hand side. The dotted lines show the relation between the iteration vectors of the two processes.*

Fourier components $u_i^T b$ for $i > 4$, then singular values smaller than σ_4 are approximated before the first four singular values are captured.

6.7.4. The Influence of Reorthogonalization in LSQR

As explained in §6.5 and illustrated above by means of the Ritz plots, the consequence of the loss of orthogonality is essentially a delay of the iterations, compared to computations in exact arithmetic. This can also be illustrated by means of the error histories for LSQR with and without reorthogonalization.

Figure 6.13 shows the error histories for the first 30 iterations of LSQR with and without reorthogonalization, applied to the shaw test problem with exact right-hand side. The iterations vectors are practically identical during the first 6 iterations, and then the error history for LSQR without reorthogonalization (i.e., standard LSQR, shown as the solid line) starts to exhibit plateaus when multiple or spurious singular values of B_k start to appear. This error history is fully in accordance with the Ritz plot in the top right part of Fig. 6.10, where a new singular value of A is captured in iterations 8, 11, and 14.

With reorthogonalization, the Lanczos bidiagonalization process captures approximately one additional singular value of A in each iteration for this test

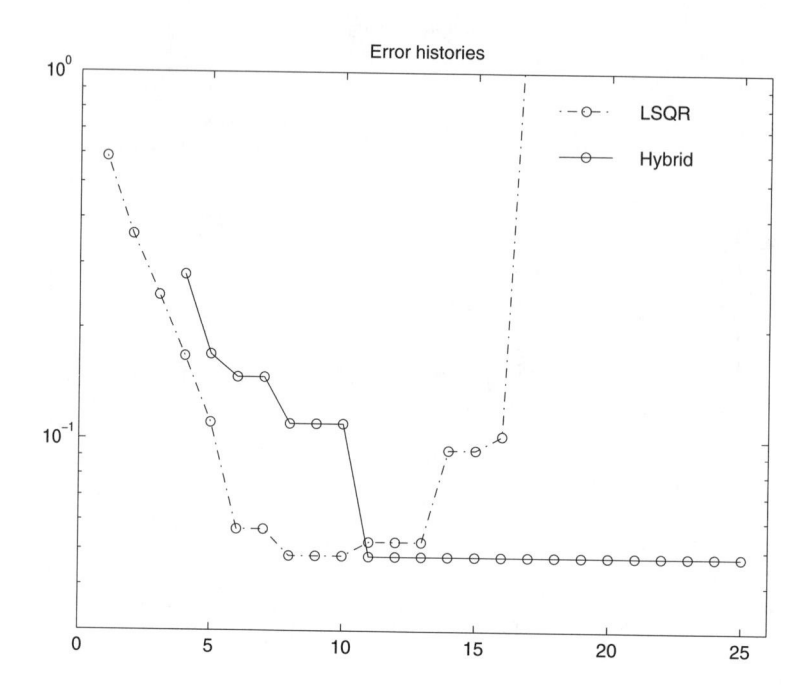

FIG. 6.14. *Error histories for LSQR and a hybrid method using TSVD in each iteration step of the Lanczos bidiagonalization process.*

problem, and the corresponding error history (shown as the dashed line in Fig. 6.13) reflects this: there are no plateaus. If $\hat{x}^{(k)}$ denotes the iteration vector computed with reorthogonalization, then $\hat{x}^{(k)} \approx x^{(k)}$ for $k = 1, \ldots, 6$, while

$$\hat{x}^{(7)} \approx x^{(8)}, \ \hat{x}^{(8)} \approx x^{(11)}, \ \hat{x}^{(9)} \approx x^{(14)}, \ \hat{x}^{(10)} \approx x^{(17)}, \ \hat{x}^{(11)} \approx x^{(26)}.$$

These relations are illustrated in Fig. 6.13 by the horizontal dotted lines.

6.7.5. A Hybrid Method in Action

The last example illustrates the use of a hybrid method, according to the terminology in §6.6. In each step of the Lanczos bidiagonalization algorithm without reorthogonalization, we use TSVD as the "inner regularization method," and the corresponding value of the truncation parameter is chosen by means of generalized cross-validation (discussed in §7.4). The test problem is the same as before, and the noise level in this example is $\|e\|_2/\|b\|_2 \approx 4 \cdot 10^{-4}$; this small value was chosen in order to fully illustrate the behavior of the hybrid method.

6.7. Iterative Regularization Methods in Action

In Figure 6.14 we compare the error history of the hybrid method (solid line) with that of the ordinary LSQR method (dash-dotted line). The hybrid method starts after $k = 4$ Lanczos steps, in order for the parameter-choice method to work properly. The hybrid method needs three more iterations than LSQR to reach the minimal error at $k = 11$. While the error of the LSQR method increases as soon as k exceeds the optimal value, the error of the hybrid method stays at its minimum once k is large enough.

7

Parameter-Choice Methods

So far, we have discussed various algorithms for computing a regularized solution. Every algorithm has its advantages and disadvantages in terms of implementation issues, filter properties, etc. However, no regularization method is complete without a method for choosing the regularization parameter, either the continuous parameter λ or the discrete parameter k, and in this chapter we discuss several parameter-choice methods and their implementations. Surveys of parameter-choice methods are rare in the literature; we are currently aware of [171] and [333].

Whenever possible, we discuss these methods in terms of a continuous regularization parameter λ, but we emphasize that similar results hold when λ is replaced by a discrete regularization parameter k. We will also restrict our discussion to the standard-form case whenever the extensions to the general case are obvious, and we use the notation $x_\lambda = x_{I_n,\lambda}$.

Most of the parameter-choice methods are based on residual norms and, in the case of the L-curve, also on the solution's "size" Ω, typically its seminorm. When solving general-form problems via a standard-form transformation, it is therefore important to recall the relations

$$\|L\,x\|_2 = \|\bar{x}\|_2 \quad \text{and} \quad \|A\,x - b\|_2 = \|\bar{A}\,\bar{x} - \bar{b}\|_2 \;;$$

cf. (2.39) and (2.45). These relations ensure that application of a norm-based parameter-choice rule to the original problem with A and b, or to the standard-form problem with $\bar{A}$ and $\bar{b}$, yields exactly the same regularization parameter.

For certain problems it may be advantageous to use more information about the residual vector than its norm. For example, the residual's autocorrelation may be useful; see [13] for an example. Such methods do not seem to have found widespread use.

We first describe some parameter-choice methods that make explicit use of the norm of the error in the right-hand side. Then we turn to methods that do not require this information, in particular, the generalized cross-validation method and the L-curve method. We finish the chapter with two extensive numerical examples that illustrate the behavior of the parameter-choice methods used in connection with some of the direct and iterative regularization methods presented in the previous two chapters.

7.1. Pragmatic Parameter Choice

In the literature on parameter-choice methods, the focus is often on the convergence of the regularized solution as the errors in the right-hand side tend to zero; see, e.g., [108], [162], and [356]. That is, given a regularization scheme for computing $x_{\text{reg}}(\lambda)$ and a parameter-choice method for computing λ, often based explicitly on the error norm $\|e\|_2 = \|b - b^{\text{exact}}\|_2$, how fast $x_{\text{reg}}(\lambda)$ converges to x^{exact} as $\|e\|_2 \to 0$ and $\lambda \to 0$ is investigated. In particular, it is desirable to develop methods whose convergence rate is as fast as possible, which leads to the study of "order optimality" of regularization methods; cf. [73] and [111, §3.2].

Another aspect of parameter-choice methods is equally important. Since the errors are "fixed" in most experiments, and since it is often the case that the experiments cannot be repeated—either due to practicalities or due to the cost of setting up the experiment—we are interested in squeezing out as much information as possible from the given data. Consequently, we want to determine the regularization parameter which balances the regularization error and the perturbation error in the computed solution for the given problem, in order to minimize the total error.

To be specific, we write the error in the regularized solution $x_{\text{reg}} = A^\# b$ as

$$\begin{aligned}
x^{\text{exact}} - x_{\text{reg}} &= A^\dagger b^{\text{exact}} - A^\# b \\
&= (A^\dagger - A^\#) b^{\text{exact}} - A^\# e \ ; \quad (7.1)
\end{aligned}$$

see also Eq. (4.40). Here, $(A^\dagger - A^\#) b^{\text{exact}}$ is the *regularization error* and $A^\# e$ is the *perturbation error*. When the discrete Picard condition is satisfied then, on the average, the regularization error decreases and the perturbation error increases as less regularization is introduced. At the optimal regularization parameter for the given problem, the two error components balance each other, and we are able to resolve all singular value decomposition (SVD) components in the regularized solution greater than the intrinsic resolution limit η_{res}; cf. (4.1) and (4.2). It is for this reason that we want to balance the two error components in x_{reg}, and it is this pragmatic point of view that is the basis for our discussion in this chapter.

From the discussion in Chapter 4 about filter factors, we see that the proper choice of the regularization parameter is a matter of choosing the right cutoff for the filter factors, i.e., the break point in the SVD/GSVD (generalized SVD) spectrum where we want the damping to set in. Let λ_{opt} denote the optimal regularization parameter that corresponds to minimization of the error $x^{\text{exact}} - x_{\text{reg}}$ (7.1). Then our goal is to derive parameter-choice algorithms that approximate this λ_{opt} as accurately and reliably as possible.

For problems with well-determined numerical rank, the choice of regularization parameter is strongly connected to the numerical ϵ-rank r_ϵ defined in §3.1. If we choose the truncation parameter $k < r_\epsilon$, then obviously we leave out too

7.1. PRAGMATIC PARAMETER CHOICE

much information and the regularization error dominates. On the other hand, if $k > r_\epsilon$ then the perturbation error may dominate, due to the division by the small singular values. Thus, it is natural to choose the regularization parameter such that precisely the largest r_ϵ SVD or GSVD components are retained in the regularized solution. Notice that r_ϵ, and thus the regularization parameter, for these problems is independent of the right-hand side.

For discrete ill-posed problems, the choice of the regularization parameter is more complicated because both the perturbation error and the regularization error are slowly varying functions of the regularization parameter. Within a certain range of regularization parameters, there is usually no particular choice which stands out as "natural" compared to the other choices.

Parameter-choice methods can roughly be divided into two classes depending on their assumptions about the error norm $\|e\|_2$. The two classes can be characterized as follows.

1. Methods based on knowledge, or a good estimate, of $\|e\|_2$.

2. Methods that do not require $\|e\|_2$, but instead seek to extract this information from the given right-hand side.

When reliable information about $\|e\|_2$ is available, then it is crucial to make use of this information, and this is the heart of the discrepancy principle and related methods. When no particular information about $\|e\|_2$ is available, then it is more difficult to devise a reliable parameter-choice method. Our experience is that it may often be advantageous to monitor several strategies and base the choice of regularization parameter on the output of all these strategies.

A fundamental result about parameter-choice methods says that if an operator is compact and infinite-dimensional, then it is not possible to construct a convergent parameter-choice strategy (i.e., satisfying $x_{\text{reg}} \to x^{\text{exact}}$ as $\|e\|_2 \to 0$) that does not make explicit use of the error norm $\|e\|_2$. See [109] or [111, Theorem 3.3] for an accessible proof.

From our pragmatic point of view, most regularization problems encountered in practice are finite-dimensional, because data are finite-dimensional, and usually the measurement setup puts a lower bound on the noise level. Hence it still makes sense to develop and use parameter-choice methods for these problems that do not make explicit use of $\|e\|_2$ (sometimes called "heuristic methods"; cf. [111, §4.5]). There are several such methods in use today, e.g., the GCV method (§7.4), and there are many years of experience saying that these methods work well in a variety of applications.

To some extent it is possible to generalize the concept of numerical rank to discrete ill-posed problems. For example, the singular value spectrum of A in a discrete ill-posed problem may resemble that of a rank-deficient problem in that there is a cluster of large singular values while the remaining singular values decay to zero. For such problems, one can consider the number of large singular values as the numerical rank of the problem. See [58] for an example in helioseismology.

Generally, the singular value spectrum for an ill-posed problem does not have such features. A general definition of numerical rank for a discrete ill-posed problem must therefore involve both the matrix and the right-hand side, because the decay of the Fourier coefficients for the right-hand side plays such an important role (see §4.5 on the discrete Picard condition). The numerical rank should also reflect the errors that are present in A and b.

The definition of an effective numerical rank given here is related to the optimal accuracy that can be obtained in the solution with the given data. Inspired by [139], we introduce the *sum of filter factors* $\rho(\lambda)$ as a function of the regularization parameter λ,

$$\rho(\lambda) = \sum_{i=1}^{n} f_i , \quad L = I_n , \qquad \rho(\lambda) = n - p + \sum_{i=1}^{p} f_i , \quad L \neq I_n . \qquad (7.2)$$

The term $n - p$ for $L \neq I_n$ is included because the unregularized component $x_0 = \sum_{i=p+1}^{n} u_i^T b \, x_i$ in the regularized solution (4.5) corresponds to $n - p$ unit filter factors. Let x_λ denote the regularized solution as a function of λ, and let λ_{opt} denote the optimal regularization parameter that minimizes the error in the solution x_λ,

$$\|x^{\mathrm{exact}} - x_{\lambda_{\mathrm{opt}}}\|_2 \leq \|x^{\mathrm{exact}} - x_\lambda\|_2 . \qquad (7.3)$$

Occasionally, one may prefer to use the seminorm $\|L \cdot \|_2$ in (7.3). Then we define the *effective numerical rank* r_{eff} of the given discrete ill-posed problem by

$$r_{\mathrm{eff}} = \rho(\lambda_{\mathrm{opt}}) . \qquad (7.4)$$

Thus, r_{eff} depends on both the underlying ill-posed problem and the errors in the given problem, and r_{eff} is related to the number of linearly independent pieces of information that can be extracted from a given problem with a given amount of errors. In general, r_{eff} is a real number which takes into account that the regularized solution generally has nonzero—but possibly very small—components for all (generalized) singular values.

The effective numerical rank r_{eff} is closely connected to the effective resolution limit η_{res} defined in (4.1). Specifically, if $\bar{r}_{\mathrm{eff}}$ is the closest integer to r_{eff}, then we have the relation

$$\eta_{\mathrm{res}} \approx |v_{\bar{r}_{\mathrm{eff}}}^T x^{\mathrm{exact}}|$$

because $\bar{r}_{\mathrm{eff}}$ is approximately equal to the number of SVD components in the regularized solution that we are able to recover.

Instead of relating the effective rank to the optimal accuracy in the solution, one can relate it to the desired accuracy in the solution. Gilliam, Lund, and

7.2. THE DISCREPANCY PRINCIPLE

Vogel [139] defined the effective rank[23] r_α as the value of $\rho(\lambda)$ that corresponds to the λ for which the expected norm of the perturbation error $A^\# e$ equals a specified tolerance α,

$$\mathcal{E}\left(\|A^\# e\|_2^2\right) = \alpha^2 \,,$$

where $A^\#$ depends on λ. Clearly, $r_\alpha \geq r_{\text{eff}}$, because the definition of r_α neglects the regularization error in x_λ which increases with λ.

We emphasize that both r_{eff} and r_α are mainly theoretical tools, because they require knowledge about x^{exact}.

7.2. The Discrepancy Principle

The most widespread $\|e\|_2$-based method is the *discrepancy principle*, usually attributed to Morozov [254]. If the ill-posed problem is consistent in the sense that $A\,x^{\text{exact}} = b^{\text{exact}}$ holds exactly, then the idea is simply to choose the regularization parameter λ such that the residual norm is equal to an a priori upper bound δ_e for $\|e\|_2$, i.e.,

$$\|A\,x_\lambda - b\|_2 = \delta_e \,, \qquad \text{where} \qquad \|e\|_2 \leq \delta_e \,. \tag{7.5}$$

The regularized solution computed by means of (7.5) corresponds to that point on the L-curve where it intersects the vertical line given by (7.5). The problem in (7.5) is identical to the least squares problem with a quadratic constraint in (5.15), and algorithms for solving (7.5) with respect to λ can be found in, e.g., [55] and [255, §26]. For a discrete regularization parameter k, one should use the smallest k for which $\|A\,x_k - b\|_2 \leq \delta_e$.

If the perturbation e has zero mean and covariance matrix $\sigma_0^2 I_m$, then the expected value of $\|e\|_2$ is $\sigma_0 \sqrt{m}$. Hence, if we assume that the exact system is consistent and that $\|e\|_2 < \|b^{\text{exact}}\|_2$, then Eq. (7.5) corresponds to choosing a regularized solution that appears on the L-curve slightly to the right of the corner which is approximately at $(\sigma_0\sqrt{m-n+p},\,\|L\,x^{\text{exact}}\|_2)$.

The *generalized discrepancy principle* [255, p. 52] also takes into account errors E in the matrix A, as well as the incompatibility measure $\delta_0^{\text{exact}} = \|b^{\text{exact}} - A^{\text{exact}}\,x^{\text{exact}}\|_2$ when the exact system is incompatible. Let δ_e and δ_E denote upper bounds for $\|e\|_2$ and $\|E\|_2$, respectively:

$$\|e\|_2 \leq \delta_e \,, \qquad \|E\|_2 \leq \delta_E \,.$$

Then the generalized discrepancy principle amounts to choosing λ such that[24]

$$\|A\,x_\lambda - b\|_2 = \delta_0^{\text{exact}} + \delta_e + \delta_E \,\|x^{\text{exact}}\|_2 \,, \tag{7.6}$$

---

[23] Gilliam, Lund, and Vogel also use the alternative term *effective number of degrees of freedom*, but here we prefer to reserve this term for the quantity $\mathcal{T}(\lambda) = m - \rho(\lambda)$; cf. (7.10).

[24] The sharper right-hand side $((\delta_0^{\text{exact}})^2 + (\delta_e + \delta_E\|x^{\text{exact}}\|_2)^2)^{1/2}$ also appears in the literature.

and this criterion can be used for all L. If the user has an a priori upper bound for $\|x_\lambda\|_2$, then this upper bound can be substituted for $\|x^{\text{exact}}\|_2$ in (7.6). Estimates for $\|E\|_2$, based solely on statistical information about the errors in A, are discussed in §3.1.

An alternative formulation to (7.6) is [255, p. 58]:

$$\|A\,x_\lambda - b\|_2 = \delta_0^{\text{exact}} + \delta_e + \Delta_{E,L}\,\|L\,x_\lambda\|_2 \,, \qquad (7.7)$$

where $\Delta_{E,L}$ is an upper bound for $\max_{L\,x \neq 0}\{\|E\,x\|_2/\|L\,x\|_2\}$, i.e., the largest generalized singular value of the pair (E, L). In particular, if $L = I_n$ then $\Delta_{E,L} = \delta_E$. The regularized solution computed by means of (7.7) corresponds to that point on the L-curve where it intersects the line given by (7.7). From a practical point of view, the approach in (7.7) is more appealing than that in (7.6) because it does not involve an a priori bound for $\|x_{\text{LS}}\|_2$ or $\|x_\lambda\|_2$.

All three formulations (7.5), (7.6), (7.7) of the discrepancy principle are based on a conservative choice of the residual norm $\|A\,x_{\text{reg}} - b\|_2$. In terms of the L-curve, they all produce regularized solutions appearing to the right of the corner, and this is particularly pronounced if $\delta_E \neq 0$. This is one explanation for the claim that the discrepancy principle leads to oversmoothing—see, e.g., [333], [211, p. 96], and [366, p. 63]—and the more inaccurate the bounds, the more oversmoothing is introduced.

Underestimation of $\|e\|_2$ is even more serious than overestimation: we see from the steep part of the L-curve that an underestimate $\delta_e < \|e\|_2$ can lead to a severely undersmoothed solution with a very large norm.

As a remedy for the oversmoothing in the white-noise case, one can replace the discrepancy principle in (7.5) with a *compensated discrepancy principle* of the form

$$\begin{aligned}\|A\,x_\lambda - b\|_2 &= \left(\|e\|_2^2 - \sigma_0^2\,\text{trace}(A\,A^{\#})\right)^{1/2} \\ &= \sigma_0\,(m - \rho(\lambda))^{1/2} = \sigma_0\,\mathcal{T}(\lambda)^{1/2}\,.\end{aligned} \qquad (7.8)$$

This compensated discrepancy principle was originally derived by Turchin [341, Eq. (3.8)]. It compensates for the fact that although $\sigma_0\sqrt{m}$ is a valid estimate for $\|e\|_2$, neither $\sigma_0\sqrt{m}$ nor $\|e\|_2$ may be a valid estimate of the residual norm $\|A\,x_\lambda - b\|_2$ (see also [333, §III.D]). This and other variants of the discrepancy principle are compiled and compared in [74, §4]. Notice that the solution of (7.8) is complicated by the fact that $A^{\#}$ depends on λ.

Assume again that e has zero mean and covariance matrix $\sigma_0^2\,I_m$. If $\|e\|_2$ or σ_0 is not known a priori, then σ_0 can be estimated by monitoring the function $\mathcal{V}$ [366, p. 68] defined by

$$\mathcal{V}(\lambda) = \frac{\|A\,x_\lambda - b\|_2^2}{\mathcal{T}(\lambda)}\,. \qquad (7.9)$$

Here, for convenience, we have defined another function $\mathcal{T}$,

$$\mathcal{T}(\lambda) = \mathrm{trace}(I_m - A\,A^{\#}) = m - (n-p) - \sum_{i=1}^{p} f_i = m - \rho(\lambda)\,, \qquad (7.10)$$

where $\rho(\lambda)$ is defined in (7.2). The quantity $\mathcal{T}$ can be considered as the *effective number of degrees of freedom* [366, pp. 63, 68]. If $\mathcal{V}$ is plotted versus λ, then on a broad scale the graph of $\mathcal{V}$ consists of a flat part and a part where $\mathcal{V}(\lambda)$ increases with λ. The plateau at the flat part is an estimate of σ_0^2. The estimate of $\|e\|_2^2$ is therefore equal to m times the value of $\mathcal{V}(\lambda)$ at the plateau.

7.3. Methods Based on Error Estimation

In this section we discuss a class of parameter-choice methods which are based on estimates for the total error (7.1) in the regularized solution x_{reg}. The goal is to minimize this error estimate and thus approximately balance the regularization and perturbation errors in x_{reg} (see §4.6 for more details about this balance).

A method developed independently by Gfrerer [138] and Raus [290] for Tikhonov regularization is based on an error estimate of the form

$$\|x^{\mathrm{exact}} - x_\lambda\|_2 \le 2\left(\|(A^\dagger - A^{\#})\,b^{\mathrm{exact}}\|_2^2 + 2\lambda^{-2}\|e\|_2^2\right)^{1/2},$$

and the right-hand side is minimized when λ satisfies the nonlinear equation $\lambda^6 (b^{\mathrm{exact}})^T (A\,A^T + \lambda^2 I_m)^{-3} b^{\mathrm{exact}} = \|e\|_2^2$. To obtain a practical algorithm, b^{exact} and $\|e\|_2$ are replaced by b and δ_e, which leads to the following nonlinear equation for computing λ:

$$\lambda^3 \left(b^T (A\,A^T + \lambda^2 I_m)^{-3} b\right)^{1/2} = \delta_e\,. \qquad (7.11)$$

See [171, §5.3] for more details as well as a numerical algorithm for solving (7.11). It is interesting to note that the left-hand side of (7.11) can also be written as $\left((b - A\,x_\lambda)^T (b - A\,x_\lambda^{(2)})\right)^{1/2}$, where $x_\lambda^{(2)}$ is the iterated Tikhonov solution after one additional iteration; cf. §6.2. Thus, the criterion (7.11) can be reformulated as

$$\|A\,x_\lambda - b\|_2 = \left(\delta_e^2 - (A\,x_\lambda - b)^T A\,(x_\lambda^{(2)} - x_\lambda)\right)^{1/2},$$

showing that the Gfrerer–Raus method is closely related to the discrepancy principle (7.5). Engl and Gfrerer [108] have analyzed the above technique and extended it to other direct and iterative regularization methods.

Hanke and Raus [174] have derived alternative upper bounds for the error $\|x^{\mathrm{exact}} - x_{\mathrm{reg}}\|_2$ that are independent of b^{exact}. For Tikhonov regularization their analysis leads to the following error estimate to be minimized:

$$\begin{aligned}
\|x^{\text{exact}} - x_\lambda\|_2 &\approx \lambda^2 \left(b^T (A\,A^T + \lambda^2 I_m)^{-3} b\right)^{1/2} \\
&= \lambda \left((b - A\,x_\lambda)^T (b - A\,x_\lambda^{(2)})\right)^{1/2}.
\end{aligned} \quad (7.12)$$

Similarly, their analysis leads to the following error estimate to be used in connection with regularizing conjugate gradient (CG) iterations:

$$\|x^{\text{exact}} - x^{(k)}\|_2 \approx \mathcal{P}_k(0)^{1/2} \|A\,x^{(k)} - b\|_2, \quad (7.13)$$

where the CG polynomial $\mathcal{P}_k$ is defined in Theorem 6.3.2. We emphasize that $\mathcal{P}_k(0)$ can be computed with almost no overhead during the CG iterations by means of the recurrence relation (6.26). Using formula (6.22) for the CG filter factors, it is easy to show that if $\theta_j^{(k)}$ are the Ritz values associated with the kth step, then

$$\mathcal{P}_k(0) = \sum_{j=1}^{k} \left(\theta_j^{(k)}\right)^{-1} = \|B_k^\dagger\|_F^2,$$

where B_k is the $(k+1) \times k$ bidiagonal matrix associated with k steps of Lanczos bidiagonalization; cf. (6.15). Thus, we can write the estimate in (7.13) as

$$\|x^{\text{exact}} - x^{(k)}\|_2 \approx \|B_k^\dagger\|_F \|A\,x^{(k)} - b\|_2 \approx \|B_k^\dagger\|_2 \|A\,x^{(k)} - b\|_2.$$

Another parameter-choice method based on error estimation is the *quasi-optimality criterion*. Here, the error estimate is given by

$$\|x^{\text{exact}} - x_\lambda\|_2 \approx \left(b^T (A\,A^T + \lambda I_m)^{-4} A\,A^T b\right)^{1/2}$$

(cf. [174, §2.1]), and minimization of this estimate leads to the problem of minimizing the function

$$\begin{aligned}
\mathcal{Q}(\lambda) &= \left\|\lambda^2 \frac{dx_\lambda}{d(\lambda^2)}\right\|_2 = \frac{1}{2} \left\|\lambda \frac{dx_\lambda}{d\lambda}\right\|_2 \\
&= \left(\sum_{i=1}^{p} \left(f_i(1-f_i) \frac{u_i^T b}{\sigma_i}\right)^2\right)^{1/2}.
\end{aligned} \quad (7.14)$$

For standard-form Tikhonov regularization it is easy to show that

$$\frac{dx_\lambda}{d(\lambda^2)} = -(A^T A + \lambda^2 I_n)^{-1} x_\lambda, \quad (7.15)$$

and therefore the second step of iterated Tikhonov regularization (cf. §6.2) leads to the solution $x_\lambda^{(2)} = x_\lambda - \lambda^2 \, dx_\lambda/d(\lambda^2)$, such that minimization of $\mathcal{Q}(\lambda)$ minimizes the correction to x_λ in this solution.

7.3. Methods Based on Error Estimation

The practical minimization of $\mathcal{Q}(\lambda)$ is complicated by the fact that this function has many local minima; see Fig. 7.1 (p. 194) for a typical example of $\mathcal{Q}(\lambda)$.

We shall now give a somewhat heuristic analysis from [184, §6.2] that relates the minimization of $\mathcal{Q}(\lambda)$ to the norms of the perturbation and regularization errors in x_λ. The filter factors satisfy the relations $f_i \approx 1$ for $\sigma_i \gg \lambda$ and $f_i \approx 0$ for $\sigma_i \ll \lambda$, and therefore we obtain

$$\mathcal{Q}(\lambda)^2 \approx \sum_{\sigma_i \geq \lambda} (1 - f_i)^2 \left(\frac{u_i^T b}{\sigma_i}\right)^2 + \sum_{\sigma_i < \lambda} f_i^2 \left(\frac{u_i^T b}{\sigma_i}\right)^2.$$

We assume that the discrete Picard condition is satisfied, and that the errors from the matrix are small such that the resolution limit in (4.1) is determined by the right-hand side errors. If i_b denotes the number of SVD components in which $u_i^T b^{\mathrm{exact}}$ dominates (cf. (4.1)), then $u_i^T b^{\mathrm{exact}} \approx u_i^T b$ and $u_i^T e < u_i^T b$ for $\sigma_i \gg \sigma_{i_b}$ (when $i \leq i_b$), while $u_i^T b^{\mathrm{exact}} \ll u_i^T b$ and $u_i^T e \approx u_i^T b$ for $\sigma_i \ll \sigma_{i_b}$ (when $i > i_b$). Using these approximations, we obtain the following relations for the regularization and perturbation errors (cf. (4.40)):

$$\|(A^\dagger - A^\#) b^{\mathrm{exact}}\|_2^2 = \sum_{i=1}^n (1 - f_i)^2 \left(\frac{u_i^T b^{\mathrm{exact}}}{\sigma_i}\right)^2 \approx \sum_{\sigma_i \geq \lambda} (1 - f_i)^2 \left(\frac{u_i^T b}{\sigma_i}\right)^2,$$

$$\|A^\# e\|_2^2 = \sum_{i=1}^n f_i^2 \left(\frac{u_i^T e}{\sigma_i}\right)^2 \approx \sum_{\sigma_i < \lambda} f_i^2 \left(\frac{u_i^T b}{\sigma_i}\right)^2.$$

Thus, if we choose $\lambda \approx \sigma_{i_b}$, then we obtain the following approximate expression for the quasi-optimality function:

$$\mathcal{Q}(\lambda)^2 \approx \|(A^\dagger - A^\#) b^{\mathrm{exact}}\|_2^2 + \|A^\# e\|_2^2.$$

Since $\|(A^\dagger - A^\#) b^{\mathrm{exact}}\|_2$ increases with λ while $\|A^\# e\|_2$ decreases with λ, it is clear that the minimizer of $\mathcal{Q}(\lambda)$ seeks to balance the regularization error and the perturbation error when λ is near its optimum value, $\lambda \approx \sigma_{i_b}$.

For a discrete regularization parameter k, we have introduced the following approach in the REGULARIZATION TOOLS package [187]. We replace λ with γ_k and use the approximations

$$\left\|\frac{dx_\lambda}{d\lambda}\right\|_2 \approx \frac{\|\Delta x_k\|_2}{\Delta \lambda}, \qquad \|\Delta x_k\|_2 = \frac{u_k^T b}{\gamma_k}, \qquad \Delta \lambda = \gamma_{k+1} - \gamma_k \approx \gamma_k$$

to obtain the following expressions for the discrete quasi-optimality criterion function:

$$\mathcal{Q}_k = \frac{u_k^T b}{\sigma_k}, \quad L = I_n, \qquad \mathcal{Q}_k = \frac{u_k^T b}{\gamma_k}, \quad L \neq I_n. \tag{7.16}$$

7.4. Generalized Cross-Validation

Generalized cross-validation (GCV) [146], [363] is a very popular and successful $\|e\|_2$-free method for choosing the regularization parameter. The GCV method is based on statistical considerations, namely, that a good value of the regularization parameter should predict missing data values. More precisely, if an arbitrary element b_i of the right-hand side b is left out, then the corresponding regularized solution should predict this observation well, and the choice of regularization parameter should be independent of an orthogonal transformation of b (including permutations of the elements of b); see [366, Chapter 4] for more details.

The GCV method is a predictive method which seeks to minimize the predictive mean-square error $\|A x_\lambda - b^{\text{exact}}\|_2$. Since b^{exact} is unknown, the GCV method works instead with the GCV function

$$\mathcal{G}(\lambda) = \frac{\|A x_\lambda - b\|_2^2}{\text{trace}(I_m - A A^\#)^2} = \frac{\mathcal{V}(\lambda)}{\mathcal{T}(\lambda)}, \qquad (7.17)$$

where $\mathcal{V}$ and $\mathcal{T}$ are defined in (7.9) and (7.10). See Fig. 7.2 (p. 196) for a typical example of the GCV function $\mathcal{G}(\lambda)$.

Wahba showed in [363] that if the discrete Picard condition is satisfied and the noise is white, then the regularization parameter λ_{GCV} which minimizes the expected value of $\mathcal{G}(\lambda)$ is near the minimizer of the expected value of the predictive mean-square error $\|A x_\lambda - b^{\text{exact}}\|_2$. More precisely, we have the relation $\lambda_{\text{GCV}} = \lambda_{\text{opt}}(1 + o(1))$, where λ_{opt} mimimizes the predictive mean-square error and $o(1) \to 0$ as $n \to \infty$. In the same paper, Wahba gives convergence rates for the expected value of $\|x^{\text{exact}} - x_{\lambda_{\text{GCV}}}\|_2$, where $x_{\lambda_{\text{GCV}}}$ is the Tikhonov solution. These results were further examined and extended by Lukas [244]. Vogel [356] has derived similar convergence results for the truncated SVD (TSVD) method.

As already mentioned above, if the errors in e are unbiased and have covariance matrix $\sigma_0^2 I_m$, then on a broad scale $\mathcal{V}$ has a plateau at approximately σ_0^2. The transition between the plateau and the increasing part of $\mathcal{V}$'s graph takes place in a (usually small) λ-interval, and the optimal λ lies in this interval. Hence, the L-curve has its characteristic corner in the same λ-interval.

The function $\mathcal{T}$ is a slowly increasing function of λ, from $m - n$ for $\lambda = 0$ to $m - n + p$ for $\lambda \to \infty$. The following theorem sheds more light on the behavior of $\mathcal{T}$.

Theorem 7.4.1. [184, Theorem 5]. *For the model distribution of generalized singular values $\gamma_i = c^{p-i}$, $i = 1, \ldots, p$, with $0 < c < 1$, $\mathcal{T}(\lambda)$ is bounded as*

$$m - n + k_\lambda - \frac{1}{1 - c^2} \leq \mathcal{T}(\lambda) \leq m - n + k_\lambda + \frac{1}{1 - c^2}, \qquad (7.18)$$

where k_λ is the number of γ_i less than λ, i.e., $\gamma_{k_\lambda} < \lambda \leq \gamma_{k_\lambda + 1}$.

7.4. GENERALIZED CROSS-VALIDATION

The GCV method seeks to locate the transition point where $\mathcal{V}(\lambda)$ changes from a very slowly varying function of λ to a more rapidly increasing function of λ, and thus to implicitly locate the L-curve's corner. But instead of working with the function $\mathcal{V}$, the GCV method uses the function $\mathcal{G}$. The denominator $\mathcal{T}$ (7.10) is a monotonically increasing function of λ, such that $\mathcal{G}$ given by (7.17) has a *minimum* in the above-mentioned transition interval. Hence, GCV replaces the problem of locating the transition point for $\mathcal{V}$ by a numerically well defined problem, namely, that of finding the minimum for the GCV function $\mathcal{G}$. Unfortunately, the unique minimum can be very flat, thus leading to numerical difficulties in computing the minimum of $\mathcal{G}(\lambda)$.

If we use Tikhonov regularization, then the denominator in (7.17) can be computed in $\mathcal{O}(n)$ operations if the bidiagonalization algorithm from §5.1.1 is used; see [103] and [171, §5.4.2] for details. Alternatively, if the filter factors are known, then they can be used to evaluate the denominator by means of the simple expressions (4.42) and (7.10) for $\|A\,x_\lambda - b\|_2$ and $\mathcal{T}(\lambda)$.

For iterative regularization methods the computation of the denominator $\mathcal{T}_k$ is more difficult because neither a canonical decomposition of A nor the filter factors $f_i^{(k)}$ are available. For regularizing CG iterations, the situation is further complicated by the fact that the matrix $A^\#$ is not unique, and therefore the GCV function $\mathcal{G}_k$ is not unique either. The following two formulas for $A^\#$ are valid:

$$A_B^\# = \hat{V}_k B_k^\dagger \hat{U}_{k+1}^T, \qquad (7.19)$$

$$A_\Sigma^\# = V \operatorname{diag}(f_1^{(k)}, \ldots, f_n^{(k)}) \Sigma^\dagger U^T, \qquad (7.20)$$

where $\hat{U}_{k+1}$ and $\hat{V}_k$ consist of the left and right Lanczos vectors from LSQR (6.15), and $f_i^{(k)}$ are the CG filter factors. Although in general $A_B^\# \neq A_\Sigma^\#$, the relation $A_B^\# b = A_\Sigma^\# b$ always holds. From (7.19)–(7.20) we obtain

$$\operatorname{trace}(I_m - A\,A_B^\#) = m - k, \qquad \operatorname{trace}(I_m - A\,A_\Sigma^\#) = m - \sum_{i=1}^n f_i^{(k)}. \qquad (7.21)$$

The first expression in (7.21) follows from the relation $A\,A_B^\# = \hat{U}_{k+1} B_k B_k^\dagger \hat{U}_{k+1}^T$, and the second expression is simply (7.10). In general, the two trace terms in (7.21) are different because in general $\sum_{i=1}^n f_i^{(k)} \neq k$. Hence, the GCV function for regularizing CG iterations is not uniquely determined.

If we use the Lanczos bidiagonalization process in infinite precision (simulated by full reorthogonalization), and if the Lanczos process captures the singular values in their natural order (cf. §6.4), then the difference between the two expressions in (7.21) is always small. In our numerical experiments we found that the maximum difference was of the order $\mathcal{O}(1)$.

A more serious complication is that the first expression in (7.21) does not hold in finite precision where the loss of orthogonality influences the CG and

Lanczos processes. The second expression, on the other hand, holds in both infinite and finite precision. Unfortunately, the CG filter factors are generally not known.

Consider again the class of problems in which all the singular values are distinct and captured by the Lanczos process in decreasing order. For these problems, a good compromise is to use the relation $\sum_{i=1}^{n} f_i^{(k)} \approx \ell_k$, where ℓ_k is the number of converged Ritz values (i.e., $\ell_k \leq k$), and then use the approximate GCV function

$$\mathcal{G}_k = \frac{\|A\,x^{(k)} - b\|_2^2}{(m - \ell_k)^2}. \qquad (7.22)$$

As long as the number of necessary CG iterations is very small compared to m, such that the behavior of $\mathcal{G}_k$ in (7.22) is mainly determined by the residual norm in the numerator, we can further approximate $\mathcal{G}_k$ by $\|A\,x^{(k)} - b\|_2^2/(m - k)^2$.

However, as k increases, the difference between ℓ_k and k also increases, and we are forced to use ℓ_k. To do this, we must either keep track of both the spurious and the multiple singular values or use reorthogonalization—either explicitly or in a selective manner; cf. [154, §9.2.4]. To reduce the storage requirement for the Lanczos vectors, it seems promising to use implicit restarts of the LSQR process as described in [40].

A different approach to computing the denominator in the GCV function, which is particularly attractive for iterative methods, is to approximate the denominator by a statistical estimate. This approach is called *Monte Carlo GCV* [140], [141]. Assume that the errors in b have covariance matrix $\sigma_0^2\,I_m$. Then the idea is to simultaneously run the iterative method on both the given right-hand side b and a random right-hand side $\tilde{b}$ whose elements have zero mean and standard deviation σ_0. If $\tilde{x}^{(k)}$ denotes the additional iteration vector corresponding to $\tilde{b}$, then

$$\tilde{b}^T(\tilde{b} - A\,\tilde{x}^{(k)}) = \sum_{i=1}^{m}(u_i^T\tilde{b})^2 - \sum_{i=1}^{n} f_i^{(k)}\,(u_i^T\tilde{b})^2$$

is an estimate of $\sigma_0^2\,\mathcal{T}_k$, and this estimate can be used in the GCV function $\mathcal{G}_k$. We emphasize that the recurrence for $\tilde{x}^{(k)}$ should use the *same* coefficients α_k and β_k as the recursion for $x^{(k)}$. This method essentially doubles the amount of work in an iterative regularization method.

The GCV method was originally designed for problems in which the regularized solution is given by $x_{\text{reg}} = A^{\#}b$ and $A^{\#}$ is independent of b. For more general regularization methods (such as the CG method) it was proposed in [301] to replace the denominator in (7.17) by $\text{trace}(I_m - A\,\mathcal{J}(x_{\text{reg}}))$, where $\mathcal{J}(x_{\text{reg}})$ is the Jacobian of x_{reg} with respect to b (note that $\mathcal{J}(x_{\text{reg}}) = A^{\#}$ if $A^{\#}$ is independent of b). The following theorem relates the trace-term to the filter factors for x_{reg}.

Theorem 7.4.2. *If $\mathcal{J}(x_{\text{reg}})$ denotes the Jacobian of x_{reg} with respect to b, and if f_i, $i = 1, \ldots, n$, are the filter factors corresponding to x_{reg}, then*

$$\text{trace}(I_m - A\,\mathcal{J}(x_{\text{reg}})) = m - \sum_{i=1}^n f_i - \sum_{i=1}^n u_i^T b \frac{\partial f_i}{\partial(u_i^T b)} \ . \tag{7.23}$$

Proof. The trace-term is invariant to a change to the SVD coordinate systems with $\beta = U^T b$ and $\xi = V^T x_{\text{reg}}$. Then

$$\mathcal{J}(\xi) = \left\{ \frac{\partial \xi}{\partial \beta} \right\} = F\,\Sigma^\dagger + G$$

where $F = \text{diag}(f_i)$ and the n columns g_i of the matrix G are given by

$$g_i = \frac{\partial (F\,\Sigma^\dagger)}{\partial \beta_i}\beta = \frac{\partial F}{\partial \beta_i}\Sigma^\dagger \beta\ , \qquad i = 1, \ldots, n\ .$$

Thus, we obtain

$$\Sigma\,\mathcal{J}(\xi) = F + \Sigma\,G = F + \text{diag}(\beta_i)\left\{\frac{\partial f_i}{\partial \beta_j}\right\}\ ,$$

which immediately leads to (7.23). □

The last term in (7.23) vanishes when $A^{\#}$ is independent of b, and we see that the difference between $\text{trace}(I_m - A\,\mathcal{J}(x_{\text{reg}}))$ and $\text{trace}(I_m - A\,A^{\#})$ is controlled by the derivatives of the filter factors, $\partial f_i/\partial(u_i^T b)$, times the corresponding Fourier coefficients $u_i^T b$. If the dependence is weak, i.e., if the derivatives are small in absolute value, then $m - \sum_{i=1}^n f_i$ is a good approximation to the trace-term.

In [301] it is also suggested that one approximate $\text{trace}(I_m - A\,\mathcal{J}(x_{\text{reg}}))$ by means of finite differences obtained by applying the CG method to the two vectors $b + d$ and $b - d$, where d is a random vector with $\|d\|_2 \ll \|b\|_2$. We have found experimentally that this approach is too sensitive to the choice of $\|d\|_2$ to be a robust parameter-choice method. More research in this topic is required.

7.5. The L-Curve Criterion

In this section we discuss another $\|e\|_2$-free parameter-choice method. It is based on the L-curve defined in §4.6 as a parametric plot of the discrete smoothing norm $\Omega(x_{\text{reg}})$ of the regularized solution, for example, the (semi)norm $\|L\,x_{\text{reg}}\|_2$, versus the corresponding residual norm $\|A\,x_{\text{reg}} - b\|_2$, with the regularization parameter λ as the parameter. As already mentioned in §4.6, the L-shaped corner of the L-curve appears for regularization parameters close to the optimal parameter that balances the regularization errors and perturbation errors in x_{reg}. This important feature is the basis for the L-curve criterion for choosing the regularization parameter, as proposed in [184] and [196].

Although the L-curve criterion is a relatively new method, it has already been used successfully in some applications; see, e.g., [45], [310], and [312]. Numerical examples that illustrate the use of the method can be found in [184] and [196].

7.5.1. Distinguishing Signal From Noise

An underlying assumption for the use of the L-curve criterion is that the residual norm $\|A\, x_{\text{reg}} - b\|_2$ and the smoothing norm $\Omega(x_{\text{reg}})$ are monotonic functions of the regularization parameter λ. This is true for Tikhonov regularization (cf. Theorem 4.6.1), and it is also trivially the case for TSVD and TGSVD—but it cannot be ensured for modified TSVD (MTSVD) or piecewise polynomial TSVD (PP-TSVD). It is also true for regularizing CG iterations (cf. §6.3) and for the Lanczos truncated total least squares (T-TLS) hybrid method from §6.6 when the solution norm is plotted versus the TLS residual norm. It is not true for the ν-method from §6.2.

Recall that if the discrete Picard condition is satisfied, then the L-curve basically consists of a vertical part and an adjacent horizontal part. The horizontal part corresponds to oversmoothed solutions where the regularization parameter is too large and the solution is dominated by regularization errors. The vertical part corresponds to underregularized solutions where the regularization parameter is too small and the solution is dominated by perturbation errors. We emphasize that this behavior does not rely on any additional properties of the problem, such as the statistical distribution of the errors.

The idea behind the L-curve criterion for choosing the regularization parameter is to choose a point on this curve at the corner between the horizontal and the vertical pieces. The rationale behind this idea is that the corner corresponds to a solution in which the regularization and perturbation errors are balanced—because the corner separates the horizontal part of the curve, where the regularization errors dominate, from the more vertical part where the perturbation errors dominate.

It is important to plot the L-curve in *log-log scale* in order to emphasize the two different parts of the curve. There is a strong intuitive justification for this. Since the singular values typically span many orders of magnitude, the behavior of the L-curve is more easily seen in such a log-log scale. In addition, the log-log scale emphasizes "flat" parts of the L-curve where the variation in either $\Omega(x_{\text{reg}})$ or $\|A\, x_{\text{reg}} - b\|_2$ is small compared to the variation in the other variable. These parts of the L-curve are often "squeezed" close to the axes in a lin-lin scale. Hence the log-log scale actually emphasizes the corner of the L-curve. One more advantage of the log-log scale is that particular scalings of x and b simply shift the L-curve vertically and horizontally.

7.5. The L-Curve Criterion

The log-log transformation has a theoretical justification as well. Consider the L-curve for TSVD, using the smoothing norm $\|\cdot\|_1$ defined in §5.7, with $\Omega(x_k) = \|x_k\|_1$ and $\|A\,x_k - b\|_1$ given by (5.75) and (5.76). As k is increased by 1, the change in $\|A\,x_k - b\|_1$ is $|u_k^T b|$, while the change in $\|x_k\|_1$ is $|u_k^T b|/\sigma_k$. Thus the slope of the kth segment of the piecewise linear interpolant is σ_k^{-1}, independent of the right-hand side for the problem. Therefore, there is no hope of distinguishing the pure signal from the noise by examining the properties of the L-curve in the lin-lin scale.

In the log-log scale, however, the slope of the kth segment is the *relative* change in $\|x_k\|_1$ divided by the *relative* change in $\|A\,x_k - b\|_1$, and these behave quite differently for signal and noise. A noiseless right-hand side $b = b^{\text{exact}}$, for which the discrete Picard condition is satisfied, has the property that the sequences $|u_k^T b^{\text{exact}}|$ and $|u_k^T b^{\text{exact}}|/\sigma_k$ both approach zero. Thus the relative change in $\|x_k\|_1$ approaches zero, while the relative change in $\|A\,x_k - b\|_1$ is finite and nonzero. Therefore, for a pure right-hand side, the L-curve in log-log coordinates becomes flat as k is increased.

A right-hand side $b = e$ consisting of pure noise gives a quite different L-curve in log-log scale. If we assume that all the error components $u_i^T e$ are roughly a constant value τ_b, then $\|x_k\|_1 \approx \tau_b/\sigma_k$ and the relative change in $\|x_k\|_1$ is approximately σ_{k-1}/σ_k. The relative change in $\|A\,x_k - b\|_1$ is $1/(m-k)$, so the slope of the piecewise linear interpolant is $(m-k)\,\sigma_{k-1}/\sigma_k$. The L-curve for the noise in log-log scale therefore has a steep slope as k increases, unlike the flat curve for the noiseless right-hand side.

The L-curves for a pure signal and for pure noise in Tikhonov regularization, using the 2-norm, have similar characteristics. Both curves are steep in lin-lin scale as $\lambda \to 0$, but only the noise curve is steep in log-log scale. This can be shown either by recalling the closeness of the TSVD and Tikhonov solutions and residual in the 2-norm, or by rather tedious computations with the 2-norm. We conclude that the log-log scale is necessary in order to properly distinguish the signal from the noise via the L-curve.

The different behavior of the L-curve in lin-lin and log-log scales is also studied in [291], but without a distinction between the different behavior of L-curves for pure signal and pure noise.

To have an *operational* definition of "corner" we define the L-curve's corner as that point on the curve

$$(\zeta(\lambda), \eta(\lambda)) = (\log \|A\,x_{\text{reg}} - b\|_2\,,\; \log \Omega(x_{\text{reg}}))$$

that has maximum curvature. Here, the curvature κ is defined as usual by

$$\kappa(\lambda) = \frac{\zeta'\eta'' - \zeta''\eta'}{((\zeta')^2 + (\eta')^2)^{3/2}}, \qquad (7.24)$$

where differentiation is with respect to λ. The curvature is a purely geometric quantity that is independent of transformations of the regularization parameter. Thus the *L-curve criterion* is, by definition, equivalent to computing the regularization parameter that maximizes the curvature in (7.24).

The convergence properties of the regularization parameter λ_L chosen by the L-curve criterion have recently been studied by Hanke [170] and Vogel [360] for the case of Tikhonov regularization.

Hanke studied the convergence of λ_L in the Hilbert space setting as the error norm $\|e\|_2 \to 0$ and showed that the L-curve criterion leads to a regularization parameter that decays too rapidly to zero. The theory as well as the numerical examples in [170] show that the undersmoothing is pronounced for small noise levels (say, $\|e\|_2/\|b^\mathrm{exact}\|_2 < 10^{-2}$) and very "smooth" solutions (i.e., with rapidly decaying coefficients $v_i^T x^\mathrm{exact}$). We note that such "smooth" solutions correspond to model problems in (4.30) with a large value of the parameter α. On the other hand, the theory predicts good performance of the L-curve criterion for less "smooth" solutions, corresponding to small values of the parameter α in the model problem (4.30).

Vogel [360] considered the behavior of λ_L as $m \to \infty$ in the Fredholm integral equation (1.2) with a discrete right-hand side. He studied a stochastic model in which the right-hand side errors have covariance matrix $\sigma_0^2 I_m$ and proved that the expected value of λ_L does not converge fast enough to zero as $m \to \infty$. Numerical results with the asymptotic noise level $\|e\|_2/\|b^\mathrm{exact}\|_2 \to 2 \cdot 10^{-2}$ as $m \to \infty$ confirm that the oversmoothing due to the difference between λ_L and the optimal λ becomes significant in this example for $n \gtrsim 4\,000$. This important result indicates that the maximum curvature may not be the optimal feature of the L-curve for selecting the regularization parameter.

7.5.2. Computational Aspects

Although the L-curve is easily defined and quite satisfying intuitively, computing the point of maximum curvature in a numerically reliable way is perhaps not as easy as it might seem. We will now summarize three cases of increasing difficulty, as discussed in [196, §5].

If the functions $\zeta(\lambda) = \log \|A\,x_\mathrm{reg} - b\|_2$ and $\eta(\lambda) = \log \Omega(x_\mathrm{reg})$ are defined by some computable formulas, and if the L-curve is twice differentiable, then it is straightforward to compute the curvature by means of (7.24). Any one-dimensional optimization routine can be used to locate the value of λ that corresponds to maximum curvature. This situation arises, e.g., when using Tikhonov regularization on a problem for which the SVD or GSVD is known. It is practical computationally, since the effort involved in such a maximization is much smaller than that for computing the SVD/GSVD.

In many situations we are limited to knowing only a finite number of points (ζ_i, η_i) on the L-curve. This is the case, e.g., for the TSVD and CG methods, and in these and other cases the underlying curve is not differentiable. In a computational sense, the L-curve then consists of a number of discrete points

7.5. THE L-CURVE CRITERION

corresponding to different values of the regularization parameter at which we have evaluated $\zeta(\lambda)$ and $\eta(\lambda)$.

We have found that in many cases, the points on a discrete L-curve are clustered, giving the L-curve fine-grained details that are not relevant for our considerations. For example, if there is a cluster of small singular values with right-hand side coefficients even smaller, then the L-curve for TSVD will have a cluster of points for the corresponding values of the truncation parameter k. This situation does not occur for Tikhonov regularization because all the components in the solution come in gradually as the filter factors change from zero to one.

For computational reasons we must define a differentiable, smooth curve associated with the discrete points in such a way that any fine-grained details are discarded while the overall shape of the L-curve is maintained; i.e., we want the approximating curve to achieve local averaging while retaining the overall shape of the L-curve. If we try to fit a pair of cubic splines to the abscissas ζ_i and ordinates η_i of the L-curve, or if we try to fit a cubic spline to η_i as a function of ζ_i, then it is difficult to approximate the corner well because dense knots are required here. This conflicts with the purpose of the fit, namely, to locate the corner.

Instead we recommend fitting a *cubic spline curve* to the discrete points of the L-curve, as described in [196, §5.2]. Such a curve has several favorable features in connection with our problem: it is twice differentiable, it can be differentiated in a numerically stable way, and it has local shape-preserving features. Yet we must be careful not to approximate the fine-grained details of clusters of points too well. Since a cubic spline curve does not intrinsically have the desired local smoothing property, we propose the following two-step algorithm for computing a cubic spline-curve approximation to a discrete L-curve.

1. Perform a local smoothing of the L-curve points, in which each point is replaced by a new point obtained by fitting a low-degree polynomial to a few neighboring points.

2. Use the new smoothed points as control points for a cubic spline curve.

We can now easily compute the corner of the spline curve by means of (7.24), and—if the regularization parameter is discrete—determine the point on the original discrete L-curve that is closest to the spline curve's corner. This strategy is implemented in routine l_curve in REGULARIZATION TOOLS [187].

In many applications, evaluating points on the L-curve is computationally demanding and one would prefer to compute as few points as possible. For such problems, with differentiable as well as nondifferentiable L-curves, we recommend an algorithm from [196, §5.3] that tries to locate the corner of the L-curve efficiently. Assume that one knows a few points on each side of the corner. Then the idea from the previous algorithm can be used to compute a sequence of new regularized solutions whose associated points on the L-curve (hopefully) approach its corner.

1. Start with a few points (ζ_i, η_i) on each side of the corner.

2. Compute an approximating three-dimensional cubic spline curve $\mathcal{S}$ for the points $(\zeta_i, \eta_i, \lambda_i)$, where λ_i is the regularization parameter that corresponds to (ζ_i, η_i).

3. Let $\mathcal{S}_2$ denote the first two coordinates of $\mathcal{S}$, such that $\mathcal{S}_2$ approximates the L-curve.

4. Compute the point on $\mathcal{S}_2$ with maximum curvature, and find the corresponding regularization parameter λ_0 from the third coordinate of $\mathcal{S}$.

5. Solve the regularization problem for $\lambda = \lambda_0$ and add the new point $(\zeta(\lambda_0), \eta(\lambda_0))$ to the L-curve.

6. Repeat from step 2 until convergence.

In step 2, it is necessary to introduce λ_i as the third coordinate of $\mathcal{S}$ because we need to associate a regularization parameter with each point on $\mathcal{S}$ (a two-dimensional spline curve with λ_i as knots does not provide this feature). Initial points for step 1 can be generated by choosing a number of regularization parameters ranging from very "large" values to very "small" values.

A different version of this algorithm is used by Kaufman and Neumaier in their "envelope guided CG algorithm" for Tikhonov regularization of large-scale ill-posed problems [221], [222]. The idea is to track the shape of the family of L-curves associated with the CG-iterates and adaptively adjust λ so as to stay as close to the corner as possible.

7.5.3. Other Aspects

Miller [249, Method 2] suggested an a priori parameter choice which is related to the L-curve. Assume that, ideally, we wish to compute a solution x^* for which

$$\|A x^* - b\|_2 \leq \delta \quad \text{and} \quad \|L x^*\|_2 \leq \alpha .$$

Then Miller showed that if we choose the Tikhonov regularization parameter λ as

$$\lambda = \delta/\alpha , \qquad (7.25)$$

then the regularized solution satisfies $\|A x_\lambda - b\|_2 \leq \sqrt{2}\,\delta$ and $\|L x_\lambda\|_2 \leq \sqrt{2}\,\alpha$. Hence, if δ and α are good estimates of $\|e\|_2$ and $\|x^{\text{exact}}\|_2$, respectively, then Miller's parameter choice (7.25) yields a solution x_λ which cannot be too far from the L-curve's corner.

7.6. Parameter-Choice Methods in Action

TABLE 7.1. *Regularization parameters.*

Method	λ
Quasi-opt.	$2.13 \cdot 10^{-2}$
GCV	$3.85 \cdot 10^{-2}$
L-curve	$1.67 \cdot 10^{-2}$
Optimal	$3.55 \cdot 10^{-2}$

Miller's parameter choice (7.25) is also related to the choice of filter factors in (5.26) if we assume that $|v_i^T x^{\text{exact}}| = \xi_o$ (i.e, independent of i), for then $\|x^{\text{exact}}\|_2 = n^{1/2} \xi_o$, $\|e\|_2 = m^{1/2} \sigma_0$, and (5.26) and (7.25) with $L = I_n$ lead to the two similar parameter choices,

$$\lambda = \sqrt{\frac{n}{m}} \frac{\|e\|_2}{\|x^{\text{exact}}\|_2} \quad \text{and} \quad \lambda = \frac{\|e\|_2}{\|x^{\text{exact}}\|_2} ,$$

respectively.

Regińska [291] relates the L-curve criterion to minimization of the function

$$\Psi_\beta(\lambda) = \|x_\lambda\|_2 \cdot \|A\,x_\lambda - b\|_2^\beta , \qquad (7.26)$$

where $\beta > 0$ is a real parameter. Intuitively, a regularization parameter λ that balances the regularization and perturbation errors in x_λ should not lead to large values of $\|x_\lambda\|_2$ or $\|A\,x_\lambda - b\|_2$. Indeed, it is proved in [291] that if the L-curve in log-log scale has maximum curvature for $\lambda = \lambda^*$, and if the tangent to the curve $(\log \|A\,x_\lambda - b\|_2, \log \|x_\lambda\|_2)$ has slope β^*, then the function $\Psi_\beta(\lambda)$ in (7.26) with $\beta = \beta^*$ has a minimum for $\lambda = \lambda^*$.

Finally, we mention that Engl and Grever [109] use the L-curve to compute the regularization parameter λ corresponding to the discrepancy principle (see §7.2).

7.6. Parameter-Choice Methods in Action

In this section we illustrate three parameter-choice methods by a numerical example using Tikhonov regularization in standard form: the quasi-optimality criterion, the GCV method, and the L-curve criterion. The test problem is identical to the second test problem in §4.8.1, namely, shaw from §1.4.3 with dimensions $m = n = 64$ and noise level $\|e\|_2 / \|b^{\text{exact}}\|_2 = 0.01$.

The left part of Fig. 7.1 shows the quasi-optimality function $\mathcal{Q}(\lambda)$ from (7.14), the GCV function $\mathcal{G}(\lambda)$ from (7.17), and the curvature $\kappa(\lambda)$ in (7.24) for the L-curve in log-log scale. The minima of the three curves are also shown by the three symbols $*$, $\times$, and $\circ$, and the corresponding values of the regularization parameter are listed in Table 7.1.

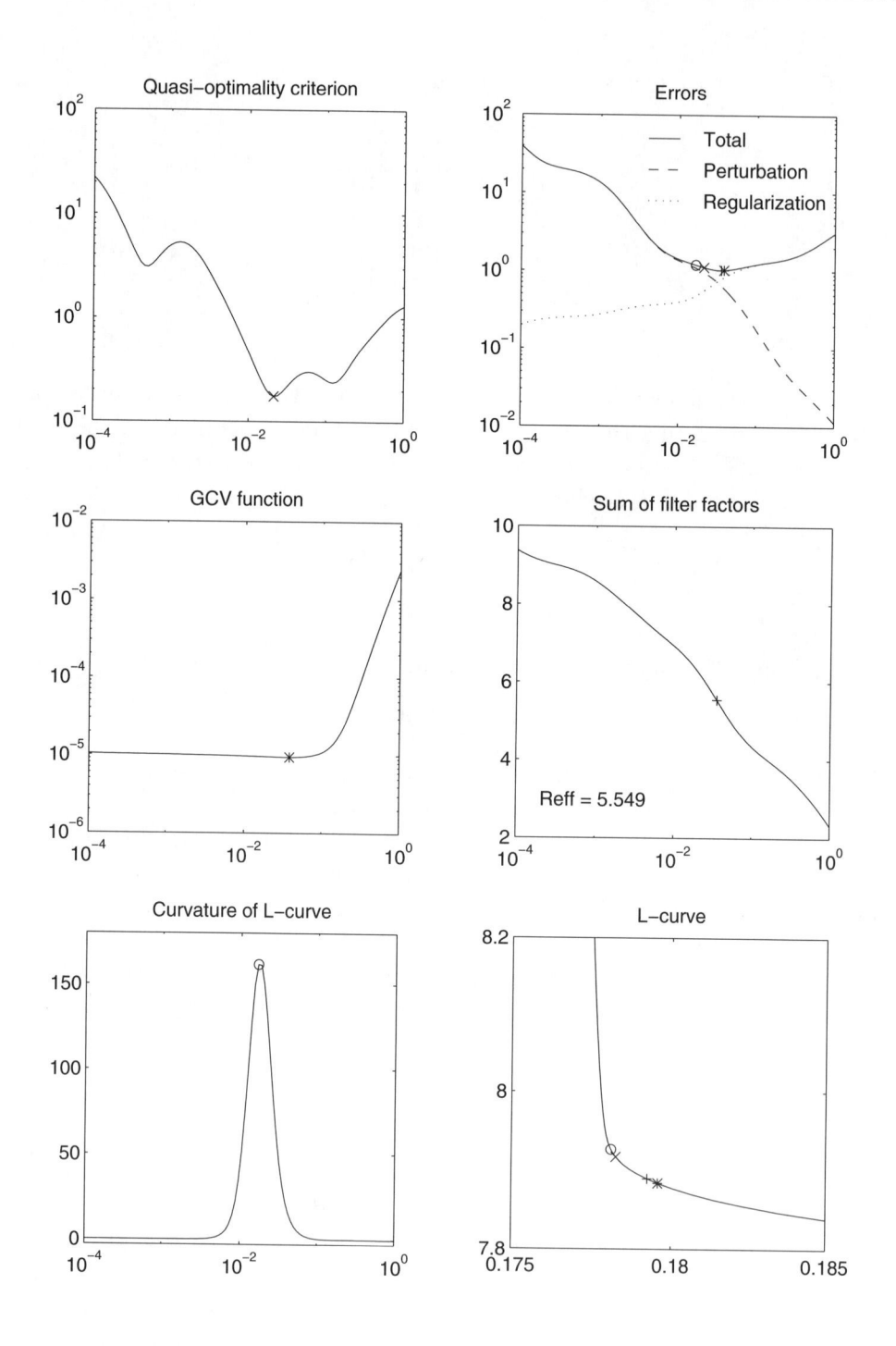

FIG. 7.1. *Left: the quasi-optimality function $\mathcal{Q}(\lambda)$, the GCV function $\mathcal{G}(\lambda)$, and the curvature $\kappa(\lambda)$ for the L-curve in log-log scale, as functions of λ. Top right: the norms of the regularization, perturbation, and total errors versus λ. Middle right: the function $\rho(\lambda) = \sum_{i=1}^{n} f_i$. Bottom right: a close-up view of the L-curve's corner.*

The top right part of Fig. 7.1 shows the norm of the perturbation error $\|A^\# e\|_2$, the norm of the regularization error $\|(A^\dagger - A^\#)b\|_2$, and the norm of the total error $\|x^{\text{exact}} - x_\lambda\|_2$ versus the regularization parameter λ. The minimum of the total error is indicated by the + symbol, and it corresponds to the optimal regularization parameter $\lambda_{\text{opt}} = 3.55 \cdot 10^{-2}$. The total error corresponding to the regularization parameters found by the above three methods are also indicated with the corresponding symbols. In this problem, the GCV method leads to a slight oversmoothing, while the quasi-optimality criterion and the L-curve method both lead to some undersmoothing where the total error is dominated by the perturbation error.

The bottom right part of Fig. 7.1 shows essentially the same information in the form of a close-up view of the L-curve's corner. We have indicated the four points on the L-curve corresponding to the values of λ found by the three parameter-choice methods plus the optimal one, using the same symbols as above. The point on the L-curve corresponding to λ_{opt} lies slightly to the right of the point with maximum curvature.

Finally, the middle right part of Fig. 7.1 shows the function $\rho(\lambda) = \sum_{i=1}^n f_i$ in (7.2), i.e., the sum of the Tikhonov filter factors. The effective numerical rank $r_{\text{eff}} = \rho(\lambda_{\text{opt}})$ defined in (7.4) is $r_{\text{eff}} \approx 5.5$ for this problem. This agrees with Fig. 4.3, where we found that approximately five SVD components can be recovered in the regularized solution for this problem with the present noise level.

To illustrate the variations of the regularization parameters found by the three methods used above, Fig. 7.2 shows the error norms for six test problems with different perturbations of the right-hand side, still with noise level 10^{-2}. In these six examples, the L-curve criterion consistently undersmooths, while the quasi-optimality criterion consistently oversmooths. The GCV method yields regularization parameters that are both too large and too small, and in one of the examples (shown in the top right part) the undersmoothing is severe.

7.7. Experimental Comparisons of the Methods

An experimental comparison of some of the parameter-choice methods discussed in this chapter was presented by Thompson et al. in [333], where some theoretical results are also presented. They considered Tikhonov regularization for image reconstruction of 64×64 and 128×128 images degraded by space-invariant blur and additive white noise with standard deviation σ_0. They compared the discrepancy principle (7.5), the compensated discrepancy principle (7.8), generalized cross-validation (7.17), and an idealized parameter-choice method based on minimization of the expected value of the predictive mean-

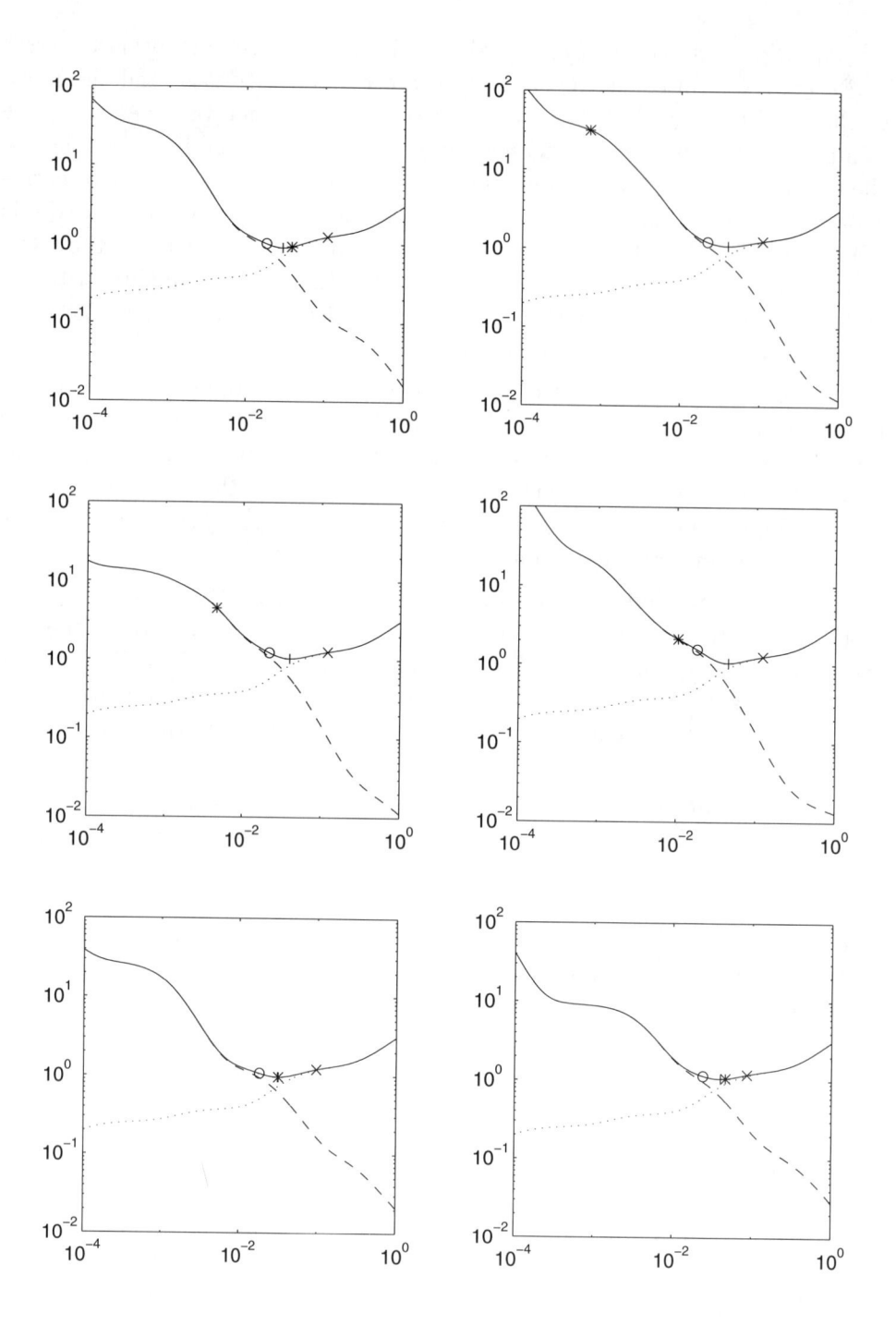

FIG. 7.2. *Norms of the perturbation errors (dashed lines), regularization errors (dotted lines), and total errors (solid lines) for six different perturbations of the right-hand side. We also indicate the total errors corresponding to regularization parameters found by the quasi-optimality criterion (crosses), the GCV method (asterisks), and the L-curve criterion (circles).*

7.7. Experimental Comparisons of the Methods

square error

$$\mathcal{E}\left(\|b^{\text{exact}} - A\,x_\lambda\|_2^2\right) = \|(I_m - A\,A_\lambda^{\#})\,b^{\text{exact}}\|_2^2 + \sigma_0^2\,\rho(\lambda)\,. \tag{7.27}$$

This criterion is, of course, only of theoretical interest since it requires knowledge of b^{exact}. The following main conclusions are made in [333].

- The results using either the true error norm $\|e\|_2$ or the estimate $\sigma_0\sqrt{m}$ are virtually identical.

- The discrepancy principle consistently oversmooths the solution.

- The compensated discrepancy principle performs well for small blur, provided σ_0 is a good noise estimate.

- In most cases, GCV and the idealized method based on (7.27) give identical results, and usually they lead to good results (with some undersmoothing).

- GCV occasionally produces a very small λ.

The occasional failure of GCV when a very small λ is computed has also been noticed by other authors [184], [189], [196], [334]. The failure occurs when the GCV function $\mathcal{G}(\lambda)$ is almost flat near the optimal λ and has a minimum for a very small λ. This situation is illustrated in Fig. 7.3, which shows the GCV function for the test problem from the top right plot of the Fig. 7.2, where GCV leads to severe undersmoothing. Other examples are shown in Fig. 5 in [184][25] and Fig. 5 in [334]. It was found experimentally in [184, §7.2] that this situation can occur, for example, when the errors in A and b are correlated.

Recently, the phenomenon has been studied in more detail for smoothing spline problems by Wahba and Wang [367]. They show empirically, supported by theory, that the probability of a very small λ goes exponentially fast to zero as the number of data points m gets large. Moreover, it is found experimentally that GCV never leads to moderately small λ. In other words, the probability distribution of λ can be described as having a mass point at zero plus a fairly concentrated distribution away from zero.

To prevent such situations, it is practical to combine the GCV method, as well as the L-curve criterion, with an upper bound $\alpha_{\max}$ on $\|L\,x_{\text{reg}}\|_2$. This upper bound prevents us from being fooled by a global minimum of the GCV function at zero, or by other corners that the L-curve may have; in the absence of other information, we seek the local minimum or the leftmost corner consistent with the bound $\alpha_{\max}$.

---

[25]The expression for $f(x)$ in Eq. (39) in [184] is incorrect; it should read $f(x) = \exp(-6(x-0.5)^2) + \exp(2(x+0.8)^2)$.

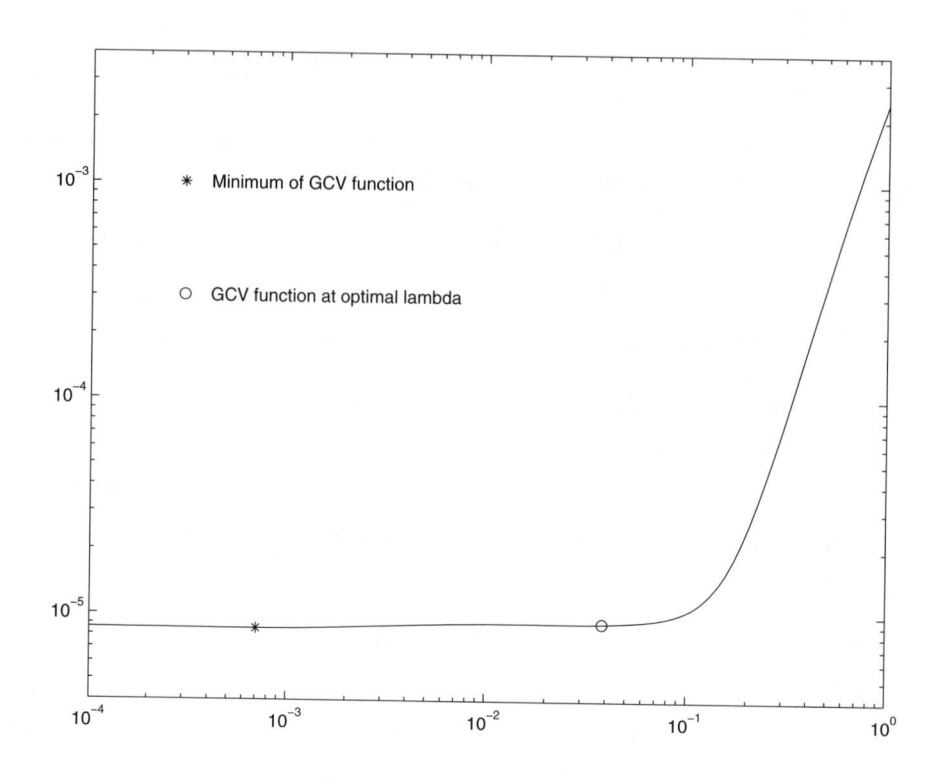

FIG. 7.3. *The GCV function $\mathcal{G}(\lambda)$ for the test problem from the top right plot of the previous figure, illustrating the occasional failure of the GCV method. The optimum value of λ is shown by the circle, and the minimum of $\mathcal{G}(\lambda)$ is shown by the asterisk. In this example, GCV leads to a severely undersmoothed solution.*

Hanke and Hansen [171, §8] made an experimental comparison of several parameter-choice methods for Tikhonov regularization, the ν-method (6.7), and regularizing CG iterations. Two test problems are considered: inverse helioseismology and image deblurring. The main conclusions are as follows.

- The discrepancy principle consistently gives the largest relative errors.

- The behavior of the regularization/parameter-choice combinations differs for the two test problems, thus illustrating the difficulties in making general conclusions.

- For the two problems considered, the following combinations of methods are most robust and, on average, yield the smallest relative errors:
    - Tikhonov regularization and GCV,
    - the ν-method and Monte Carlo GCV,
    - regularizing CG iterations and the L-curve criterion.

7.7. Experimental Comparisons of the Methods

Below, we elaborate on some of the results from [171, §8] and present new results as well. Our tests are performed in MATLAB using the REGULARIZATION TOOLS package (see §8 and [187]). Instead of the small test problems from Chapter 1, we use two larger test problems arising in helioseismology and image deblurring which are more realistic and therefore give a better impression of the behavior of the methods. Our results clearly illustrate how difficult it is to make general conclusions about regularization and parameter-choice methods.

7.7.1. Inverse Helioseismology

Our first test problem is the inverse helioseismology test problem from [171, §§2.1 and 8.1]. The underlying model is a Fredholm integral equation of the form (1.2) with a discrete right-hand side, involving m functionals k_i on the unknown function f. This function describes the internal rotation of the sun as a function of radius. The functions k_i are made up of linear combinations of eigenfunctions that describe nonradial stellar oscillations, according to some stellar model. The right-hand side consists of measured frequency shifts (due to the rotation) of the observed oscillations of the surface of the sun.

The coefficient matrix A in this problem is 212×100, L is an approximation to the second derivative operator of size 98×100, and the exact solution x^{exact} represents a realistic solution. The norm of the exact right-hand side $b^{\text{exact}} = A\,x^{\text{exact}}$ is $\|b^{\text{exact}}\|_2 = 6.0$, and white noise e with standard deviation $\sigma_0 = 10^{-2}$ is added to get $b = b^{\text{exact}} + e$. Hence, the expected noise-to-signal ratio is

$$\mathcal{E}(\|e\|_2) / \|b^{\text{exact}}\|_2 = \sigma_0 \sqrt{m} / \|b^{\text{exact}}\|_2 = 2.4 \cdot 10^{-2} \,.$$

Since it is impossible to recover the solution near the center of the sun, represented by the elements of x with small indices, we follow [171, §8.1] and "chop off" the first 12 elements of x when comparing the computed solutions. Hence, for this problem we define the relative error η_{reg} in the regularized solution x_{reg} as (using MATLAB notation):

$$\eta_{\text{reg}} = \frac{\|x^{\text{exact}}(13{:}100) - x_{\text{reg}}(13{:}100)\|_2}{\|x^{\text{exact}}(13{:}100)\|_2}. \tag{7.28}$$

To compare the various parameter-choice methods, we generated 500 test problems, and for each test problem we computed the regularization parameter by means of GCV, the L-curve criterion, and the discrepancy principle using the following three values for δ_e in (7.5):

$$\mathcal{E}(\|e\|_2) = \sigma_0 \sqrt{m}\,, \qquad 1.05 \cdot \sigma_0 \sqrt{m}\,, \qquad \text{and} \qquad 0.95 \cdot \sigma_0 \sqrt{m}\,.$$

In this way we can study the sensitivity of the discrepancy principle to the quality of the error estimate. For comparison, for each test problem we also

TABLE 7.2. *Average and maximum relative errors for Tikhonov regularization applied to the inverse helioseismology test problem.*

Parameter-choice method	Average relative error	Maximum relative error
"Optimal"	$3.4 \cdot 10^{-2}$	$5.1 \cdot 10^{-2}$
GCV	$3.8 \cdot 10^{-2}$	$1.4 \cdot 10^{-1}$
L-curve criterion	$4.9 \cdot 10^{-2}$	$7.0 \cdot 10^{-2}$
Discrepancy principle	$5.1 \cdot 10^{-2}$	>1
Ditto $+5\%$	$5.7 \cdot 10^{-2}$	>1
Ditto -5%	$5.9 \cdot 10^{-2}$	>1

computed the optimal regularization parameter, λ_{opt} or k_{opt}, that minimizes the relative error in (7.28).

Histograms of the relative errors for Tikhonov regularization are shown in Fig. 7.4. Notice that the abscissa axis shows the *logarithm* of the relative errors and that all our histograms are *normalized* to be independent of the sample size; i.e., each stave represents the percentage of data in the particular bin. The corresponding average and maximum values of the relative errors are listed in Table 7.2. The average values are computed as 10^μ, where μ is the median of $\log \eta_{reg}$.

When Tikhonov regularization is applied to this test problem, GCV leads to regularized solutions with the smallest average value of the relative errors, but with a minor tail to the right of the central distribution. The L-curve criterion, on the other hand, leads to larger errors, but with no tail in the distribution. The discrepancy principle leads to yet larger errors and a significant tail that stretches beyond 1 (i.e., some of the relative errors are greater than 1). A slight overestimate of $\|e\|_2$ by 5% increases the average error slightly and virtually eliminates the tail, while a slight underestimate by 5% dramatically increases the tail.

Still, for Tikhonov regularization, Fig. 7.5 shows the relations between the regularization parameter λ chosen by the particular parameter-choice method and the corresponding relative error η_{reg}. For clarity, for each test problem we normalize the computed regularization parameter with λ_{opt}; i.e., we plot η_{reg} versus λ/λ_{opt}. The number in the upper left corner of each subplot is the number of results for which $\lambda < 10^{-2}\lambda_{opt}$. Independently of the parameter-choice method, we see that the average errors have a minimum for $\lambda = \lambda_{opt}$ and that undersmoothing leads to significantly greater errors than oversmoothing. This is clear from the shape of the L-curve.

To obtain more insight into the behavior of the ordinary and compensated discrepancy principles, we compared these two methods for two variants of the right-hand side, one using the exact $\|e\|_2$ (computed as the norm of the actual

7.7. Experimental Comparisons of the Methods

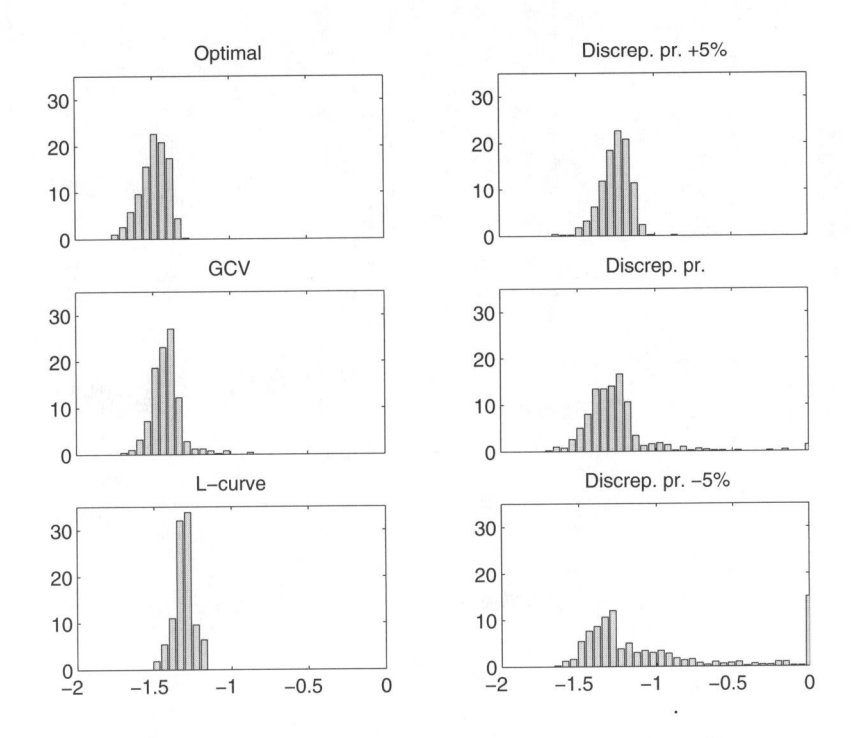

FIG. 7.4. *Normalized histograms of the relative errors η_{reg} for Tikhonov regularization applied to the inverse helioseismology test problem.*

perturbation), and one using the estimate $\sigma_0\sqrt{m}$. The results are shown in Fig. 7.6. Both methods require very good estimates of $\|e\|_2$ in order to be robust. When such good estimates are available, the discrepancy principle still oversmooths, while the compensated discrepancy principle yields results that are close to optimal, with an average relative error of $3.7 \cdot 10^{-2}$.

Combining the information in Figs. 7.4–7.6, we make the following conclusions for Tikhonov regularization.

1. The GCV method, on average, leads to a slight oversmoothing which accounts for the increased average error, compared to the optimal results. Occasionally, GCV undersmooths, leading to larger errors that constitute the histogram's tail.

2. The L-curve criterion consistently oversmooths—there is no $\lambda < \lambda_{\text{opt}}$. Hence, the average error is greater than that for GCV, but the histogram has no tail.

3. Using the estimate $\sigma_0\sqrt{m}$ of $\|e\|_2$, both the ordinary and compensated discrepancy principles oversmooth as well as undersmooth. The large

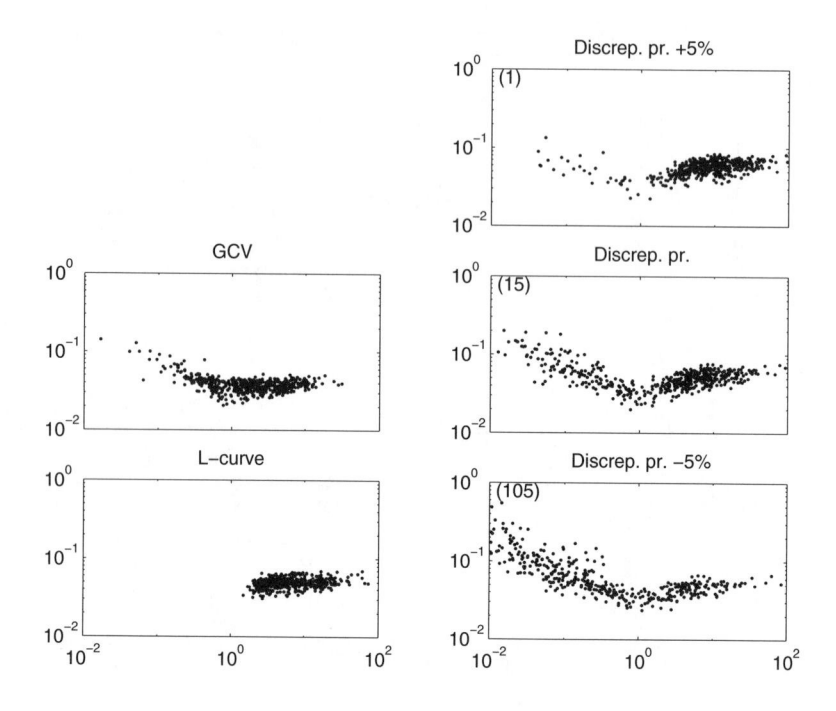

FIG. 7.5. *Relative errors η_{reg} versus normalized regularization parameters $\lambda/\lambda_{\mathrm{opt}}$ for Tikhonov regularization applied to the inverse helioseismology test problem.*

number of instances of undersmoothing leads to a large tail in the histogram.

4. Using the exact value of $\|e\|_2$, the discrepancy principle consistently oversmooths while the compensated discrepancy principle yields relative errors comparable to GCV.

5. The L-curve criterion is more *robust* than GCV, in the sense that the L-curve criterion never leads to large errors, while GCV occasionally does.

6. Both discrepancy principles are less robust, except when a very good estimate of $\|e\|_2$ is available.

The results for regularizing CG iterations applied to the inverse helioseismology test problem, using LSQR with a maximum of 15 iterations, are shown in Figs. 7.7 and 7.8. The abscissa axis in Fig. 7.8 is k/k_{opt} and therefore "reversed" in comparison with Fig. 7.5. The number in the upper right corner of each subplot for the discrepancy principle is the number of tests for which the residual norm $\|b - A\,x^{(k)}\|_2$ failed to fall below the right-hand side of (7.5). The corresponding average and maximum errors are listed in Table 7.3. On average, the optimal solution required 5.3 iterations. When so few iterations are

7.7. Experimental Comparisons of the Methods

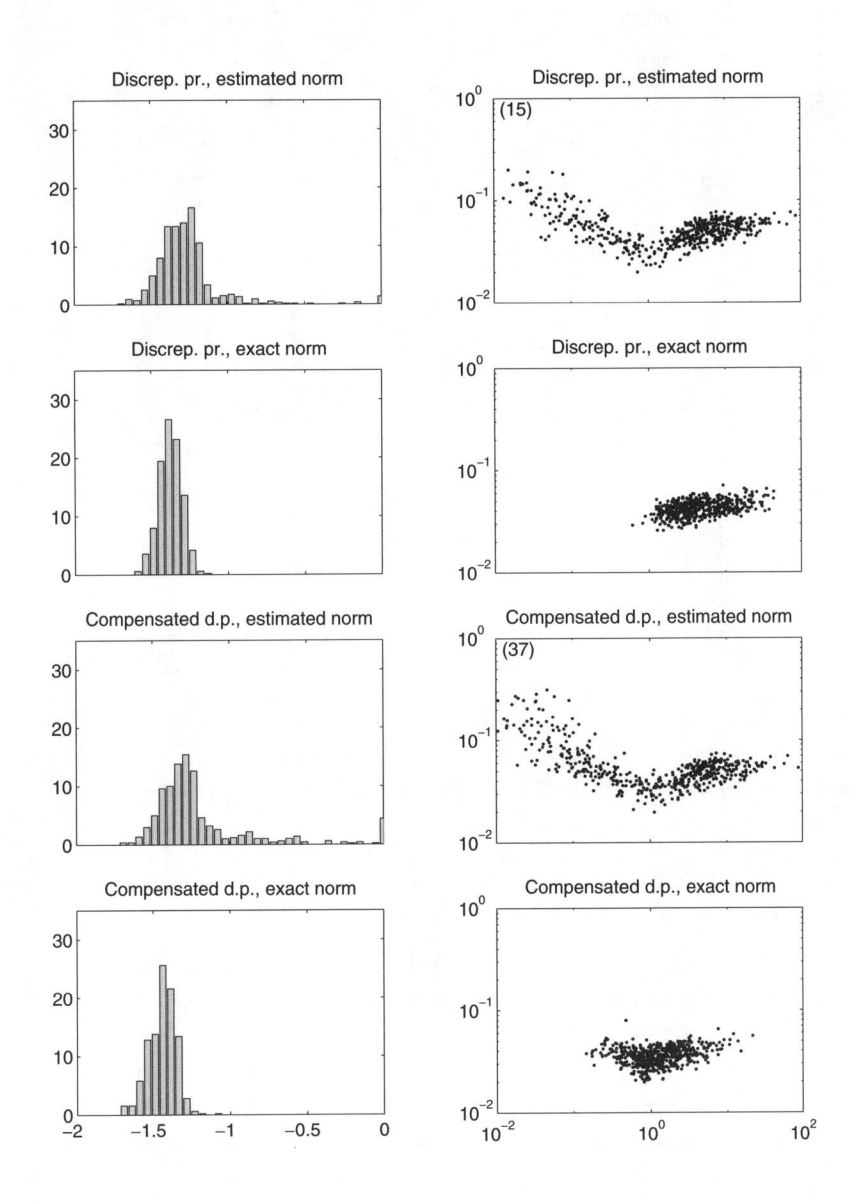

FIG. 7.6. *Results for the ordinary and compensated discrepancy principles, using either the exact or the estimated error norm, for Tikhonov regularization applied to the inverse helioseismology test problem.*

sufficient, relative to n, then LSQR is a computationally attractive alternative to Tikhonov regularization even for matrices that are not sparse or structured.

LSQR was used without any reorthogonalization, and the denominator in the GCV function was approximated by $(m - k)^2$, where k is the iteration

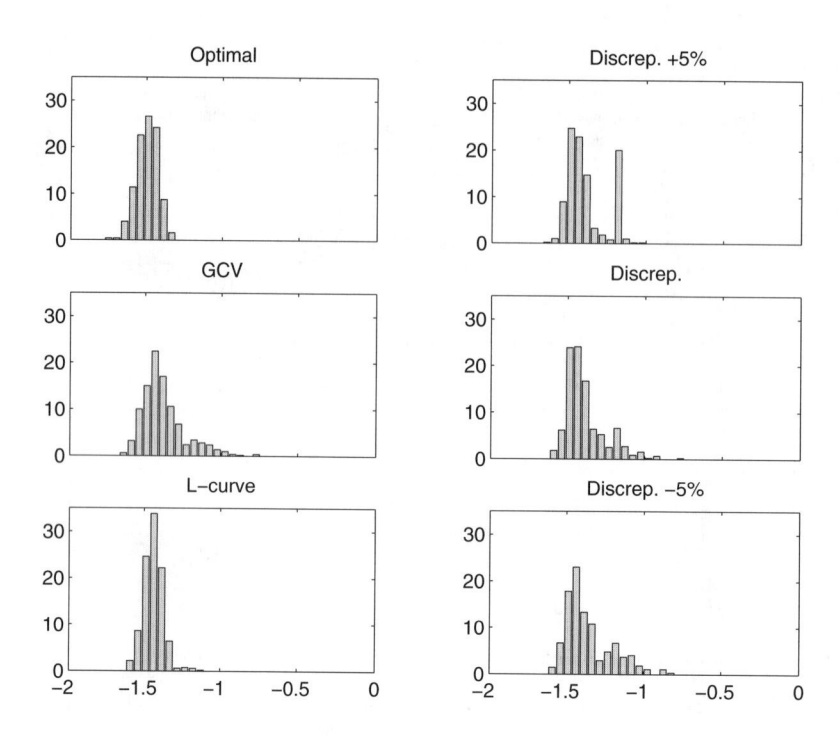

FIG. 7.7. *Normalized histograms of the relative errors η_{reg} for regularizing CG iterations (LSQR) applied to the inverse helioseismology test problem.*

TABLE 7.3. *Average and maximum relative errors for regularizing CG iterations (LSQR) applied to the inverse helioseismology test problem.*

Parameter-choice method	Average relative error	Maximum relative error
"Optimal"	$3.2 \cdot 10^{-2}$	$4.6 \cdot 10^{-2}$
GCV	$3.8 \cdot 10^{-2}$	$1.8 \cdot 10^{-1}$
L-curve criterion	$3.7 \cdot 10^{-2}$	$7.5 \cdot 10^{-2}$
Discrepancy principle	$3.8 \cdot 10^{-2}$	$1.8 \cdot 10^{-1}$
Ditto +5%	$3.7 \cdot 10^{-2}$	$9.9 \cdot 10^{-2}$
Ditto −5%	$3.9 \cdot 10^{-2}$	$1.5 \cdot 10^{-1}$

number. The deviation of this approximate denominator from the correct one is minor, since $k \ll m$.

We note in passing that the CG error estimate (7.13) for this test problem is monotonically increasing with k and therefore not useful here.

For regularizing CG iterations applied to the inverse helioseismology test problem, we make the following conclusions.

7.7. EXPERIMENTAL COMPARISONS OF THE METHODS

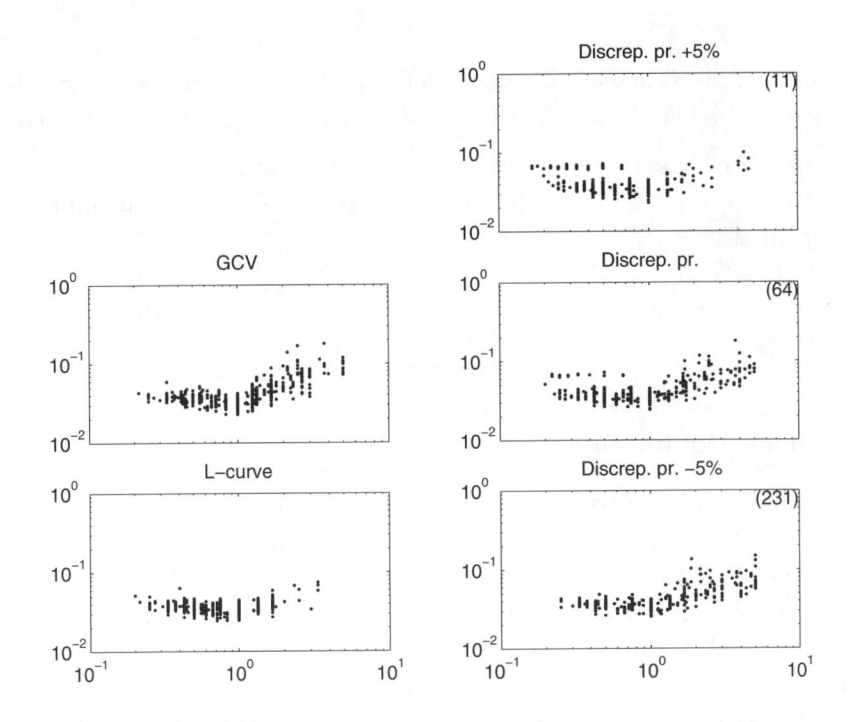

FIG. 7.8. *Relative errors η_{reg} versus normalized regularization parameters k/k_{opt} for regularizing CG iterations (LSQR) applied to the inverse helioseismology test problem.*

1. All three parameter-choice methods lead to average errors of approximately the same size.

2. Neither the discrepancy principle nor GCV are very robust.

3. The L-curve criterion is significantly more robust and, additionally, leads to the smallest relative errors.

Hence, for this regularization method and this problem, the L-curve criterion is the method of choice.

Overall, we conclude for the inverse helioseismology test problem that both Tikhonov regularization and regularizing CG iterations are capable of producing regularized solutions with relative errors typically in the range .03–.05. The L-curve criterion is always the most robust method, and for regularizing CG iterations it also produces the smallest errors, while GCV produces the smallest errors for Tikhonov's method. Both the ordinary and compensated discrepancy principles are only useful in practice if very good noise estimates are available.

TABLE 7.4. *Average and maximum relative errors and average number of matrix multiplications for LSQR and RHYBRID applied to the image deblurring test problem.*

Parameter-choice method	Average relative error	Maximum relative error	Average no. mat. mult.
"Optimal" LSQR	.39	.40	58
LSQR + GCV	.44	.52	89
LSQR + L-curve crit.	.40	.48	76
RHYBRID	.45	.57	320

7.7.2. Image Deblurring

This test problem is a first-kind Fredholm integral equation of the form (1.1). It is similar to the image deblurring test problems used in [171, §8.2] and [265], and it is included in the REGULARIZATION TOOLS package [187] as function blur. The test image is 16×16 and is shown in Fig. 7.10. The matrix A is 256×256 and it models Gaussian spatially invariant blur with point spread function $\exp\left(0.1\left((i-1)^2 + (j-1)^2\right)\right)$. Small values of the point spread function are replaced by zero, and the resulting matrix A is a block banded Toeplitz matrix with banded Toeplitz blocks. Both bandwidths are 7; i.e., only pixels within a distance 4 contribute to the blurring.

Iterative regularization methods are particularly suited for matrices with this structure, and we used LSQR as well as the RHYBRID algorithm, which implements Lanczos bidiagonalization with full reorthogonalization, implicit restarts, and GCV as the parameter-choice method; see [40] for details. The MATLAB function rhybrid was kindly provided by Eric Grimme. We used standard-form regularization ($L = I_n$), the norm of the right-hand side is $\|b^{\text{exact}}\|_2 = 6.3 \cdot 10^3$, and we added Poisson noise with Poisson parameter equal to 9.5, leading to $\mathcal{E}(\|e\|_2) = 160$ and noise-to-signal ratio $2.5 \cdot 10^{-2}$. Poisson noise is a realistic type of noise in many image processing applications where the data are bound to be nonnegative integers. See [171, §8.2] for more details.

We computed the relative error as $\eta_{\text{reg}} = \|x^{\text{exact}} - x_{\text{reg}}\|_2 / \|x^{\text{exact}}\|_2$, and the results are summarized in Table 7.4 and Fig. 7.9. The number of matrix multiplications in LSQR is $2(k+1)$, and for RHYBRID this number was found by adding a counter to the rhybrid code. Both GCV with denominator $(m-k)^2$ and the L-curve criterion, on average, produce too many iterations and therefore lead to undersmoothing for this problem. LSQR with the L-curve criterion is quite robust, and the average errors are close to the optimal errors, while LSQR with GCV is less robust and leads to larger average errors. RHYBRID is very robust, at the expense of average errors larger than those of LSQR with the L-curve criterion. Moreover, the average number of matrix multiplications is much larger than for LSQR (but fine tuning of RHYBRID to the particular

7.7. Experimental Comparisons of the Methods

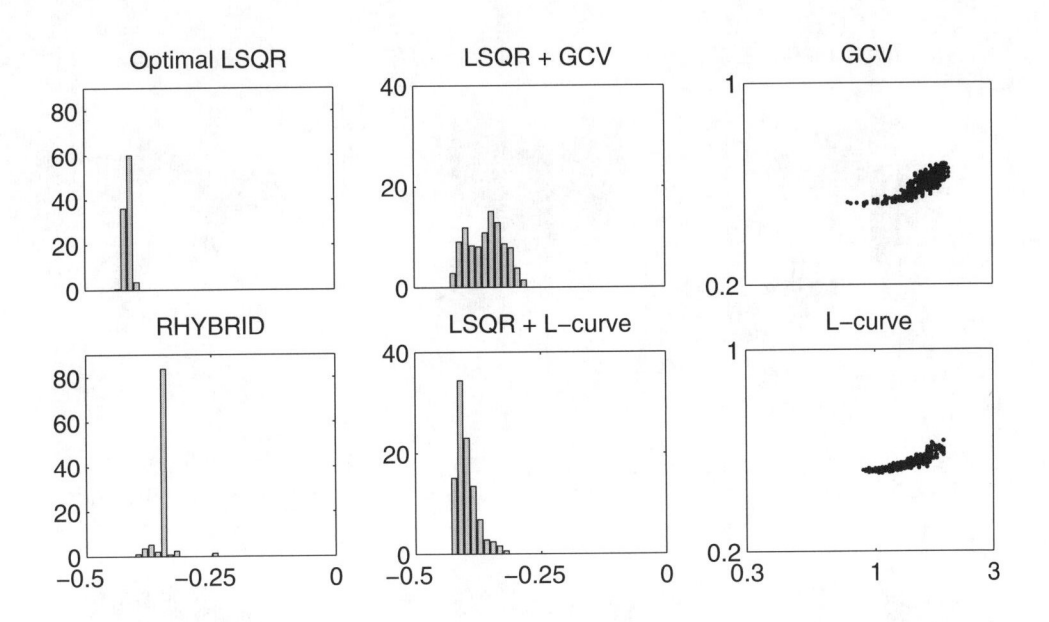

FIG. 7.9. *Error histograms and error/regularization parameter plots for iterative regularization methods applied to the image deblurring test problem.*

problem may reduce this number).

Examples of the reconstructed images for one particular test are shown in Fig. 7.10. The gray scale is fixed in all five plots, and negative pixels are displayed as zeros. The effect of taking too many iterations, which is the case for both GCV and the L-curve criterion, is clearly seen as unwanted oscillations in the reconstructed image. Although the 2-norm of the residual of the image from the RHYBRID method is larger than that of the image from LSQR with the L-curve criterion—64.4 compared to 38.3—the RHYBRID image may appear more appealing. This illustrates the difficulties of using the 2-norm for the residual in connection with image processing.

We remark that although the size of our test problem is very small compared to real applications, we can still make valid conclusions about the deblurring capabilities of the methods because the blurring in our test problem is local and spatially invariant.

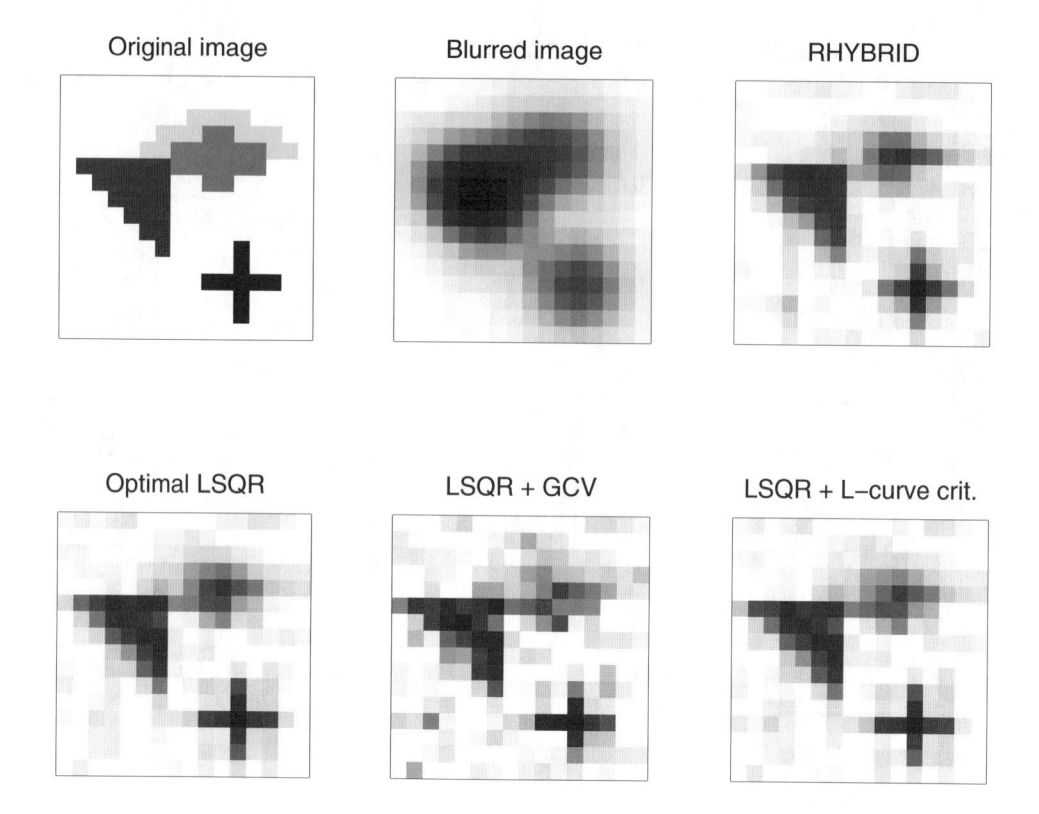

FIG. 7.10. *The original, the blurred, and the reconstructed images for the image deblurring test problem.*

8

Regularization Tools

The software package REGULARIZATION TOOLS [187], Version 3.0, consists of 53 MATLAB routines for analysis and solution of discrete ill-posed problems. By means of this package, the user can experiment with different regularization strategies, compare them, and draw conclusions that would otherwise require a major programming effort. In addition to the analysis and solution routines, the package also includes 12 test problems. The package, as well as the corresponding manual [186], is available from Netlib in the file `numeralgo/na4`, as well as from the author.

The purpose of REGULARIZATION TOOLS is to provide the user with easy-to-use routines, based on numerically robust and efficient algorithms, for doing experiments with regularization of discrete ill-posed problems. By means of this package, the user can experiment with different regularization strategies, compare them, and draw conclusions that would otherwise require a major programming effort. For discrete ill-posed problems, which are indeed difficult to treat numerically, such an approach is superior to a single black-box routine.

There have been other attempts to write general software for discrete ill-posed problems, all of them implementing Tikhonov regularization. They are listed here in chronological order.

- What appears to be the first package was written in the object-oriented language Simula by Eldén [99], and it concentrated on the use of bidiagonalization (§5.1.1) and standard-form transformation (§2.3.1) to solve the Tikhonov problem in its three incarnations (5.1), (5.14), and (5.15).

- The Fortran program *CONTIN* by Provencher [288] uses the quadrature method to discretize the integral equation (1.2). Equality and inequality constraints can be imposed on the discretized solution. The regularization parameter is chosen either by solving $\mathcal{V}(\lambda) = \sigma_0^2$ (cf. (7.9)) or via a confidence region estimate. The linear systems are solved by means of orthogonal transformations and SVD, and the inequality constraints are handled by means of the LDP least distance programming routine from [231].

- The Fortran package *ARIES* by Drake [94] uses least squares collocation and a B-spline basis (whose order can be chosen by the user) to discretize the integral equation (1.1), and cross-validation to choose the regularization parameter. The linear systems are solved by means of SVD.

- In the Fortran package *F1REGU* by te Riele [332], standard collocation and ⊓ basis functions are used to discretize the integral equation (1.1). There is no built-in parameter-choice method; the regularization parameter λ must be provided by the user. The matrix $A^T A + \lambda^2 L^T L$ is explicitly formed, followed by Cholesky factorization.

- The Fortran package *GCVPACK*, developed by Wahba and her coworkers [20], uses thin plate smoothing splines as basis functions for problems in one or several variables. The parameter-choice method is GCV, and QR factorizations and SVD are used for solving the linear systems.

- The Fortran programs *FTIKREG* and *NLREG* by Weese [369], [370] solve linear and nonlinear problems, discretized by means of the rectangle quadrature method, and optionally with nonnegativity constraints on the solution. The regularization parameter is computed by minimization of the estimate $\|(I_n - \Xi) x_\lambda\|_2^2 + \sigma_0^2 \sum_{i=1}^n f_i^2/\sigma_i^2$ of the squared norm of the total error (7.1). The unconstrained linear problem is solved by means of the author's "homemade GSVD," the constrained linear problem is solved via an active set method, and the nonlinear problem is solved via Gauss–Newton's method.

- The book [339] contains a package of Fortran routines which, in addition to Tikhonov regularization for general problems and convolution problems, solve the least squares problem with inequality constraints; cf. §5.1.4. Tikhonov regularization with the Sobolev norm $\Omega(x)^2 = \|x\|_2^2 + \|L_1 x\|_2^2$ is implemented as follows. First the Cholesky factor C of $I_n + L_1^T L_1$ is computed; then $A C^{-1}$ is formed and the bidiagonalization $A C^{-1} = \bar{U} \bar{B} \bar{V}^T$ is computed; and finally the tridiagonal matrix $\bar{B}^T \bar{B} + \lambda I_n$ is explicitly formed for each value of λ.

The packages [66] and [316], which both minimize the linear functional $w^T x$ subject to $\|A x - b\|_2 \leq \delta$ and $x_{\min} \leq x \leq x_{\max}$ (cf. Eqs. (5.65)–(5.66)) should also be mentioned. This problem arises, e.g., in gravimetry and geomagnetism.

The philosophy behind the package REGULARIZATION TOOLS is modularity and regularity between the routines—in fact, one of the inspirations was a paper by Natterer [260] where a "numerical analyst's toolkit for ill-posed problems" is suggested. Many routines in the package are based on the SVD of A or the GSVD of (A, L). For example, the Tikhonov solution x_λ is computed by means of the SVD/GSVD expansions (4.3) and (4.5). The SVD/GSVD is also used to compute the L-curve via (4.41) and (4.42), the GCV function $\mathcal{G}$ defined in (7.17), and the T-TLS solution from §3.2.2, as well as to solve the

8. REGULARIZATION TOOLS

inequality-constrained problems in (5.14) and (5.15). This is not necessarily the best approach in a given large-scale application, but it is well suited for MATLAB and for this package.

The following six tables, grouped by subject area, list all 53 MATLAB routines in the package together with a short description of their purposes. More details about all the routines, as well as a tutorial introduction to the package, can be found in the manual [186].

ANALYSIS ROUTINES	
fil_fac	Computes filter factors for some regularization methods
gcv	Plots the GCV function and computes its minimum
lagrange	Plots the Lagrange function $\|A\,x - b\|_2^2 + \lambda^2\,\|L\,x\|_2^2$ and its derivative
l_corner	Locates the L-shaped corner of the L-curve
l_curve	Computes the L-curve, plots it, and computes its corner
picard	Plots the (generalized) singular values, the Fourier coefficients for the right-hand side, and a possibly smoothed curve of the solution's Fourier-coefficients
plot_lc	Plots an L-curve
quasiopt	Plots the quasi-optimality function and computes its minimum

REGULARIZATION ROUTINES	
cgls	Computes the least squares solution based on k steps of the CG algorithm
discrep	Minimizes the solution (semi)norm subject to an upper bound on the residual norm (discrepancy principle)
dsvd	Computes a damped SVD/GSVD solution
lsqi	Minimizes the residual norm subject to an upper bound on the (semi)norm of the solution
lsqr	Computes the least squares solution based on k steps of the LSQR algorithm
maxent	Computes the maximum entropy regularized solution
mtsvd	Computes the modified TSVD solution
nu	Computes the solution based on k steps of Brakhage's iterative ν-method
pcgls	Same as cgls, but for general-form regularization
plsqr	Same as lsqr, but for general-form regularization
pnu	Same as nu, but for general-form regularization
tgsvd	Computes the TGSVD solution
tikhonov	Computes the Tikhonov regularized solution
tsvd	Computes the TSVD solution

TEST PROBLEMS	
baart	First-kind Fredholm integral equation
blur	Image deblurring test problems
deriv2	Computation of second derivative
foxgood	Severely ill posed test problem
heat	Inverse heat equation
ilaplace	Inverse Laplace transformation
parallax	Stellar parallax problem with real observations
phillips	Phillips' "famous" test problem
shaw	One-dimensional image deblurring model
spikes	Test problem with a "spiky" solution
ursell	Integral equation with no square integrable solution
wing	Test problem with a discontinuous solution

STANDARD-FORM TRANSFORMATION	
gen_form	Transforms a standard-form solution back into the general-form setting
std_form	Transforms a general-form problem into one in standard form

UTILITY ROUTINES	
bidiag	Bidiagonalization of a matrix by Householder transformations
cgsvd	Computes the compact GSVD of a matrix pair
csvd	Computes the compact SVD of an $m \times n$ matrix
get_l	Produces an $(n-d) \times n$ matrix which is the discrete approximation to the dth-order derivative operator
lanc_b	Performs k steps of the Lanczos bidiagonalization process with/without reorthogonalization
regutm	Generates random test matrices for regularization methods

AUXILIARY ROUTINES	
app_hh_l	Applies a Householder transformation from the left
gen_hh	Generates a Householder transformation
heb_new	Newton–Raphson iteration with Hebden's rational approximation, used in lsqi
lsolve	Inversion with A-weighted generalized inverse of L
ltsolve	Inversion with transposed A-weighted generalized inverse of L
newton	Newton–Raphson iteration, used in discrep
pinit	Initialization for treating general-form problems
pythag	Computes $\sqrt{a^2+b^2}$ without over/underflow
regudemo	Tutorial introduction to REGULARIZATION TOOLS
spleval	Computes points on a spline or spline curve

8. REGULARIZATION TOOLS

There are 12 built-in test problems in REGULARIZATION TOOLS, Version 3.0. Eleven of them are taken from the literature (cf. the manual pages in [186] for references) while the remaining, spikes, is "cooked up" for this package. Three of them are described in §1.4. All of them have in common that they are easy to generate, and they share the characteristic features of discrete ill-posed problems discussed in Chapter 4.

All the test problems are derived from discretizations of a Fredholm integral equation of the first kind:

$$\int_a^b K(s,t)\, f(t)\, dt = g(s)\, , \qquad c \leq s \leq d\, .$$

Both discretization techniques mentioned in §1.3 are used: the quadrature method and the Galerkin method with orthonormal basis functions.

In the quadrature method, the integral is approximated by a weighted sum,

$$\int_a^b K(s,t)\, f(t)\, dt \approx I_n(s) = \sum_{i=1}^n w_j\, K(s,t_j)\, f(t_j)\, .$$

In particular, for the midpoint rule $w_j = (b-a)/n$ and $t_j = (j-\frac{1}{2})(b-a)/n$, $j = 1, \ldots, n$. Collocation in the n points $s_1, \ldots, s_n$ then leads to the requirements $I_n(s_i) = g(s_i)$, $i = 1, \ldots, n$. Hence, we obtain a system of linear algebraic equations $A\,x = b$ with elements given by $a_{ij} = w_j\, K(s_i, t_j)$ and $b_i = g(s_i)$ for $i,j = 1, \ldots, n$; cf. Eq. (1.11). If the solution f is known then we represent it by the vector x with elements $x_j = f(t_j)$, $j = 1, \ldots, n$.

In the Galerkin method, we choose the following orthonormal ⊓ functions as basis functions:

$$\psi_i(s) = \begin{cases} h_s^{-(1/2)}, & s \in [s_{i-1}, s_i], \\ 0, & \text{elsewhere,} \end{cases} \qquad \phi_i(t) = \begin{cases} h_t^{-(1/2)}, & t \in [t_{i-1}, t_i], \\ 0, & \text{elsewhere,} \end{cases}$$

in which $h_s = (d-c)/n$, $h_t = (b-a)/n$, and $s_i = i h_s$, $t_i = i h_t$, $i = 0, \ldots, n$. Then the Galerkin method leads to a system of linear equations $A\,x = b$ with elements given by $a_{ij} = \int_a^b K(s,t)\, \phi_i(s)\, \psi_j(t)\, ds\, dt$ and $b_i = \int_c^d g(s)\, \phi_i(s)\, ds$, $i, j = 1, \ldots, n$; cf. Eq. (1.12). Similarly, we represent the solution f by the vector x with elements $x_j = \int_a^b \phi_j(t)\, f(t)\, dt$, $j = 1, \ldots, n$.

We stress that for both methods the product $A\,x$ is, in general, *different* from the vector b due to the discretization errors. Table 8.1 gives an overview of the 12 test problems, while graphs of x for $n = 100$ are given in the individual manual pages in [186].

In addition to these fixed test problems, we also provide a function regutm, implementing the algorithm in §4.7, for generating "random" test matrices with an increasing number of sign changes in the left and right singular vectors u_i and v_i as i increases.

Some extensions to REGULARIZATION TOOLS are also available from the author. For example, we provide the 212 × 100 test problem from helioseismology used in [171] and §7.7.1.

TABLE 8.1. *Test problems in* REGULARIZATION TOOLS, *Version* 3.0.

Test problem	Discretization	$Ax = b$
baart	Galerkin	no
blur	–	yes
deriv2	Galerkin	yes
foxgood	Quadrature	no
heat	Quadrature	yes
ilaplace	Quadrature	yes
parallax	Galerkin	x not known
phillips	Galerkin	no
shaw	Quadrature	yes
spikes	"Cooked up"	yes
ursell	Galerkin	no x exists
wing	Galerkin	no

Bibliography

[1] R. C. Allen, W. R. Boland, V. Faber, and G. M. Wing, *Singular values and condition numbers of Galerkin matrices arising from linear integral equations of the first kind*, J. Math. Anal. Appl., 109 (1985), pp. 564–590.

[2] R. C. Allen, W. R. Boland, and G. M. Wing, *Numerical experiments involving Galerkin and collocation methods for linear integral equations of the first kind*, J. Comput. Phys., 49 (1983), pp. 465–477.

[3] U. Amato and W. Hughes, *Maximum entropy regularization of Fredholm integral equations of the first kind*, Inverse Problems, 7 (1991), pp. 793–808.

[4] E. Anderson, Z. Bai, C. Bischof, J. Demmel, J. Dongarra, J. Du Croz, A. Greenbaum, S. Hammarling, A. McKenney, O. Ostrouchow, and D. Sorensen, *LAPACK Users' Guide*, Second Edition, SIAM, Philadelphia, 1995.

[5] R. S. Anderssen, *On the use of linear functionals for Abel-type integral equations in applications*, in R. S. Anderssen, F. R. de Hoog, and M. A. Lukas (Eds.), The Application and Numerical Solution of Integral Equations, Sijthoff and Noordhoff, Leyden, the Netherlands, 1980, pp. 119–134.

[6] R. S. Anderssen, *The linear functional strategy for improperly posed problems*, in J. R. Cannon and U. Hornung (Eds.), Inverse Problems, Birkhäuser, Boston, 1986, pp. 11–30.

[7] R. S. Anderssen and P. M. Prenter, *A formal comparison of methods proposed for the numerical solution of first kind integral equations*, J. Austral. Math. Soc. Ser. B, 22 (1981), pp. 488–500.

[8] A. L. Andrew, *Eigenvalues and singular values of certain random matrices*, J. Comput. Appl. Math., 30 (1990), pp. 165–171.

[9] H. C. Andrews and B. R. Hunt, *Digital Image Restoration*, Prentice-Hall, Englewood Cliffs, NJ, 1977.

[10] G. Anger, *Inverse Problems in Differential Equations*, Akademie-Verlag, Berlin, 1990.

[11] C. M. Aulick and T. M. Gallie, *Isolating error effects in solving ill-posed problems*, SIAM J. Alg. Disc. Meth., 4 (1983), pp. 371–376.

[12] O. Axelsson, *Iterative Solution Methods*, Cambridge University Press, Cambridge, UK, 1994.

[13] M. L. Baart, *The use of auto-correlation for pseudo-rank determination in noisy ill-conditioned linear least-squares problems*, IMA J. Numer. Anal., 2 (1982), pp. 241–247.

[14] E. Babolian and L. M. Delves, *An augmented Galerkin method for first kind Fredholm equations*, J. IMA, 24 (1979), pp. 157–174.

[15] G. E. Backus and J. F. Gilbert, *The resolving power of gross earth data*, Geophys. J. R. Astron. Soc., 16 (1968), pp. 169–205.

[16] Z. Bai and J. Demmel, *Computing the generalized singular value decomposition*, SIAM J. Sci. Comput., 14 (1993), pp. 1464–1486.

[17] C. T. H. Baker, *The Numerical Treatment of Integral Equations*, Clarendon Press, Oxford, UK, 1977.

[18] C. T. H. Baker, L. Fox, D. F. Miller, and K. Wright, *Numerical solution of Fredholm integral equations of first kind*, Comput. J., 7 (1964), pp. 141–148.

[19] R. Barakat and G. Newsam, *Remote sensing of the refractive index structure parameter via inversion of Tatarski's integral equation for both spherical and plane wave situations*, Radio Science, 19 (1984), pp. 1041–1056.

[20] D. M. Bates, M. J. Lindstrom, G. Wahba, and B. S. Yandell, *GCVPACK – routines for generalized cross validation*, Comm. Statist. Simulation Comput., 16 (1987), pp. 263–297.

[21] J. Baumeister, *Stable Solution of Inverse Problems*, Vieweg, Braunschweig, Germany, 1987.

[22] J. A. Belward, *Further studies of the application of constrained minimization methods to Fredholm integral equations of the first kind*, IMA J. Numer. Anal., 5 (1985), pp. 125–139.

[23] M. W. Berry, *Large-scale sparse singular value computations*, Internat. J. Supercomput. Appl., 6 (1992), pp. 13–49.

[24] M. W. Berry and G. H. Golub, *Estimating the largest singular values of large sparse matrices via modified moments*, Numer. Algorithms, 1 (1991), pp. 353–374.

[25] M. W. Berry and A. Sameh, *An overview of parallel algorithms for the singular value and symmetric eigenvalue problems*, J. Comput. Appl. Math., 27 (1989), pp. 191–213.

[26] M. Bertero, P. Boccacci, G. J. Brakenhoff, F. Malfanti, and H. T. M. van der Voort, *Three-dimensional image restoration and super-resolution in fluorescence confocal microscopy*, J. Microscopy, 157 (1990), pp. 3–20.

[27] M. Bertero and C. De Mol, *SVD for linear inverse problems*, in [250], pp. 341–348.

[28] M. Bertero, C. De Mol, and E. R. Pike, *Linear inverse problems with discrete data: I. General formulation and singular system analysis*, Inverse Problems, 1 (1985), pp. 301–330.

[29] M. Bertero, C. De Mol, and E. R. Pike, *Applied inverse problems in optics*, in [110], pp. 291–313.

[30] M. Bertero, T. A. Poggio, and V. Torre, *Ill-posed problems in early vision*, Proc. IEEE, 76 (1988), pp. 869–889.

[31] C. H. Bischof and P. C. Hansen, *Structure preserving and rank-revealing QR-factorizations*, SIAM J. Sci. Stat. Comput., 12 (1991), pp. 1332–1350.

[32] C. H. Bischof and P. C. Hansen, *A block algorithm for computing rank-revealing QR factorizations*, Numer. Algorithms, 2 (1992), pp. 371–392.

[33] C. H. Bischof and G. Quintana-Ortí, *Computing rank-revealing QR factorizations of dense matrices*, ACM Trans. Math. Software, to appear.

[34] C. H. Bischof and G. M. Shroff, *On updating signal subspaces*, IEEE Trans. Signal Process., SP-40 (1992), pp. 96–105.

[35] Å. Björck, *A bidiagonalization algorithm for solving large and sparse ill-posed systems of linear equations*, BIT, 28 (1988), pp. 659–670.

[36] Å. Björck, *Numerical Methods for Least Squares Problems*, SIAM, Philadelphia, 1996.

[37] Å. Björck and L. Eldén, *Methods in Numerical Algebra for Ill-Posed Problems*, Report LiTH-R-33-1979, Dept. of Mathematics, Linköping University, Sweden, 1979.

[38] Å. Björck and T. Elfving, *Accelerated projection methods for computing pseudoinverse solutions of systems of linear equations*, BIT, 19 (1979), pp. 145–163.

[39] Å. Björck, T. Elfving, and Z. Strakos, *Stability of conjugate gradient and Lanczos methods for linear least squares problems*, SIAM J. Matrix Anal. Appl., to appear.

[40] Å. Björck, E. Grimme, and P. M. Van Dooren, *An implicit shift bidiagonalization algorithm for ill-posed systems*, BIT, 34 (1994), pp. 510–534.

[41] A. W. Bojanczyk, L. M. Ewerbring, F. T. Luk, and P. Van Dooren, *An accurate product SVD algorithm*, Signal Processing, 25 (1991), pp. 189–201. This paper can also be found in [343].

[42] A. W. Bojanczyk and J. M. Lebak, *Downdating a ULLV decomposition of two matrices*, in J. G. Lewis (Ed.), Proceedings of the Fifth SIAM Conference on Applied Linear Algebra, SIAM, Philadelphia, 1994, pp. 261–265.

[43] H. Brakhage, *On ill-posed problems and the method of conjugate gradients*, in [110], pp. 165–175.

[44] J. A. Cadzow and D. M. Wilkes, *Enhanced rational signal modeling*, Signal Processing, 25 (1991), pp. 171–188.

[45] A. S. Carasso, *Overcoming Hölder discontinuity in ill-posed continuation problems*, SIAM J. Numer. Anal., 31 (1994), pp. 1535–1557.

[46] Y. Censor, *Row-action methods for huge and sparse systems and their applications*, SIAM Review, 23 (1981), pp. 444–464.

[47] R. H. Chan, J. G. Nagy, and R. J. Plemmons, *FFT-based preconditioners for Toeplitz-block least squares problems*, SIAM J. Numer. Anal., 30 (1993), pp. 1740–1768.

[48] T. F. Chan, *An improved algorithm for computing the singular value decomposition*, ACM Trans. Math. Software, 8 (1982), pp. 72–83.

[49] T. F. Chan, *On the existence and computation of LU-factorizations with small pivots*, Math. Comp., 42 (1984), pp. 535–547.

[50] T. F. Chan, *Rank revealing QR-factorizations*, Linear Algebra Appl., 88/89 (1987), pp. 67–82.

[51] T. F. Chan and P. C. Hansen, *Computing truncated SVD least squares solutions by rank revealing QR factorizations*, SIAM J. Sci. Stat. Comput., 11 (1990), pp. 519–530.

[52] T. F. Chan and P. C. Hansen, *Some applications of the rank revealing QR factorization*, SIAM J. Sci. Stat. Comput., 13 (1992), pp. 727–741.

[53] T. F. Chan and P. C. Hansen, *Low-rank revealing QR factorizations*, Numer. Linear Algebra Appl., 1 (1994), pp. 33–44.

[54] T. F. Chan and D. E. Foulser, *Effectively well-conditioned linear systems*, SIAM J. Sci. Stat. Comput., 9 (1988), pp. 963–969.

[55] T. F. Chan, J. Olkin, and D. W. Cooley, *Solving quadratically constrained least squares using black box unconstrained solvers*, BIT, 32 (1992), pp. 481–495.

[56] S. Chandrasekaran and I. Ipsen, *On rank-revealing QR factorizations*, SIAM J. Matrix Anal. Appl., 15 (1994), pp. 592–622.

[57] S. Chandrasekaran and I. Ipsen, *On the sensitivity of solution components in linear systems of equations*, SIAM J. Matrix Anal. Appl., 16 (1995), pp. 93–112.

[58] J. Christensen-Dalsgaard, P. C. Hansen, and M. Thompson, *GSVD analysis of helioseismic inversions*, Month. Not. R. Astr. Soc., 264 (1993), pp. 541–465.

[59] J. Christensen-Dalsgaard, J. Schou, and M. Thompson, *A comparison of methods for inverting helioseismic data*, Month. Not. R. Astr. Soc., 242 (1990), pp. 353–369.

[60] S. Christiansen and P. C. Hansen, *An analysis of the boundary collocation method based on the effective condition number*, J. Comput. Appl. Math., 54 (1994), pp. 15–36.

[61] J. A. Cochran, *The Analysis of Linear Integral Equations*, McGraw-Hill, New York, 1972.

[62] D. Colton and R. Kress, *Integral Equation Methods for Scattering Theory*, Wiley, New York, 1983.

[63] R. Courant and D. Hilbert, *Methods of Mathematical Physics, Vol. II*, Interscience, New York, 1953.

[64] M. G. Cox, *Data approximation by splines in one and two independent variables*, in A. Iserles and M. J. D. Powell (Eds.), The State of the Art in Numerical Analysis, Clarendon Press, Oxford, UK, 1987, pp. 111–138.

[65] I. J. D. Craig and J. C. Brown, *Inverse Problems in Astronomy*, Adam Hilger, Bristol, UK, 1986.

[66] M. Cuer and R. Bayer, *A Package of Routines for Linear Inverse Problems*, Report, Dép. Physique Mathématique, Université des Sciences et Techniques du Languedoc, Montpellier, France, 1979.

[67] J. K. Cullum, *The effective choice of the smoothing norm in regularization*, Math. Comp., 33 (1979), pp. 149–170.

[68] J. K. Cullum, *Peaks and Plateaus in Lanczos Methods for Solving Nonsymmetric Systems of Equations $Ax = b$*, Report RC 18084, IBM T. J. Watson Research Center, Yorktown Heights, NY, 1992. Portions of this report are published in [69].

[69] J. K. Cullum, *Peaks, plateaus, numerical instabilities in a Galerkin minimal residual pair of methods for solving $Ax = b$*, Appl. Numer. Math., 19 (1995), pp. 255–278.

[70] J. K. Cullum and A. Greenbaum, *Relations between Galerkin and norm-minimizing iterative methods for solving linear systems*, SIAM J. Matrix Anal. Appl., 17 (1996), pp. 223–247.

[71] J. K. Cullum and R. A. Willoughby, *Lanczos Algorithms for Large Symmetric Eigenvalue Problems*, Birkhäuser, Boston, 1985.

[72] J. J. M. Cuppen, *A Numerical Solution of the Inverse Problem of Electrocardiography*, Ph.D. Thesis, Dept. of Mathematics, University of Amsterdam, 1983.

[73] A. R. Davies, *Optimality in regularization*, in M. Bertero and E. R. Pike (Eds.), Inverse Problems in Scattering and Imaging, Adam Hilger, Bristol, UK, 1992, pp. 393–410.

[74] A. R. Davies and M. F. Hassan, *Optimality in the regularization of ill-posed inverse problems*, in P. C. Sabatier (Ed.), Inverse Problems: An Interdisciplinary Study, Academic Press, London, UK, 1987.

[75] M. E. Davison, *A singular value decomposition for the Radon transform in n-dimensional space*, Numer. Funct. Anal. Optim., 3 (1981), pp. 321–340.

[76] A. Dax, *On regularized least norm problems*, SIAM J. Optim., 2 (1992), pp. 602–618.

[77] A. Dax, *On row relaxation methods for large constrained least squares problems*, SIAM J. Sci. Comput., 14 (1993), pp. 570–584.

[78] A. Dax and L. Eldén, *Iterative Improvement of Regularized Solutions via Bidiagonalization*, Report MAT-R-1996-12, Dept. of Mathematics, Linköping University, Sweden, 1996; J. Numer. Linear Algebra Appl., to appear.

[79] F. R. de Hoog, *Review of Fredholm equations of the first kind*, in R. S. Anderssen, F. R. de Hoog, and M. A. Lukas (Eds.), The Application and Numerical Solution of Integral Equations, Sijthoff and Noordhoff, Leyden, the Netherlands, 1980, pp. 119–134.

[80] L. M. Delves and J. L. Mohamed, *Computational Methods for Integral Equations*, Cambridge University Press, Cambridge, UK, 1985.

[81] L. M. Delves and J. Walsh (Eds.), *Numerical Solution of Integral Equations*, Clarendon Press, Oxford, 1974.

[82] J. W. Demmel, *Applied Numerical Linear Algebra*, SIAM, Philadelphia, 1997.

[83] J. Demmel, M. T. Heath, and H. A. van der Vorst, *Parallel numerical linear algebra*, Acta Numerica, 2 (1993), pp. 111–199.

[84] C. De Mol, *A critical survey of regularized inversion methods*, in M. Bertero and E. R. Pike (Eds.), Inverse Problems in Scattering and Imaging, Adam Hilger, Bristol, UK, 1992, pp. 345–370.

[85] G. Demoment, *Image reconstruction and restoration: Overview of common estimation structures and problems*, IEEE Trans. Acoust. Speech Signal Process., ASSP-37 (1987), pp. 2024–2036.

[86] B. L. R. De Moor, *Generalizations of the OSVD: Structure, properties and applications*, in [343], pp. 83–98.

[87] B. L. R. De Moor, *On the structure and geometry of the product singular value decomposition*, Linear Algebra Appl., 168 (1992), pp. 95–136.

[88] B. L. R. De Moor, *The singular value decomposition and long and short spaces of noisy matrices*, IEEE Trans. Signal Process., 41 (1993), pp. 2826–2838.

[89] B. L. R. De Moor, *Structured total least squares and L_2 approximation problems*, Linear Algebra Appl., 188/189 (1993), pp. 163–205.

[90] B. L. R. De Moor and G. H. Golub, *The restricted singular value decomposition: Properties and applications*, SIAM J. Matrix Anal. Appl., 12 (1991), pp. 401–425.

[91] B. L. R. De Moor and H. Zha, *A tree of generalizations of the ordinary singular value decomposition*, Linear Algebra Appl., 147 (1991), pp. 469–500.

[92] E. F. Deprettere (Ed.), *SVD and Signal Processing. Algorithms, Applications and Architectures*, North-Holland, Amsterdam, 1988.

[93] J. J. Dongarra, J. R. Bunch, C. B. Moler, and G. W. Stewart, *Linpack Users' Guide*, SIAM, Philadelphia, 1979.

[94] J. B. Drake, *ARIES: A Computer Program for the Solution of First Kind Integral Equations with Noisy Data*, Report K/CSD/TM-43, Dept. of Computer Science, Oak Ridge National Laboratory, Oak Ridge, TN, October 1983.

[95] A. Edelman, *Eigenvalues and Condition Numbers of Random Matrices*, Ph.D. Thesis, Report 89-7, Dept. of Mathematics, MIT, Cambridge, MA, May 1989.

[96] B. Eicke, A. K. Louis and R. Plato, *The instability of some gradient methods for ill-posed problems*, Numer. Math., 58 (1990), pp. 129–134.

[97] M. P. Ekstrom and R. L. Rhodes, *On the application of eigenvector expansions to numerical deconvolution*, J. Comput. Phys., 14 (1974), pp. 319–340.

[98] L. Eldén, *Algorithms for the regularization of ill-conditioned least squares problems*, BIT, 17 (1977), pp. 134–145.

[99] L. Eldén, *A Program for Interactive Regularization*, Report LiTH-MAT-R-79-25, Dept. of Mathematics, Linköping University, Sweden, 1979.

[100] L. Eldén, *A weighted pseudoinverse, generalized singular values, and constrained least squares problems*, BIT, 22 (1982), pp. 487–501.

[101] L. Eldén, *An efficient algorithm for the regularization of ill-conditioned least squares problems with triangular Toeplitz matrix*, SIAM J. Sci. Stat. Comput., 5 (1984), pp. 229–236.

[102] L. Eldén, *An algorithm for the regularization of ill-conditioned, banded least squares problems*, SIAM J. Sci. Stat. Comput., 5 (1984), pp. 237–254.

[103] L. Eldén, *A note on the computation of the generalized cross-validation function for ill-conditioned least squares problems*, BIT, 24 (1984), pp. 467–472.

[104] L. Eldén, *Algorithms for the computation of functionals defined on the solution of a discrete ill-posed problem*, BIT, 30 (1990), pp. 466–483.

[105] T. Elfving, *Some Numerical Results Obtained with Two Gradient Methods for Solving the Linear Least Squares Problem*, Report LiTH-MAT-R-75-5, Dept. of Mathematics, Linköping University, Sweden, 1978.

[106] T. Elfving, *An algorithm for maximum entropy image reconstruction from noisy data*, Math. Comput. Modelling, 12 (1989), pp. 729–745.

[107] H. W. Engl, *Regularization methods for the stable solution of inverse problems*, Surveys Math. Indust., 3 (1993), pp. 71–143.

[108] H. W. Engl and H. Gfrerer, *A posteriori parameter choice for general regularization methods for solving linear ill-posed problems*, Appl. Numer. Math., 4 (1988), pp. 395–417.

[109] H. W. Engl and W. Grever, *Using the L-curve for determining optimal regularization parameters*, Numer. Math., 69 (1994), pp. 25–31.

[110] H. W. Engl and C. W. Groetsch (Eds.), *Inverse and Ill-Posed Problems*, Academic Press, London, 1987.

[111] H. W. Engl, M. Hanke, and A. Neubauer, *Regularization of Inverse Problems*, Kluwer, Dordrecht, the Netherlands, 1996.

[112] W. A. Essah and L. M. Delves, *The numerical solution of first kind integral equations*, J. Comp. Appl. Math., 27 (1989), pp. 363–387.

[113] L. M. Ewerbring and F. T. Luk, *Canonical correlations and generalized SVD: Applications and algorithms*, in Real Time Signal Processing XI, Proc. SPIE, Vol. 977, 1988, pp. 206–222; also published in J. Comp. Appl. Math., 27 (1989), pp. 37–52.

[114] V. Faber, A. Manteuffel, A. B. White Jr., and G. M. Wing, *Asymptotic behavior of singular values and singular functions of certain convolution operators*, Comput. Math. Appl., 12A (1986), pp. 733–747.

[115] V. Faber and G. M. Wing, *Asymptotic behavior of singular values of convolution operators*, Rocky Mountain J. Math., 16 (1986), pp. 567–574.

[116] V. Faber and G. M. Wing, *Singular values of fractional integral operators: A unification of theorems of Hille, Tamakin, and Chang*, J. Math. Anal. Appl., 120 (1986), pp. 745–760.

[117] V. Faber and G. M. Wing, *Effective bounds for the singular values of integral operators*, J. Integral Equations Appl., 1 (1988), pp. 55–64.

[118] K. V. Fernando and S. J. Hammarling, *A product induced singular value decomposition (ΠSVD) for two matrices and balanced realisation*, in B. N. Datta, C. R. Johnson, M. A. Kaashoek, R. J. Plemmons, and E. D. Sontag (Eds.), Linear Algebra in Signals, Systems, and Control, SIAM, Philadelphia, 1987, pp. 128–140.

[119] R. D. Fierro, *Perturbation analysis for two-sided (or complete) orthogonal decompositions*, SIAM J. Matrix Anal. Appl., 17 (1996), pp. 383–400.

[120] R. D. Fierro and J. R. Bunch, *Collinearity and total least squares*, SIAM J. Matrix Anal. Appl., 15 (1994), pp. 1167–1181.

[121] R. D. Fierro and J. R. Bunch, *Bounding the subspaces from rank revealing two-sided orthogonal decompositions*, SIAM J. Matrix Anal. Appl., 16 (1995), pp. 743–759.

[122] R. D. Fierro and J. R. Bunch, *Orthogonal projections and total least squares*, Numer. Lininer Algebra Appl., 2 (1995), pp. 135–153.

[123] R. D. Fierro and J. R. Bunch, *Perturbation theory for orthogonal projection methods with application to least squares and total least squares*, Linear Algebra Appl., 234 (1996), pp. 71–96.

[124] R. D. Fierro, G. H. Golub, P. C. Hansen, and D. P. O'Leary, *Regularization by truncated total least squares*, SIAM J. Sci. Comput., 18 (1997), pp. 1223–1241.

[125] R. D. Fierro and P. C. Hansen, *Low-rank revealing two-sided orthogonal decompositions*, Numer. Algorithms, 15 (1997), pp. 37–55.

[126] R. D. Fierro and P. C. Hansen, *Accuracy of TSVD solutions computed from rank-revealing decompositions*, Numer. Math., 70 (1995), pp. 453–471.

[127] A. A. Figueiras-Vidal, D. Docampo-Amoedo, J. R. Casar-Corredera, and A. Artes-Rodriguez, *Adaptive iterative algorithms for spiky deconvolution*, IEEE Trans. Acoust. Speech Signal Process., 38 (1990), pp. 1462–1466.

[128] H. E. Fleming, *Equivalence of regularization and truncated iteration in the solution of ill-posed image reconstruction problems*, Linear Algebra Appl., 130 (1990), pp. 133–150.

[129] G. E. Forsythe, M. A. Malcolm, and C. B. Moler, *Computer Methods for Mathematical Computations*, Prentice-Hall, Englewood Cliffs, NJ, 1977.

[130] L. V. Foster, *Rank and null space calculations using matrix decomposition without column interchanges*, Linear Algebra Appl., 74 (1986), pp. 47–71.

[131] L. V. Foster, *The probability of large diagonal elements in the QR factorization*, SIAM J. Sci. Stat. Comput., 11 (1990), pp. 531–544.

[132] J. N. Franklin, *Well-posed stochastic extensions to ill-posed linear problems*, J. Math. Anal. Appl., 31 (1970), pp. 682–716.

[133] J. N. Franklin, *Minimum principles for ill-posed problems*, SIAM J. Math. Anal., 9 (1978), pp. 638–650.

[134] B. R. Frieden, *Estimating occurrence laws with maximum probability, and the transition to entropic estimators*, in [311], pp. 133–169.

[135] W. Gander, *Least squares with a quadratic constraint*, Numer. Math., 36 (1981), pp. 291–307.

[136] B. S. Garbow, J. M. Boyle, J. J. Dongarra, and C. B. Moler, *Matrix Eigensystem Routines: EISPACK Guide Extension*, Springer-Verlag, New York, 1977.

[137] J. A. George and M. T. Heath, *Solution of sparse linear least squares problems using Givens rotations*, Linear Algebra Appl., 34 (1980), pp. 69–83.

[138] H. Gfrerer, *An a posteriori parameter choice for ordinary and iterated Tikhonov regularization of ill-posed problems leading to optimal convergence rates*, Math. Comp., 49 (1987), pp. 507–522.

[139] D. S. Gilliam, J. R. Lund, and C. R. Vogel, *Quantifying information content for ill-posed problems*, Inverse Problems, 6 (1990), pp. 725–736.

[140] D. A. Girard, *A fast 'Monte-Carlo cross-validation' procedure for large least squares problems with noisy data*, Numer. Math., 56 (1989), pp. 1–23.

[141] D. A. Girard, *Asymptotic optimality of fast randomized versions of GCV and C_L in ridge regression and regularization*, Ann. Statist., 19 (1991), pp. 1950–1963.

[142] V. B. Glasko, *Inverse Problems of Mathematical Physics*, American Institute of Physics, New York, NY, 1998 (translation of Russian book from 1984).

[143] A. A. Goldstein, *Convex programming in Hilbert space*, Bull. AMS, 70 (1964), pp. 709–710.

[144] G. H. Golub, *Numerical methods for solving linear least squares problems*, Numer. Math., 7 (1965), pp. 206–216.

[145] G. H. Golub, P. C. Hansen, and D. P. O'Leary, *Tikhonov Regularization and Total Least Squares*, Technical Report IMM-REP-1997-15, Dept. of Mathematical Modelling, Technical University of Denmark, Lyngby, Denmark, 1997; submitted to SIAM J. Matrix Anal. Appl.

[146] G. H. Golub, M. T. Heath, and G. Wahba, *Generalized cross-validation as a method for choosing a good ridge parameter*, Technometrics, 21 (1979), pp. 215–223.

[147] G. H. Golub, A. Hoffman, and G. W. Stewart, *A generalization of the Eckart–Young–Mirsky matrix approximation theorem*, Linear Algebra Appl., 88/89 (1987), pp. 317–327.

[148] G. H. Golub and W. Kahan, *Calculating the singular values and pseudoinverse of a matrix*, SIAM J. Numer. Anal. Ser. B, 2 (1965), pp. 205–224.

[149] G. H. Golub, V. Klema, and G. W. Stewart, *Rank Degeneracy and Least Squares*, Report TR-456, Computer Science Dept., University of Maryland, College Park, MD, June 1976.

[150] G. H. Golub, F. T. Luk, and M. L. Overton, *A block Lanczos method for computing the singular values and corresponding singular vectors of a matrix*, ACM Trans. Math. Software, 7 (1981), pp. 149–169.

[151] G. H. Golub and C. Reinsch, *Singular value decomposition and least squares solutions*, Numer. Math., 14 (1970), pp. 403–420.

[152] G. H. Golub and P. Van Dooren, *Numerical Linear Algebra, Digital Signal Processing and Parallel Algorithms*, NATO ASI Series, Springer-Verlag, Berlin, 1991.

[153] G. H. Golub and C. F. Van Loan, *An analysis of the total least squares problem*, SIAM J. Numer. Anal., 17 (1980), pp. 883–893.

[154] G. H. Golub and C. F. Van Loan, *Matrix Computations*, Third Edition, the Johns Hopkins University Press, Baltimore, MD, 1996.

[155] G. H. Golub and J. M. Varah, *On a characterization of the best ℓ_2-scaling of a matrix*, SIAM J. Numer. Anal., 11 (1974), pp. 472–479.

[156] G. H. Golub and U. von Matt, *Quadratically constrained least squares and quadratic problems*, Numer. Math., 59 (1991), pp. 561–580.

[157] J. Graves and P. M. Prenter, *Numerical iterative filters applied to first kind Fredholm integral equations*, Numer. Math., 30 (1978), pp. 281–299.

[158] A. Greenbaum and Z. Strakos, *Predicting the behavior of finite precision Lanczos and conjugate gradient computations*, SIAM J. Matrix Anal. Appl., 13 (1992), pp. 121–137.

[159] C. W. Groetsch, *Generalized Inverses of Linear Operators*, Marcel Dekker, New York, 1977.

[160] C. W. Groetsch, *The Theory of Tikhonov Regularization for Fredholm Equations of the First Kind*, Research Notes in Mathematics 105, Pitman, Boston, 1984.

[161] C. W. Groetsch, *Inverse Problems in the Mathematical Sciences*, Vieweg, Wiesbaden, Germany, 1993.

[162] C. W. Groetsch and C. R. Vogel, *Asymptotic theory of filtering for linear operator equations with discrete noisy data*, Math. Comp., 49 (1987), pp. 499–506.

[163] M. Gu and S. C. Eisenstat, *Efficient algorithms for computing a strong rank-revealing QR factorization*, SIAM J. Sci. Comput., 17 (1996), pp. 848–869.

[164] J. Hadamard, *Lectures on Cauchy's Problem in Linear Partial Differential Equations*, Yale University Press, New Haven, CT, 1923.

[165] S. Hammarling, *Parallel algorithms for singular value problems*, in [152], pp. 173–187.

[166] M. Hanke, *Accelerated Landweber iterations for the solution of ill-posed equations*, Numer. Math., 60 (1991), pp. 341–373.

[167] M. Hanke, *Regularization with differential operators: An iterative approach*, Numer. Funct. Anal. Optim., 13 (1992), pp. 523–540.

[168] M. Hanke, *Iterative solution of underdetermined linear systems by transformation to standard form*, in Proceedings Numerical Mathematics in Theory and Practice, Dept. of Mathematics, University of West Bohemia, Plzeň, Czech Republic, 1993, pp. 55–63.

[169] M. Hanke, *Conjugate Gradient Type Methods for Ill-Posed Problems*, Pitman Research Notes in Mathematics 327, Longman, Harlow, UK, 1995.

[170] M. Hanke, *Limitations of the L-curve method in ill-posed problems*, BIT, 36 (1996), pp. 287–301.

[171] M. Hanke and P. C. Hansen, *Regularization methods for large-scale problems*, Surveys Math. Indust., 3 (1993), pp. 253–315.

[172] M. Hanke and J. G. Nagy, *Restoration of atmosperically blurred images by symmetric indefinite conjugate gradient techniques*, Inverse Problems, 12 (1996), pp. 157–173.

[173] M. Hanke, J. G. Nagy, and R. J. Plemmons, *Preconditioned iterative regularization methods for ill-posed problems*, in L. Reichel, A. Ruttan, and R. S. Varga (Eds.), Numerical Linear Algebra, de Gruyter, Berlin, Germany, 1993, pp. 141–163.

[174] M. Hanke and T. Raus, *A general heuristic for choosing the regularization parameter in ill-posed problems*, SIAM J. Sci. Comput., 17 (1996), pp. 956–972.

[175] P. C. Hansen, *Detection of near-singularity in Cholesky and LDL^T Factorizations*, J. Comp. Appl. Math., 19 (1987), pp. 293–299.

[176] P. C. Hansen, *The truncated SVD as a method for regularization*, BIT, 27 (1987), pp. 534–553.

[177] P. C. Hansen, *The 2-norm of random matrices*, J. Comp. Appl. Math., 23 (1988), pp. 117–120.

[178] P. C. Hansen, *Computation of the singular value expansion*, Computing, 40 (1988), pp. 185–199.

[179] P. C. Hansen, *Regularization, GSVD and truncated GSVD*, BIT, 29 (1989), pp. 491–504.

[180] P. C. Hansen, *Perturbations bounds for discrete Tikhonov regularization*, Inverse Problems, 5 (1989), pp. L41–L44.

[181] P. C. Hansen, *Truncated SVD solutions to discrete ill-posed problems with ill-determined numerical rank*, SIAM J. Sci. Stat. Comput., 11 (1990), pp. 503–518.

[182] P. C. Hansen, *Relations between SVD and GSVD of discrete regularization problems in standard and general form*, Linear Algebra Appl., 141 (1990), pp. 165–176.

[183] P. C. Hansen, *The discrete Picard condition for discrete ill-posed problems*, BIT, 30 (1990), pp. 658–672.

[184] P. C. Hansen, *Analysis of discrete ill-posed problems by means of the L-curve*, SIAM Review, 34 (1992), pp. 561–580.

[185] P. C. Hansen, *Numerical tools for analysis and solution of Fredholm integral equations of the first kind*, Inverse Problems, 8 (1992), pp. 849–872.

[186] P. C. Hansen, *Regularization Tools. A Matlab Package for Analysis and Solution of Discrete Ill-Posed Problems*, Manual and Tutorial, Technical Report, Dept. of Mathematical Modelling, Technical University of Denmark, 1998. The software is published in [187] and is available from Netlib as na4 in the directory numeralgo, as well as from the author.

[187] P. C. Hansen, *Regularization Tools: A Matlab package for analysis and solution of discrete ill-posed problems*, Numer. Algorithms, 6 (1994), pp. 1–35.

[188] P. C. Hansen, *The Backus-Gilbert method: SVD analysis and fast implementation*, Inverse Problems, 10 (1994), pp. 895–904.

[189] P. C. Hansen, *Test matrices for regularization methods*, SIAM J. Sci. Comput., 16 (1995), pp. 506–512.

[190] P. C. Hansen, *Rank deficient prewhitening by QSVD and QUTV*, BIT, to appear.

[191] P. C. Hansen and S. Christiansen, *An SVD analysis of linear algebraic equations derived from first kind integral equations*, J. Comput. Appl. Math., 12–13 (1985), pp. 341–357.

[192] P. C. Hansen and H. Gesmar, *Fast orthogonal decomposition of ill-conditioned Toeplitz matrices*, Numer. Algorithms, 4 (1993), pp. 151–166.

[193] P. C. Hansen and M. Hanke, *A Lanczos algorithm for computing the largest quotient singular values in regularization problems*, in [250], pp. 131–138.

[194] P. C. Hansen and S. H. Jensen, *FIR filter representations of reduced-rank noise reduction*, IEEE Trans. Signal Process., to appear.

[195] P. C. Hansen and K. Mosegaard, *Piecewise polynomial solutions without a priori break points*, Numer. Linear Algebra Appl., 3 (1996), pp. 513–524.

[196] P. C. Hansen and D. P. O'Leary, *The use of the L-curve in the regularization of discrete ill-posed problems*, SIAM J. Sci. Comput., 14 (1993), pp. 1487–1503.

[197] P. C. Hansen and D. P. O'Leary, *Regularization algorithms based on total least squares*, in S. Van Huffel (Ed.), Recent Advances in Total Least Squares Techniques and Errors-in-Variables Modeling, SIAM, Philadelphia, 1997, pp. 127–137.

[198] P. C. Hansen, D. P. O'Leary, and G. W. Stewart, unpublished results, presented at the XII Householder Symposium, Lake Arrowhead, CA, 1993.

[199] P. C. Hansen, T. Sekii, and H. Shibahashi, *The modified truncated SVD method for regularization in general form*, SIAM J. Sci. Stat. Comput., 13 (1992), pp. 1142–1150.

[200] P. S. K. Hansen, P. C. Hansen, S. D. Hansen, and J. Aa. Sørensen, *Noise reduction of speech signals using the rank-revealing ULLV decomposition*, in G. Ramponi, G. L. Sicuranza, S. Carrato, and S. Marsi (Eds.), Signal Processing VIII: Theories and Applications, EUSIPCO, Trieste, Italy, 1996, pp. 967–970.

[201] R. J. Hanson, *A numerical method for solving Fredholm integral equations of the first kind using singular values*, SIAM J. Numer. Anal., 8 (1971), pp. 616–622.

[202] R. F. Harrington, *Field Computations by Moment Methods*, Macmillan, New York, 1868.

[203] S. Haykin, *Adaptive Filter Theory*, Third Edition, Prentice-Hall, Englewood Cliffs, NJ, 1996.

[204] M. T. Heath, *The Numerical Solution of Ill-Conditioned Systems of Linear Equations*, Report ORNL-4957 UC-32, Oak Ridge National Laboratory, Oak Ridge, TN, July 1974.

[205] M. T. Heath, A. J. Laub, C. C. Paige, and R. C. Ward, *Computing the singular value decomposition of a product of two matrices*, SIAM J. Sci. Stat. Comput., 7 (1986), pp. 1147–1159.

[206] M. R. Hestenes and E. Stiefel, *Methods of conjugate gradients for solving linear systems*, J. Res. Nat. Bur. Standards, 49 (1952), pp. 409–436.

[207] N. J. Higham, *A survey of condition number estimation for triangular matrices*, SIAM Review, 29 (1987), pp. 575–596.

[208] N. J. Higham, *Analysis of the Cholesky decomposition of a semi-definite matrix*, in M. G. Cox and S. Hammarling (Eds.), Reliable Numerical Computation, Clarendon Press, Oxford, 1990, pp. 161–185.

[209] N. J. Higham, *Algorithm 694: A collection of test matrices in Matlab*, ACM Trans. Math. Software, 17 (1991), pp. 289–305. Version 3.0 is described in N. J. Higham, *The Test Matrix Toolbox for Matlab*, Numerical Analysis Report No. 237, Dept. of Mathematics, University of Manchester, September 1995. Available from ftp.mathworks.com in file pub/contrib/linalg/testmatrix.sh.

[210] J. W. Hilgers, *On the equivalence of regularization and certain reproducing kernel Hilbert space approaches for solving first kind problems*, SIAM J. Numer. Anal., 13 (1976), pp. 172–184 and 15 (1978), p. 1301 (erratum).

[211] B. Hofmann, *Regularization for Applied Inverse and Ill-Posed Problems*, Teubner, Stuttgart, Germany, 1986.

[212] B. Hofmann, *Regularization of nonlinear problems and the degree of ill-posedness*, in G. Anger, R. Gorenflo, H. Jochmann, H. Moritz, and W. Webers (Eds.), Inverse Problems: Principles and Applications in Geophysics, Technology, and Medicine, Akademie Verlag, Berlin, 1993.

[213] Y. P. Hong and C.-T. Pan, *The rank revealing QR decomposition and SVD*, Math. Comp., 58 (1992), pp. 213–232.

[214] T. A. Hua and R. F. Gunst, *Generalized ridge regression: A note on negative ridge parameters*, Comm. Statist. Theory Methods, 12 (1983), pp. 37–45.

[215] T.-M. Hwang, W.-W. Lin and E. K. Yang, *Rank revealing LU factorizations*, Linear Algebra Appl., 175 (1992), pp. 115–141.

[216] *IMSL MATH/LIBRARY User's Manual*, Version 3.0, IMSL Inc., Houston, TX, 1996.

[217] D. D. Jackson, *The use of a priori data to resolve non-uniqueness in linear inversion*, Geophys. J. R. Astron. Soc., 57 (1979), pp. 137–157.

[218] S. H. Jensen, P. C. Hansen, S. D. Hansen, and J. Aa. Sørensen, *Reduction of broad-band noise in speech by truncated QSVD*, IEEE Trans. Audio Speech Process., 3 (1995), pp. 439–448.

[219] S. Kaczmarz, *Angenäherte Auflösung von Systemen Linearer Gleichungen*, Bull. Acad. Polon. Sci. Lett. A, 35 (1937), pp. 355–357.

[220] A. K. Katsaggelos, J. Biermond, R. W. Schafer, and R. M. Merserreau, *A regularized iterative image restoration algorithm*, IEEE Trans. Signal Process., 39 (1991), pp. 914–929.

[221] L. Kaufman and A. Neumaier, *PET regularization by envelope guided conjugate gradients*, IEEE Trans. Medical Imaging, 15 (1996), pp. 385–389.

[222] L. Kaufman and A. Neumaier, *Regularization of ill-posed problems by envelope guided conjugate gradients*, J. Comp. Graph. Statistics, to appear.

[223] J. B. Keller, *Inverse problems*, Amer. Math. Monthly, 83 (1976), pp. 107–118.

[224] J. T. King, *Multilevel algorithms for ill-posed problems*, Numer. Math., 61 (1992), pp. 311–334.

[225] J. T. King and D. Chillingworth, *Approximation of generalized inverses by iterated regularization*, Numer. Funct. Anal. Optim., 1 (1979), pp. 499–513.

[226] A. Kirsch, *An Introduction to the Mathematical Theory of Inverse Problems*, Springer-Verlag, New York, 1996.

[227] R. Kress, *Linear Integral Equations*, Springer-Verlag, Berlin, 1989.

[228] C. Lanczos, *An iteration method for the solution of the eigenvalue problem of linear differential and integral operators*, J. Res. Nat. Bur. Standards, 45 (1950), pp. 255–282.

[229] J. Larsen, H. Lund-Andersen, and B. Krogsaa, *Transient transport across the blood-retina barrier*, Bull. Math. Biol., 45 (1983), pp. 749–758.

[230] R. M. Larsen and P. C. Hansen, *Efficient implementation of the SOLA mollifier method*, Astron. Astrophys. Suppl. Ser., 121 (1997), pp. 587–598.

[231] C. L. Lawson and R. J. Hanson, *Solving Least Squares Problems*, Prentice-Hall, Englewood Cliffs, NJ, 1974; reprinted by SIAM, Philadelphia, 1995.

[232] S. Levy and P. K. Fullager, *Reconstruction of a sparse spike train from a portion of its spectrum and application to high-resolution deconvolution*, Geophysics, 46 (1981), pp. 1235–1243.

[233] F. Li and R. J. Vaccaro, *Analytical performance prediction of subspace-based algorithms for DOA estimation*, in [343], pp. 243–260.

[234] R.-C. Li, *Bounds on perturbations of generalized singular values and of associated subspaces*, SIAM J. Matrix Anal. Appl., 14 (1993), pp. 195–234.

[235] P. Linz, *Uncertainty in the solution of linear operator equations*, BIT, 24 (1984), pp. 92–101.

[236] P. Linz, *A new numerical method for ill-posed problems*, Inverse Problems, 10 (1994), pp. L1–L6.

[237] F. Lorenzelli, P. C. Hansen, T. F. Chan, and K. Yao, *A systolic implementation of the Chan/Foster RRQR algorithm*, IEEE Trans. Signal Process., 42 (1994), pp. 2205–2208.

[238] A. K. Louis, *Convergence of the conjugate gradient method for compact operators*, in [110], pp. 177–183.

[239] A. K. Louis, *Inverse und schlecht gestellte Probleme*, Teubner, Stuttgart, Germany, 1989.

[240] A. K. Louis and P. Maass, *A mollifier method for linear operator equations of the first kind*, Inverse Problems, 6 (1990), pp. 427–440.

[241] F. T. Luk and S. Qiao, *A new matrix decomposition for signal processing*, Automatica, 30 (1994), pp. 39–43.

[242] F. T. Luk and S. Qiao, *A symmetric rank-revealing Toeplitz matrix decomposition*, J. VLSI Signal Proc., 14 (1996), pp. 19–28.

[243] F. T. Luk and S. Qiao, *An adaptive algorithm for interference cancelling in array processing*, in F. T. Luk (Ed.), Advanced Signal Processing Algorithms, Architectures, and Implementations VI, SPIE Proceedings, Vol. 2846, 1996, pp. 151–161.

[244] M. A. Lukas, *Asymptotic optimality of generalized cross-validation for choosing the regularization parameter*, Numer. Math., 66 (1993), pp. 41–66.

[245] A. MacLeod, *Finite-dimensional regularization with nonidentity smoothing matrices*, Linear Algebra Appl., 111 (1988), pp. 191–207.

[246] D. W. Marquardt, *Generalized inverses, ridge regression, biased linear estimation, and nonlinear estimation*, Technometrics, 12 (1970), pp. 591–612.

[247] R. Mathias and G. W. Stewart, *A block QR algorithm and the singular value decomposition*, Linear Algebra Appl., 182 (1993), pp. 91–100.

[248] W. Menke, *Geophysical Data Analysis: Discrete Inverse Theory*, Academic Press, San Diego, 1989.

[249] K. Miller, *Least squares methods for ill-posed problems with a prescribed bound*, SIAM J. Math. Anal., 1 (1970), pp. 52–74.

[250] M. Moonen and B. De Moor (Eds.), *SVD and Signal Processing, III. Algorithms, Architectures and Applications*, Elsevier, Amsterdam, the Netherlands, 1995.

[251] M. S. Moonen, G. H. Golub, and B. L. R. De Moor (Eds.), *Linear Algebra for Large Scale and Real-Time Applications*, NATO ASI Series, Kluwer, Dordrecht, the Netherlands, 1993.

[252] M. Moonen, P. Van Dooren, and J. Vandewalle, *A singular value decomposition updating algorithm for subspace tracking*, SIAM J. Matrix Anal. Appl., 13 (1992), pp. 1015–1038.

[253] M. Moonen, P. Van Dooren, and F. Vanpoucke, *On the QR algorithm and updating the SVD and the URV decomposition in parallel*, Linear Algebra Appl., 188/189 (1993), pp. 549–568.

[254] V. A. Morozov, *On the solution of functional equations by the method of regularization*, Soviet Math. Dokl., 7 (1966), pp. 414–417.

[255] V. A. Morozov, *Methods for Solving Incorrectly Posed Problems*, Springer-Verlag, New York, 1984.

[256] D. A. Muiro, *The Mollification Method and the Numerical Solution of Ill-Posed Problems*, Wiley, New York, 1993.

[257] *NAG Fortran Library Manual, Mark* 17, NAG Ltd., Oxford, 1995.

[258] J. G. Nagy and D. P. O'Leary, *Restoring Images Degraded by Spatially-Variant Blur*, Report CS-TR-3426, Dept. of Computer Science, University of Maryland, College Park, MD, February 1994; SIAM J. Sci. Comput., to appear.

[259] F. Natterer, *The Mathematics of Computerized Tomography*, Wiley, New, York, 1986.

[260] F. Natterer, *Numerical treatment of ill-posed problems*, in G. Talenti (Ed.), Inverse Problems, Lecture Notes in Mathematics 1225, Springer-Verlag, Berlin, 1986, pp. 142–167.

[261] A. S. Nemirovskii, *The regularizing properties of the adjoint gradient method in ill-posed problems*, U.S.S.R. Comput. Math. and Math. Phys., 26 (1986), pp. 7–16; English translation of Zh. Vychisl. Mat. Mat. Fiz., 26 (1986), pp. 332–347.

[262] G. Nolet, *Solving or resolving inadequate and noisy tomographic systems*, J. Comp. Phys., 61 (1985), pp. 463–482.

[263] G. Nolet (Ed.), *Seismic Tomography*, Kluwer, Dordrecht, the Netherlands, 1987.

[264] D. M. O'Brien and J. N. Holt, *The extension of generalized cross-validation to a multi-parameter class of estimators*, J. Austral. Math. Soc. Ser. B, 22 (1981), pp. 501–514.

[265] D. P. O'Leary, *The SVD in image restoration*, in [250], pp. 315–322.

[266] D. P. O'Leary and B. W. Rust, *Confidence intervals for inequality-constrained least squares problems, with applications to ill-posed problems*, SIAM J. Sci. Stat. Comput., 7 (1986), pp. 473–489.

[267] D. P. O'Leary and J. A. Simmons, *A bidiagonalization-regularization procedure for large-scale regularization of ill-posed problems*, SIAM J. Sci. Stat. Comput., 2 (1981), pp. 474–489.

[268] J. Ory and R. G. Pratt, *Are our parameter estimators biased? The significance of finite-difference regularization operators*, Inverse Problems, 11 (1995), pp. 397–424.

[269] Y. Osaki and H. Shibahashi (Eds.), *Progress of Seismology of the Sun and Stars*, Springer-Verlag, Berlin, 1990.

[270] C. C. Paige, *Some aspects of generalized QR factorizations*, in M. Cox and S. Hammarling (Eds.), Reliable Numerical Computation, Clarendon Press, Oxford, 1990.

[271] C. C. Paige and M. S. Saunders, *Solution of sparse indefinite systems of linear equations*, SIAM J. Numer. Anal., 2 (1975), pp. 617–629.

[272] C. C. Paige and M. S. Saunders, *Towards a generalized singular value decomposition*, SIAM J. Numer. Anal., 18 (1981), pp. 398–405.

[273] C. C. Paige and M. A. Saunders, *LSQR: An algorithm for sparse linear equations and sparse least squares*, ACM Trans. Math. Software, 8 (1982), pp. 43–71.

[274] C. C. Paige and M. A. Saunders, *Algorithm 583. LSQR: Sparse linear equations and least squares problems*, ACM Trans. Math. Software, 8 (1982), pp. 195–209.

[275] C. C. Paige and M. Wei, *History and generality of the CS decomposition*, Linear Algebra Appl., 108/109 (1994), pp. 303–326.

[276] C.-T. Pan and P. T. P. Tang, *Bounds on singular values revealed by QR factorizations*, in [250], pp. 157–165.

[277] R. L. Parker, *Understanding inverse theory*, Ann. Rev. Earth Planet. Sci., 5 (1977), pp. 35–64.

[278] R. L. Parker, *Geophysical Inverse Theory*, Princeton University Press, Princeton, NJ, 1994.

[279] B. N. Parlett, *The Symmetric Eigenvalue Problem*, Prentice-Hall, Englewood Cliffs, NJ, 1980; reprinted by SIAM, Philadelphia, 1997.

[280] D. L. Phillips, *A technique for the numerical solution of certain integral equations of the first kind*, J. Assoc. Comput. Mach., 9 (1962), pp. 84–97.

[281] M. Piana and M. Bertero, *Projected Landweber method and preconditioning*, Inverse Problems, 13 (1997), pp. 441–464.

[282] D. J. Pierce, *A Sparse URL Rather than a URV Factorization*, Report MEA-TR-203, Boeing Computer Services, Seattle, WA, November 1992.

[283] D. J. Pierce and J. G. Lewis, *Sparse multifrontal rank revealing QR factorization*, SIAM J. Matrix Anal. Appl., 18 (1997), pp. 159–180.

[284] J. E. Pierce and B. W. Rust, *Constrained least squares interval estimation*, SIAM J. Sci. Stat. Comput., 6 (1985), pp. 670–683.

[285] F. P. Pijpers and M. J. Thompson, *Faster formulations of the optimally localized averages method for helioseismic inversions*, Astron. Astrophys., 262 (1992), pp. L33–L36.

[286] F. P. Pijpers and M. J. Thompson, *The SOLA method for helioseismic inversion*, Astron. Astrophys., 281 (1994), pp. 231–240.

[287] W. H. Press, S. A. Teukolsky, W. T. Vetterling, and B. P. Flannery, *Numerical Recipes*, Second Edition, Cambridge University Press, Cambridge, UK, 1992.

[288] S. W. Provencher, *A constrained regularization method for inverting data represented by linear algebraic or integral equations*, Comput. Phys. Comm., 27 (1982), pp. 213–227.

[289] S. Qiao, *Approximating the PSVD and QSVD*, in [250], pp. 149–155.

[290] T. Raus, *The principle of the residual in the solution of ill-posed problems with nonselfadjoint operator*, Uchen. Zap. Tartu Gos. Univ., 75 (1985), pp. 12–20 (in Russian).

[291] T. Regińska, *A regularization parameter in discrete ill-posed problems*, SIAM J. Sci. Comput., 17 (1996), pp. 740–749.

[292] L. Reichel and W. B. Gragg, *Algorithm 686: Fortran subroutines for updating the QR decomposition*, ACM Trans. Math. Software, 16 (1990), pp. 369–377.

[293] G. R. Richter, *Numerical solution of integral equations of the first kind with nonsmooth kernels*, SIAM J. Numer. Anal., 15 (1978), pp. 511–522.

[294] J. D. Riley, *Solving systems of linear equations with a positive definite, symmetric, but possibly ill-conditioned matrix*, Math. Tables Aids Comput., 9 (1955), pp. 96–101.

[295] L. I. Rudin, S. Osher, and E. Fatemi, *Nonlinear total variation based noise removal algorithms*, Phys. D, 60 (1992), pp. 259–268.

[296] B. W. Rust and W. R. Burrus, *Mathematical Programming and the Numerical Solution of Linear Equations*, Elsevier, New York, 1972.

[297] B. W. Rust and D. P. O'Leary, *Confidence intervals for discrete approximations to ill-posed problems*, J. Comp. Graph. Stat., 3 (1994), pp. 67–96.

[298] H. Rutishauser, *Once again: The least squares problem*, Linear Algebra Appl., 1 (1968), pp. 479–488.

[299] C. Sánchez-Avila, *An adaptive regularization method for deconvolution of signals with edges by convex projections*, IEEE Trans. Signal Process., 42 (1994), pp. 1849–1851.

[300] R. J. Santos, *Equivalence of regularization and truncated iteration for general ill-posed problems*, Linear Algebra Appl., 236 (1996), pp. 25–33.

[301] R. J. Santos and A. R. De Pierro, *Generalized Cross-Validation Applied to the Conjugate Gradients Method as a Stopping Rule*, Report, Universidade Federal de Minas Gerais, Belo Horizonte, Brazil, 1995.

[302] F. Santosa, Y.-H. Pao, W. W. Symes, and C. Holland (Eds.), *Inverse Problems of Acoustic and Elastic Waves*, SIAM, Philadelphia, 1984.

[303] F. Santosa and W. W. Symes, *Linear inversion of band-limited reflection seismograms*, SIAM J. Sci. Stat. Comput., 7 (1986), pp. 1307–1330.

[304] T. K. Sarkar and S. M. Rao, *A simple technique for solving E-field integral equations for conducting bodies at internal resonances*, IEEE Trans. Antennas Propagation, 30 (1982), pp. 1250–1254.

[305] M. A. Saunders, *Solution of sparse rectangular systems using LSQR and CRAIG*, BIT, 35 (1995), pp. 588–604.

[306] J. A. Scales and A. Gersztenkorn, *Robust methods in inverse theory*, Inverse Problems, 4 (1988), pp. 1071–1091.

[307] L. L. Scharf, *The SVD and reduced rank signal processing*, Signal Process., 25 (1991), pp. 113–133. This paper can also be found in [343].

[308] L. L. Scharf and D. W. Tufts, *Rank reduction for modeling stationary signals*, IEEE Trans. Acoust. Speech Signal Process., 35 (1987), pp. 350–355.

[309] C. B. Shaw, Jr., *Improvement of the resolution of an instrument by numerical solution of an integral equation*, J. Math. Anal. Appl., 37 (1972), pp. 83–112.

[310] L. Simcik and P. Linz, *Qualitative Regularization: Resolving Non-Smooth Solutions*, Report CSE-94-12, Dept. of Computer Science, University of California, Davis, CA, 1994.

[311] C. R. Smith and W. T. Grandy (Eds.), *Maximum-Entropy and Bayesian Methods in Inverse Problems*, D. Reidel, Boston, 1985.

[312] R. C. Smith, K. L. Bowers, and C. R. Vogel, *Numerical recovery of material parameters in Euler-Bernoulli beam models*, ICASE Report 91-14, NASA Langley Research Center, February 1991; J. Math. Systems, Estimation, and Control, to appear.

[313] F. Smithies, *The eigenvalues and singular values of integral equations*, Proc. London Math. Soc., 43 (1937), pp. 255–279.

[314] F. Smithies, *Integral Equations*, Cambridge University Press, Cambridge, UK, 1958.

[315] W. Squire, *The solution of ill-conditioned linear systems arising from Fredholm equations of the first kind by steepest descents and conjugate gradients*, Int. J. Numer. Meth. Eng., 10 (1976), pp. 607–617.

[316] P. B. Stark, program sbvq, *Strict bounds on linear functionals of bounded variables subject to a quadratic misfit criterion on linear data constraints*, Fortran program available by anonymous ftp from ftp://stat-ftp.berkeley.edu/pub/users/stark/code, 1988.

[317] G. W. Stewart, *A note on the perturbation of singular values*, Lin. Alg. Appl., 28 (1979), pp. 213–216.

[318] G. W. Stewart, *On efficient generation of orthogonal matrices with an application to condition estimation*, SIAM J. Numer. Anal., 17 (1980), pp. 403–409.

[319] G. W. Stewart, *Rank degeneracy*, SIAM J. Sci. Stat. Comput., 5 (1984), pp. 403–413.

[320] G. W. Stewart, *On the invariance of perturbed null vectors under column scaling*, Numer. Math., 44 (1984), pp. 61–65.

[321] G. W. Stewart, *Collinearity and least squares regression*, Statistical Science, 2 (1987), pp. 68–100.

[322] G. W. Stewart, *An updating algorithm for subspace tracking*, IEEE Trans. Signal Proc., 40 (1992), pp. 1535–1541.

[323] G. W. Stewart, *Updating a rank-revealing ULV decomposition*, SIAM J. Matrix Anal. Appl., 14 (1993), pp. 494–499.

[324] G. W. Stewart, *On the early history of the singular value decomposition*, SIAM Review, 35 (1993), pp. 551–566.

[325] G. W. Stewart, *Determinating rank in the presence of error*, in [251], pp. 275–291.

[326] G. W. Stewart and J. Sun, *Matrix Perturbation Theory*, Academic Press, Boston, 1990.

[327] O. N. Strand, *Theory and methods related to the singular-function expansion and Landweber's iteration for integral equations of the first kind*, SIAM J. Numer. Anal., 11 (1974), pp. 798–825.

[328] E. de Sturler and H. A. van der Vorst, *Reducing the effect of global communication in GMRES(m) and CG on parallel distributed memory computers*, Appl. Numer. Math., 18 (1995), pp. 441–459.

[329] B. F. Swindel, *Geometry of ridge regression illustrated*, Amer. Statist., 35 (1981), pp. 12–15.

[330] A. A. Tal, *Numerical Solution of Fredholm Integral Equations of the First Kind*, Report TR-66-34, Computer Science Center, University of Maryland, College Park, MD, December 1966.

[331] A. Tarantola, *Inverse Problem Theory*, Elsevier, Amsterdam, 1987.

[332] H. J. J. te Riele, *A program for solving first kind Fredholm integral equations by means of regularization*, Comput. Phys. Comm., 36 (1985), pp. 423–432.

[333] A. M. Thompson, J. C. Brown, J. W. Kay, and D. M. Titterington, *A study of methods for choosing the smoothing parameter in image restoration by regularization*, IEEE Trans. Pattern Anal. Machine Intell., 13 (1991), pp. 3326–3339.

[334] A. M. Thompson, J. W. Kay, and D. M. Titterington, *A cautionary note about crossvalidatory choice*, J. Statist. Comput. Simulation, 33 (1989), pp. 199–216.

[335] A. J. Thorpe and L. L. Scharf, *Data adaptive rank-shaping methods for solving least squares problems*, IEEE Trans. Signal Process., 43 (1995), pp. 1591–1601.

[336] A. N. Tikhonov, *Solution of incorrectly formulated problems and the regularization method*, Soviet Math. Dokl., 4 (1963), pp. 1035–1038; English translation of Dokl. Akad. Nauk. SSSR, 151 (1963), pp. 501–504.

[337] A. N. Tikhonov and V. Y. Arsenin, *Solutions of Ill-Posed Problems*, Winston, Washington, D.C., 1977.

[338] A. N. Tikhonov and A. V. Goncharsky (Eds.), *Ill-Posed Problems in the Natural Sciences*, MIR, Moscow, 1987.

[339] A. N. Tikhonov, A. V. Goncharsky, V. V. Stepanov, and A. G. Yagola, *Numerical Methods for the Solution of Ill-Posed Problems*, Kluwer, Dordrecht, the Netherlands, 1995.

[340] D. M. Titterington, *Common structure of smoothing techniques in statistics*, Int. Stat. Rev., 53 (1985), pp. 141–170.

[341] V. F. Turchin, *Solution of the Fredholm equation of the first kind in a statistical ensemble of smooth functions*, U.S.S.R. Comput. Math. and Math. Phys., 7 (1967), pp. 79–96; English translation of Zh. Vychisl. Mat. Mat. Fiz., 7 (1867), pp. 1270–1284.

[342] S. Twomey, *On the numerical solution of Fredholm integral equations of the first kind by inversion of the linear system produced by quadrature*, J. Assoc. Comput. Mach., 10 (1963), pp. 97–101.

[343] R. Vaccaro (Ed.), *SVD and Signal Processing, II. Algorithms, Analysis and Applications*, Elsevier, Amsterdam, 1991.

[344] A. van der Sluis, *Condition numbers and equilibration of matrices*, Numer. Math., 14 (1969), pp. 14–23.

[345] A. van der Sluis and H. A. van der Vorst, *The rate of convergence of conjugate gradients*, Numer. Math., 48 (1986), pp. 543–560.

[346] A. van der Sluis and H. A. van der Vorst, *SIRT- and CG-type methods for iterative solution of sparse linear least-squares problems*, Linear Algebra Appl., 130 (1990), pp. 257–302.

[347] P. M. Van Dooren, *Structured linear algebra problems in digital signal processing*, in [152], pp. 361–384.

[348] S. Van Huffel and J. Vandewalle, *The Total Least Squares Problem—Computational Aspects and Analysis*, SIAM, Philadelphia, 1991.

[349] S. Van Huffel and H. Zha, *An efficient total least squares algorithm based on a rank-revealing two-sided orthogonal decomposition*, Numer. Algorithms, 4 (1993), pp. 101–133.

[350] C. F. Van Loan, *Generalizing the singular value decomposition*, SIAM J. Numer. Anal., 13 (1976), pp. 76–83.

[351] C. F. Van Loan, *Computational Frameworks for the Fast Fourier Transform*, SIAM, Philadelphia, 1992.

[352] J. M. Varah, *On the numerical solution of ill-conditioned linear systems with applications to ill-posed problems*, SIAM J. Numer. Anal., 10 (1973), pp. 257–267.

[353] J. M. Varah, *A practical examination of some numerical methods for linear discrete ill-posed problems*, SIAM Review, 21 (1979), pp. 100–111.

[354] J. M. Varah, *Pitfalls in the numerical solution of linear ill-posed problems*, SIAM J. Sci. Stat. Comput., 4 (1983), pp. 164–176.

[355] V. V. Voevodin, *The method of regularization*, U.S.S.R. Comp. Math. and Math. Phys., 9 (1969), pp. 228–232; English translation of Zh. Vychisl. Mat. Mat. Fiz., 9 (1969), pp. 673–675.

[356] C. R. Vogel, *Optimal choice of the truncation level for the truncated SVD solution of linear first kind integral equations when data are noisy*, SIAM J. Numer. Anal., 23 (1986), pp. 109–117.

[357] C. R. Vogel, *Solving Ill-Conditioned Linear Systems Using the Conjugate Gradient Method*, Report, Dept. of Mathematical Sciences, Montana State University, Bozeman, 1987.

[358] C. R. Vogel, *An overview of numerical methods for nonlinear ill-posed problems*, in [110], pp. 231–245.

[359] C. R. Vogel, *Total Variation Regularization for Ill-Posed Problems*, Report, Dept. of Mathematical Sciences, Montana State University, Bozeman, 1993.

[360] C. R. Vogel, *Non-convergence of the L-curve regularization parameter selection method*, Inverse Problems, 12 (1996), pp. 535–547.

[361] C. R. Vogel and M. E. Oman, *Iterative methods for total variation denoising*, SIAM J. Sci. Comput., 17 (1996), pp. 227–238.

[362] C. R. Vogel and J. G. Wade, *Iterative SVD-based methods for ill-posed problems*, SIAM J. Sci. Comput., 15 (1994), pp. 736–754.

[363] G. Wahba, *Practical approximate solutions to linear operator equations when the data are noisy*, SIAM J. Numer. Anal., 14 (1977), pp. 651–667.

[364] G. Wahba, *Ill-Posed Problems: Numerical and Statistical Methods for Mildly, Moderately and Severely Ill-Posed Problems with Noisy Data*, Technical Report 595, Dept. of Statistics, University of Wisconsin, Madison, 1980.

[365] G. Wahba, *Three topics in ill-posed problems*, in [110], pp. 37–51.

[366] G. Wahba, *Spline Models for Observational Data*, CBMS-NSF Regional Conference Series in Applied Mathematics, Vol. 59, SIAM, Philadelphia, 1990.

[367] G. Wahba and Y. Wang, *Behavior near zero of the distribution of GCV smoothing parameter estimates*, Statist. Probab. Lett., 25 (1995), pp. 105–111.

[368] P. Å. Wedin, *On angles between subspaces of a finite dimensional inner product space*, in B. Kågstrøm and Axel Ruhe (Eds.), Matrix Pencils, Lecture Notes in Mathematics 973, Springer-Verlag, Berlin, 1983, pp. 263–285.

[369] J. Weese, *A reliable and fast method for the solution of Fredholm integral equations of the first kind based on Tikhonov regularization*, Comput. Phys. Comm., 69 (1992), pp. 99–111.

[370] J. Weese, *A regularization method for nonlinear ill-posed problems*, Comput. Phys. Comm., 77 (1993), pp. 429–440.

[371] M. Wei, *Algebraic relations between the total least squares and least squares problems with more than one solution*, Numer. Math., 62 (1992), pp. 123–148.

[372] G. M. Wing, *Condition numbers of matrices arising from the numerical solution of linear integral equations of the first kind*, J. Integral Equations, 9 (Suppl.) (1985), pp. 191–204.

[373] G. M. Wing and J. D. Zahrt, *A Primer on Integral Equations of the First Kind*, SIAM, Philadelphia, 1991.

[374] H. Zha, *The restricted singular value decomposition of matrix triplets*, SIAM J. Matrix Anal. Appl., 12 (1991), pp. 172–194.

[375] H. Zha, *The product-product singular value decomposition of matrix triplets*, BIT, 31 (1991), pp. 711–726.

[376] H. Zha, *A numerical algorithm for computing the restricted singular value decomposition of matrix triplets*, Linear Algebra Appl., 168 (1992), pp. 1–25.

[377] H. Zha, *Computing the generalized singular values/vectors of large sparse or structured matrix pairs*, Numer. Math., 72 (1996), pp. 391–417.

[378] H. Zha and P. C. Hansen, *Regularization and the general Gauss-Markov linear model*, Math. Comp., 55 (1990), pp. 613–624.

Index

a priori estimate, 13, 74, 75, 103
ART, 141
averaging kernel
    definition, 79, 116
    examples, 91, 93
A-weighted generalized inverse
    definition, 39
    efficient multiplication, 41–42

Babolian and Delves, 122
Backus–Gilbert method
    definition, 118
    example, 130
    Lanczos algorithm, 119
    SVD analysis, 119
basic solution
    via RRQR, 59
    via SVD, 50
bidiagonalization
    algorithm, 101–102
    norm relations, 102

CGLS algorithm, 143
collinearity, 4
collocation, 12, 16
column pivoting, 32, 51
condition number, 22
confidence intervals, 123
conjugate gradients (CG)
    CG polynomial, 148
    CGLS implementation, 143
    convergence rate, 153–154
    error function, 141
    filter factors, 146–149, 154–157
    Krylov subspace characterization, 124
    link to Lanczos method, 146
    LSQR implementation, 143
    residual norm, 124
    solution norm, 141
    three-term recurrence, 148
convexity, 106
covariance matrix, 80
Craig's algorithm, 144
curvature of L-curve, 189

damped SVD, 120
deblurring, 15, 206
decay
    Fourier coefficients, 9, 70, 81, 90, 98
    generalized singular values, 24, 26
    singular values, 8, 20, 26, 90, 98, 152
degree of ill-posedness, 8
deregularization, 55, 115
derivative operator, 12
discontinuous solution, 16, 95–97, 121, 131
discrepancy principle
    compensated, 180, 200–202
    generalized, 179
    ordinary, 179, 193, 200–202
    sensitivity, 199–202
discrete ill-posed problem

characterization, 2, 20–21, 24–25, 69–71
    model problem, 81
discrete Picard condition, 82
discrete smoothing norm, 12–13, 75
discretization methods, 11, 213

Eckart–Young–Mirsky theorem, 53
effective number of degrees of freedom, 181, 184
effective numerical rank, 178, 195
effective resolution limit, 71, 83, 91, 97, 178
effectively well conditioned, 4
error estimates, 181–183
error history, 164, 165, 171
    plateaus, 171
error levels, 70
extreme residual norms, 85

filter factors, 72, 82
    CG method, 146, 154–157
    damped SVD, 120
    examples, 91–93, 157, 158
    iterated Tikhonov regularization, 140
    Landweber iteration, 138
    O'Brien and Holt's method, 108
    Rutishauser's method, 107
    Tikhonov regularization, 73, 100
    TSVD/TGSVD methods, 109
    T-TLS method, 112
Franklin's method, 107
Fredholm integral equation
    discretization, 11, 213
    generic form, 5
    with discrete right-hand side, 6

Galerkin method, 11, 15, 43, 213
Gauss–Markov linear model
    definition, 48, 108
    perturbation bound, 109
    regularized version, 108
GCVPACK, 210

generalized cross-validation (GCV)
    and CG, 185–186
    example, 194
    for general methods, 186
    function, 184
    Monte Carlo GCV, 186
    nonuniqueness, 185
    occasional failure, 197
    using bidiagonalization, 185
generalized singular values, 22
    decay, 24
generalized SVD (GSVD)
    algorithms and software, 29
    definition, 22
    relation to SVD, 23–25, 42
Givens rotations, 76, 102

Hadamard, 4
Hankel matrix, 14
helioseismology, 199
Hestenes and Stiefel, 141
hybrid method, 163, 172

ill-posed probem, 4
image deblurring, 15, 206
inequality constraints, 106
influence matrix, 79
inner product, 7
inner regularization, 163
inverse Laplace transformation, 10
inverse problems, 4–5
iterated Tikhonov regularization, 106, 140

Kaczmarz's method, 141
kernel, 5, 6
Krylov subspace
    definition, 145
    other bases, 149, 150

Lagrange polynomial, 151
Lanczos bidiagonalization
    examples, 167–170
    finite-precision aspects, 157–162, 167–170

for Backus–Gilbert method, 119
for GSVD, 42
for T-TLS, 163
implicit restart, 164
LSQR algorithm, 143
reorthogonalization, 164, 171
starting vector, 149, 160, 168
Lanczos vectors, 143
Landweber iteration, 138–140
LAPACK, 29, 101
L-curve
    analysis, 83–88
    criterion
        computational aspects, 190–192
        convergence properties, 190
        definition, 190
        example, 194
    definition, 83
    discrete, 190–191
    examples, 94, 95, 129, 132, 194
    for Tikhonov regularization, 84
    location of corner, 87, 189
    log-log scale, 188
    properties, 85
least squares problems, 21
least squares solution, 21
least squares TSVD, 54
linear functionals, 122, 210
LSQR algorithm, 143

matrix quotient, 23
maximum entropy regularization, 121
mildly ill posed, 8
Miller's method, 192
minimum variance estimator, 101
minimum variance TSVD, 54
MINRES, 142
model problem, 81, 97, 111, 153
moderately ill posed, 8
modified regularization matrix, 77, 95
modified TSVD (MTSVD), 51, 110

mollifier methods
    Backus–Gilbert formulation, 118
    definition, 116
    target function formulation, 116
monotonicity, 106
Monte Carlo GCV, 186

noise level, 97
nonnegativity, 106
norm in SVD basis, 124
null space, 23
numerical ϵ-rank, 30, 46, 49, 176
numerical null space
    definition, 48
    example, 66
    via RRQR, 31, 60
    via URV/ULV, 34, 61
numerical range
    definition, 48
    example, 67
    via RRQR, 60
    via subset selection, 50
    via URV/ULV, 61
ν-method, 140

O'Brien and Holt, 108
optimally localized averaging, 116
order optimality, 176
oscillating matrices, 88
oscillations, 3, 8, 25, 92
oversmoothing, 85, 128

parallel aspects, 99, 136
perturbation error, 85, 176, 194, 196
Picard condition, 9
piecewise polynomial solution, 121
point spread function, 16, 206
PP-TSVD method
    definition, 121
    example, 131
pragmatic parameter-choice methods, 176
preconditioning, 136
prewhitening, 55, 62, 67

product SVD (PSVD), 26
pseudoinverse, 21
pseudorank, 46

quadratic constraints, 105
quadrature method, 11, 213
quasi-optimality criterion
    definition, 182
    example, 194
quotient SVD, 23

rank reduction, 54
rank-deficient problem
    characterization, 2, 46
        treatment via rank-revealing decompositions, 58
        treatment via SVD, 49
rank-revealing LU decomposition, 33
rank-revealing QR decomposition
    algorithms, 38
    approximate subspaces, 60
    approximate TSVD solution, 59
    definition, 31
    existence, 31
    tightness bounds, 32–33
regularization
    by discretization, 69
    characterization of methods, 123
    of discrete ill-posed problem, 71
    of ill-posed problem, 10
    smoothing of exact solution, 78
regularization error, 79, 85, 176, 194, 196
regularization matrix L, 12, 73, 94
regularized inverse, 78, 100
regularized TLS, 114
relative decay rate, 82–83, 97, 111
reorthogonalization, 145, 164, 171
resolution matrix
    definition, 78
    example, 130
restricted SVD (RSVD), 27–28, 44, 109
RHYBRID algorithm, 206

ridge regression, 101
ridge trace, 111, 127
Riemann–Lebesgue lemma, 6
Ritz plot, 167–169
Ritz polynomial, 146
Ritz values
    characterization, 146, 150
    convergence, 149–153
    examples, 166–170
Rutishauser's method, 106

scaling of matrices, 47
second derivative, 14
semiconverence, 135
seminorm, 13, 74–78
severely ill posed, 9
sign changes, 20, 25, 88
singular functions, 7
singular value decomposition (SVD)
    algorithms and software, 29
    definition, 19–20
    link to eigenvalue decompositions, 20
    relation to SVE, 20
singular value expansion (SVE), 6
    algorithm, 43
    approximation results, 43
    expression for solution, 7
    fundamental relation, 7
singular values
    decay, 20
    of kernel, 7
    of matrix, 20
    perturbation bounds, 46
sinusoids in noise, 13
size of solution, 12, 124–125, 175
smoothing effect, 6, 8, 21
smoothing norm, 11
smoothness of kernel, 8
Sobolev norm, 75–77
spurious singular values, 160, 168–170
standard-form problem, 38, 137

standard-form transformation, 38
   as "preconditioner," 137
   explicit, 40
   implicit, 41
   norm relations, 40
subset selection
   comparison of methods, 60, 63
   via RRQR, 58
   via SVD, 50
subspace angle, 45
SVD–GSVD relation, 24, 42
symmetric rank-revealing decomposition, 35

target function, 116
test problems
   discontinuous solution, 16
   in REGULARIZATION TOOLS, 213
   in TEST MATRIX TOOLBOX, 88
   inverse helioseismology, 199
   one-dimensional image deblurring, 15
   random matrices, 88
   second derivative, 14
   sinusoids in noise, 13
   two-dimensional image deblurring, 206
thin SVD, 20
Tikhonov regularization
   formulation, 11, 100
   geometric perspective, 127
   perturbation bounds, 104–105
   regularization error, 81
   regularized inverse, 100, 110
   scaled residual, 100
   statistical setting, 101
total least squares (TLS), 52
total variation, 120
truncated GSVD (TGSVD) method
   filter factors, 109
   for discrete ill-posed problems, 109
   for rank-deficient problems, 51
   matrix, 55
   perturbation bounds, 57
   regularized inverse, 110
   relation to Tikhonov regularization, 110–111
truncated GSVD (TGSVD) solution, 51
truncated SVD (TSVD) method
   filter factors, 109
   filtering (in signal proc.), 56
   for discrete ill-posed problems, 109
   for rank-deficient problems, 50
   geometric perspective, 127
   matrix, 49
   perturbation bounds, 56
   relation to Tikhonov regularization, 110–111
truncated SVD (TSVD) solution
   definition, 50
   numerical example, 64
   via RRQR, 59
   via URV/ULV, 61
truncated TLS (T-TLS) method
   definition, 52
   filter factors, 112
   for discrete ill-posed problems, 112
   for rank-deficient problems, 52
   perturbation bounds, 57

ULLV decomposition, 36, 62
undersmoothing, 85, 128
URV/ULV decompositions
   algorithms, 38
   approximate subspaces, 61, 63
   approximate TSVD solution, 61, 63
   definitions, 33
   tightness bounds, 34

Wiener filtering, 125

zero-crossings, 8